Managing Groundwater for Irrigated Agriculture

Managing Groundwater *for* Irrigated Agriculture

The Relevance of Biodrainage Model

A.S. Kapoor

RAWAT PUBLICATIONS
Jaipur | New Delhi | Bangalore | Hyderabad | Guwahati | Kolkata

ISBN 978-81-316-0624-7

Published by
Prem Rawat for **Rawat Publications**
Satyam Apts., Sector 3, Jawahar Nagar, Jaipur 302 004 (India)
Phone: 0141 265 1748 / 7006 Fax: 0141 265 1748
E-mail: info@rawatbooks.com
Website: www.rawatbooks.com

New Delhi Office
4858/24, Ansari Road, Daryaganj, New Delhi 110 002
Phone: 011 2326 3290

Also at *Bangalore, Hyderabad, Guwahati* and *Kolkata*

Typeset by *d'lines* DTP SERVICES, Jaipur
Printed at Chaman Enterprises, New Delhi

To

my granddaughters

ANUBHA, ADITI, ANUBHOOTI and **AVNI**

Contents

Preface

Agriculture is widely practised to feed human population which is growing at a fast pace. Forests are needed to sustain ecology and environment. Land use, for agriculture and forestry, has to be maintained in appropriate balance. In some regions of the world, too much land has already been diverted for agriculture use resulting in an imbalance.

There are wide variations in climatic conditions in different parts of the world. Leaving aside extreme conditions, agriculture and forestry are feasible in most climatic conditions where water is available. Natural precipitation is the source of all fresh water. Where it occurs uniformly and in plenty, vegetation grows readily and easily. Where it is scanty, arid conditions prevail. In such regions agriculture needs the support of irrigation.

Development projects are undertaken to store water in reservoirs by constructing dams and are used for power generation and irrigation. Irrigation provides crops and plants with water needed for their growth, but at the same time some water percolates down into the ground by seepage from the conveyance system and during application of irrigation water in the field. This disturbs the pre-existing natural groundwater balance and the groundwater table starts rising. This rise continues till a new balance is achieved by additional withdrawals from the groundwater equal to the recharge from irrigation. In the absence of sufficient withdrawals, the rise of groundwater table continues and when it comes up to near the ground surface, it causes waterlogging, rendering the land unsuitable for further agricultural use. Moreover, the salts contained in groundwater are transported into the pore space of top soils by capillary action and concentration of salts in this manner causes soil salinisation.

Waterlogging and salinity have become major hazards for irrigated agricultural lands and have raised doubts about sustainability of irrigation, particularly in dry arid regions. This threat is generally met by providing subsurface drainage under fields, by installing perforated pipes to collect water and conveying drainage water to sea, rivers, lakes or evaporation tanks. But discharge of saline drainage water in natural rivers, lakes, etc. pollutes them and is objectionable.

Biodrainage, in which the property of plants and trees to transpire water is used to remove excess groundwater, can be a biological option to provide the needed drainage. Trees and plants also remove minerals to some extent from the soil. They act like biological pumps and can help in achieving water balance as well as salt balance to mitigate the problem of waterlogging and salinisation.

Waterlogging and salinity have occurred at some places where trees and natural perennial vegetation were cleared for development of agriculture systems as in Australia. Biodrainage through tree plantations can be an appropriate remedy under such conditions. Apprehensions of progressive increase in soil salinity under the trees and doubts regarding their long-term survival are misconceived.

In some regions like the Aral Sea Basin in Central Asia, salinity in river waters is increasing because of return flow of saline drainage water from irrigated lands. Biodrainage can help in combating this problem.

The book describes the prevailing conditions and status of irrigation, drainage and forestry in different regions of the world. It presents the relevant aspects of groundwater salinity, plant physiology, uptake of water and minerals by crops and trees. Some case studies of drainage by plantations are also presented. An attempt has been made to formulate principles of planning and design of biodrainage systems. While fields for future study and research have been identified, the diverse management aspects of biodrainage are also described.

In recent years, research and field trials were carried out in several experimental stations in different parts of India to ascertain feasibility of biodrainage through trees. The results are encouraging. These are described in Chapter 7.

Biodrainage is a promising and environment-friendly option, which would be socially acceptable and a least cost measure to mitigate the problems of waterlogging and salinity.

The author hopes that the propositions and suggestions made in the book will find acceptance and provide stimulus for further study and research to bridge the gaps in knowledge.

This book draws heavily upon my earlier published book *Biodrainage: A Biological Option for Controlling Waterlogging and Salinity*. It incorporates information, data and results of research studies that became available since the publication of the this work.

Acknowledgements

Many people have provided valuable assistance in preparing this book. The Indira Gandhi Nahar Project personnel helped in making field investigations and in the collection of samples and data. The Agriculture Department laboratories carried out tests on soil and water samples as well as chemical analysis of biomass produce. The Forest Department made available valuable information relating to the forestry work. The Indira Gandhi Nahar Board (IGNB) staff rendered invaluable assistance in typing, editing and printing the typescript. The Computer Section of the Agriculture Department prepared the illustrative figures. I am very grateful to all of them. M.K. Sogani and Phool Chand Sharma of IGNB, in particular, provided great support.

Liberal use has been made of relevant information and data published in many papers and books, a reference list of which has been given after each chapter. The contribution that such information/data has made to the theme of this book is thankfully acknowledged.

Following the recommendations of the 8th ICID International Drainage Workshop at New Delhi (2000) and suggestions contained in IPTRID–FAO knowledge synthesis report (2002), the Ministry of Water Resources, India organised several research programmes through Agricultural Universities, Forest Institutes and Research Stations to test the efficacy of biodrainage through trees under different climatic and physiographic conditions. The reported results are incorporated in Chapter 7. The author is thankful to ICID, IPTRID, MOWR (India), and very many scientists and researchers for the invaluable information and data produced and made available to the author.

Overview

▼ INTRODUCTION

The land and water resources on the surface of earth which, in the past, were quite bountiful are no longer so. The needs of the growing population are putting an increasing pressure on the natural resources. Quite often the harnessing and exploitation of natural resources is done at the cost of the environment. Land degradation, water pollution and deforestation are the direct results of human activities, which if not checked, can cause irreparable and irreversible damage to the environment. We owe a responsibility to the future generations to protect and preserve natural resources in a healthy condition for them. The Earth Summit of Rio de Janeiro (1992) stressed that economic productivity, social responsibility and environmental protection belong together as an inseparable whole, for guaranteeing the equitable development opportunities for all countries and preserving the natural resources for the coming generations.

Potable water and usable land are not unlimited. Good quality water, in particular, is in short supply in most parts of the world and is scarce in dry arid regions. Natural waterflows need to be harnessed and used judiciously to avoid damage to flora and fauna and to prevent land degradation due to waterlogging and salinisation. The disciplined use of water in a regulated manner can help meet all the human needs and, at the same time, protect and preserve land and environment.

Agriculture is not possible without water. Crops must be supplied with water in the required quantities they need it. Natural precipitation rarely occurs to match with agricultural requirements. Therefore, it is necessary to harness and store water and ensure its supply for agriculture by practising irrigation. Presently, irrigation facility is available to about 18 per cent of the total cropped area of 1474 million ha in the world, and contributes about one-third to the world's total food produce. To meet the growing future human requirements of agriculture produce, land productivity must be improved and for this, it would be necessary to increase and extend irrigation facilities.

The practice of irrigation has not been without adverse consequences. Water seeping from the canals and the fields percolates down to the groundwater table, causing disturbance in the previously existing water balance. The rise of the groundwater table, if unchecked can ultimately result in its coming up near the ground surface. The root zone of the plants gets filled up with water leaving no air in the soil pores for the plant roots to breathe. Most plants and crops cannot grow under such conditions of waterlogging. All groundwaters contain some mineral salts. Under waterlogged conditions, water evaporating from the ground leaves behind the dissolved salts on the ground surface and in the soil pores. This results in salinisation of lands which then become totally unproductive. The reclamation of such lands is quite difficult. About 20 per cent of the world's irrigated area of 260 million ha is already affected by secondary salinisation and the malady is on the rise.

The Sumerian empire flourished about 4000 years ago in Mesopotamia, in the plains of the rivers Tigris and Euphrates, on the basis of a highly developed irrigation system. Later, large-scale salinisation, however, rendered the farm lands unproductive and this contributed to the collapse of the empire. In California's Imperial Valley, drainage water from irrigated lands is discharged into the Salton Sea whose salinity is on the increase. In San Joaquin Valley, drainage water from irrigated lands on discharge into the Kesterson Reservoir, caused problems of toxicity and discovery of selenium in the biota. Waterlogging, salinisation and drainage problems are widespread in Australia, India, Pakistan and Egypt. Mismanaged irrigation and drainage policies in the Aral Sea Basin, Central Asia, have caused one of the greatest catastrophes in history. Hundreds of thousands of hectares of fertile land are being removed from agricultural use. Agrochemical minerals and waste water have made water unfit for consumption. Out of the 20 species of fish, only two are left in small numbers. Worsening environmental conditions have caused a socio-economic crisis in the region.

In Pakistan, large investments have been made on Salinity Control and Reclamation Projects (SCARP). An expenditure of Rs 2.1 billion has been incurred on drainage measures. It is estimated that an investment of Rs 20 billion would be needed for another 2 million ha. However, the results so far have not been satisfactory.

DRAINAGE MEASURES CURRENTLY PRACTISED AND THEIR SHORTCOMINGS

The subsurface drainage system, in which perforated pipes are buried in trenches is the most commonly practised system of overcoming the waterlogging problem. The pipes are encased in filters and laid at sufficient depths in the fields (1–2 m). The percolating subsurface water collects into these drains and flows out of the area through a network of sub-collector, collector and main drains. This checks the rise of the groundwater table into the root zone, prevents waterlogging and enables sustainable agriculture.

It is estimated that an increase in waterlogging and salinisation of irrigated lands is occurring at the annual rate of about 0.5 million ha while the horizontal drainage facility is being provided to an additional area of about 0.1 to 0.2 million ha each year. Subsurface drainage pipe systems have been installed in more than 35 countries.

It is necessary to make a distinction between the need for drainage in wet and humid regions from that in dry arid areas. In wet humid regions, the groundwater is generally non-saline and the objective of drainage measures is to simply remove excess water from the ground surface and the soil in the root zone. Normal surface and subsurface drainage measures are effective and adequate. The use of groundwater by pumping (vertical drainage) can be equally effective and is widely practised. In dry arid regions, irrigation is necessary to raise crops. The groundwater table is quite deep before the introduction of irrigation and the groundwater is generally saline. Where and when the groundwater table rises and comes up near the ground surface, it becomes necessary to provide subsurface drainage. Vertical drainage (extraction of groundwater by pumping) is not feasible because the water is saline and unfit for agriculture. Horizontal drainage (by subsurface perforated pipelines) can be effectively practised but several constraints have to be overcome before this can be done.

First, the cost of providing horizontal drainage is quite high. It can be economically viable and affordable only where the land is used to raise remunerative commercial crops, vegetables or horticulture. The market prices of food grains, which meet the primary needs of the human population, do not permit high investments on subsurface drainage without direct or indirect subsidies to the farmers. In the developing countries, the largest areas under cultivation are under food grains. The investment costs on providing horizontal subsurface drainage systems on such a large scale thus cannot be met.

Second, the subsurface drainage systems need proper care in operation and maintenance. The pipes and drains need periodical cleaning. Silting and weed growth in open drains are common problems. The drainage water has to be pumped where adequate land gradient is not available for the gravity flow. This requires energy and proper arrangements for the upkeep of pumping units, which is not easy to achieve under the conditions prevailing in most developing countries. Already, the systems installed in Central Asia, Pakistan and other countries are facing this problem and managers are searching for solutions.

Third and most important, the drainage water which is saline and polluted is required to be disposed off in a harmless manner. This is feasible where an outlet to the sea is available in which the drains can discharge the saline polluted water. But there are large landlocked irrigated areas which do not have an easy access to the sea and, agricultural drainage water has thus to be discharged into rivers or lakes. This causes pollution and degradation of fresh water bodies. The detrimental effect towards environment has been recognised and subsurface drainage has been restricted, even prohibited in many locations in the developed countries. The problems of toxicity and discovery of selenium in the biota, due to drainage water discharged into the Kesterson Reservoir in San Joaquin Valley, USA is a widely reported and researched case.

As a consequence of the above-mentioned constraints and other local reasons, the annual rate of installation of subsurface drainage that was about 3,00,000 hectares during the 1980s came down to about one-half in 1990s. In some places (for instance, Germany and UK), drained areas were converted back to marshlands. In several countries like Netherlands, Switzerland and Germany, the investments on drainage are restricted to the rehabilitation of ancient systems.

In this context, alternative options must be thought of and tried, otherwise it may not be possible to practise irrigation, particularly in the dry arid regions, on a sustainable basis. What happened in Mesopotamia cannot be allowed to recur. The need to preserve the environment and to discourage measures that cause environmental degradation can no longer be overlooked.

EVOLUTION OF BIODRAINAGE METHODOLOGY

Tree plantations were raised on the Indira Gandhi Nahar Project (IGNP), India, along the canals to protect the canal embankments from wind erosion and prevent deposit of windblown sand into the canal. They were quite effective in achieving these objectives. In addition, they brought about the unintended benefit of significantly depressing the groundwater table which had come up very high in adjoining areas. A study was undertaken to assess whether tree plantations could be used as a drainage measure. The results were very encouraging and the concepts and methodology of biodrainage, described in this book, were developed.

The two important characteristics of trees on which the concept and methodology of biodrainage are based are:

Tree Water Use (TWU)

All trees uptake water from the subsoil through their root systems, most of which is transpired into the atmosphere. This characteristic of trees can be used to remove excess water from the soils. Properly planned tree plantations can uptake and remove enough quantity of water from the shallow water table to check its rise and provide protection against waterlogging. They can also be used to reclaim the already waterlogged lands. This methodology is termed as 'biodrainage'.

The rate of transpiration by trees depends upon external and internal factors. The important external factors are humidity of air, temperature, wind velocity, intensity of light, atmospheric pressure and available water supply. The important internal factors that influence the rate of transpiration are stomatal frequency, structural peculiarities of leaf, leaf area, amount of spongy tissue, inter-cellular spaces, orientation of leaf and the extent of the root system.

The reported results of the capacity of trees to grow and transpire water show great variance. It is best to correlate the Tree Water Use (TWU) rate with the rate of evapotranspiration form crops (ET_o) and the rate of evaporation from free water surface (A_{pan}). Crop evapotranspiration (ET_o) is defined as the rate of evapotranspiration from an extensive surface of 8 to 15 cm tall green grass cover of uniform height, actively growing, completely shading the ground and not short of water (FAO Paper No. 24). A_{pan} is the observed rate of evaporation form the water surface in a pan of standard size and is generally 1.15 to 1.20 times of ET_o.

The published annual tree water use values range from 0.6 A_{pan} for irrigated globules with full canopy cover in Westerns Australia (Marshall and Chester 1991) to 1.9 A_{pan} for *Eucalyptus camaldulensis* (Morris and Wafner 1987). In non-stressed environments, literature reports TWU for eucalyptus between 1.2 to 1.5 times of ET_o (Strible 1975; Sharma 1984).

In a study in the desert area of Rajasthan, India, annual TWU with a density of 1900 trees/ha was estimated as 3446 mm which is about 1.2 A_{pan}. In a field experiment at CSSRI, Karnal, India, annual TWU of individual eucalyptus trees grown in lysimetres was observed to be 2.59 times of ET_o. A study in Pakistan showed that TWU per day by two and a half month seedlings of different plant species under luxury water conditions ranged between 129 litres to 174 litres (Bhutta N.N. et al. 2000). Reported estimates of annual TWU for Tamarix plants in south western USA range between 2200 mm to 2800 mm.

On the basis of reported studies and experiments, a conservative estimate can be made that TWU from eucalyptus and other high water-consuming tree species under non-saline ($Ece<4$ ds/m) conditions can be expected to be not less than A_{pan}. The tree water use (TWU) decreases with an increase in the soil water salinity. In an experiment at CSSRI, Karnal, India, it was found that the TWU of eucalyptus was 81, 64, 63 and 53 per cent at groundwater salinity levels of 3, 6, 9 and 12 ds/m respectively as compared to the TWU at groundwater salinity of 0.4 ds/m.

Mineral Absorption by Plants

The plant biomass is mostly made up of carbon, oxygen and hydrogen which elements are obtained by the plant from the atmosphere. The plants also contain nitrogen and some minerals which are received from soil water through the plant roots. There are several theories to explain the process and mechanism of salt absorption by plant roots and movement of minerals within the plant body. Definite findings are yet to emerge. But there is general consensus that source of all minerals like calcium (Ca^{++}), sodium (Na^{+}), potassium (K^{+}), magnesium (Mg^{++}), sulphur (S^{-}) etc., found in plant biomass is the soil water.

The mineral content in biomass varies from plant specie to specie and depends upon the prevailing conditions of soil, water and climate under which the plants grow.

The mineral content also varies in different parts of the plant body. Plant physiology literature mentions an average mineral content of about 5 per cent in biomass on dry weight basis.

The mineral content in plants is generally determined by ash analysis or by the wet digestion method. In the diacid wet digestion method, the plant sample is washed with hydrochloric acid, treated and heated with nitric acid and then with perchloric acid. The mineral element (Na^+, K^+, Ca^{++}, Mg^{++}...) content in the sample is then determined by standard methods. Some of the elements like chlorine, sulphur etc., volatilise and escape on heating or on treatment by acids and are not fully estimated in the analysis. The chemical analysis of plant biomass, therefore, gives lower values of mineral content than the actuals.

Mineral compounds (of Na, Ca, Mg etc.) in soil solutions are broken up into elemental forms as ions on absorption by plant roots. The commonly found salts in soil solution are bicarbonates, chlorides and sulphates of calcium, sodium and magnesium. On an overall average basis, the weight of cation elements (Ca, Na, Mg...) in the mineral salt compound is about 33 per cent. The total weight of mineral salt compounds absorbed by a plant is, therefore, about three times the total weight of cation elements found in the plant.

Mineral element content in biomass of plants and trees has been determined and reported by a number of researchers. The overall average figure is about 5 per cent. Sample of tree plant and major crops grown in the IGNP, Rajasthan, India, were collected and analysed (1998) which showed cation element content in utilisable biomass produce from eucalyptus plantations to be not less than 2 per cent and that in the crops not less than 3.33 per cent. The mineral salt absorption by the tree plantations and crops at three times would be about 6 and 10 per cent of the total dry biomass produce respectively.

The best course would be to collect tree and plant samples from the site and location under consideration and determine mineral element contents by analysis in the laboratory.

▼ CASE STUDIES USED FOR EVOLVING AND TESTING BIODRAINAGE

(i) Indira Gandhi Nahar Project (IGNP), Rajasthan, India

The irrigation project is spread over an area of 1.87 million ha to utilise 9.2 billion cubic metres (bcm) of river water. The project construction is nearly complete. Irrigation was commenced four decades back and the annual irrigation in 2000 was about one million ha. The depth of groundwater table in general, was more than 40 metres before the commencement of irrigation. Some areas, particularly in the upper reaches and along the canals in the lower reaches got waterlogged after some time.

Water in the main canal under stage II of the project was released in a phased manner in the years 1975 (up to 189 km), 1983 (up to 293 km) and 1986 (up to 445 km). Soon after the release of water in the main canal, it was found that seepage water collected at the natural ground surface on both sides of the canal at many places. A survey conducted in 1991 indicated that the total number of locations where pools of water were permanently formed was 127, covering an area of 900 hectares.

Tree plantation work was undertaken along the canals primarily to protect the canal embankments from being blown away by high winds and to prevent sand deposits inside the canals. A 1524 m long and 261 m wide strip along the left side of the main canal was selected for case study. Plantation work in this strip was carried out during the years 1987 to 1994. The density of planted trees was about 1900 trees/ha and the important species planted were *Eucalyptus camaldulensis* and *Acacia nilotica.*

The plantation trees on growth had a profound impact on the groundwater table conditions. By the year 1997, the overall water pool area had reduced from 900 ha to 20 ha. In the reach selected for case study, the water pool area which was 25 ha (1988) disappeared totally by the year 1994. At one section the groundwater level under the plantation went down by 15.7 m in a period of six years (1991–1997). At a distance of 500 m, the draw-down was 8 m. At another section, the groundwater went down by 14 m under the plantation during the period of 6 years. At a distance of 100 m, the draw-down was 10.4 m and at a distance of 300 m draw-down was 7.8 m.

The plantation acted as bio-pump drawing out water at the rate of 3.93 m^3 per hr per ha. The Tree Water Use was estimated as 3446 mm per year.

The following conclusions were drawn from the study:

- The tree water use was 3446 mm/year which is about 1.2 A_{pan}.
- The draw-down of the groundwater table underneath the plantations was 15 m. The draw-down impact was quite significant up to a distance of 500 m.
- No abnormal increase in salinity levels in soils and groundwater was observed underneath the plantations.

(ii) The Aral Sea Basin, Central Asia

A programme to utilise the waters of rivers Amu Darya and Syr Darya for irrigation was undertaken in the 1960s and an area of 8.0 m ha was irrigated in the year 2000. The withdrawal of water from the rivers resulted in progressively reduced inflow into the Aral sea. This caused shrinking of the surface area of the sea, increase in the salinity of sea water resulting in extinction of many fish species, health and environmental hazards to the inhabiting population. The ill-effects of extensive irrigation include waterlogging and salinisation of soils. The quality of flow waters in the rivers is progressively deteriorating. One of the important causes of deterioration of the quality of river waters, is discharge of saline drainage water from the irrigated fields into the rivers.

The problem is proposed to be overcome by cutting down on the volume of water that is excessively used for irrigation and by extension and reconstruction of the horizontal subsurface and open drainage systems. The success of this approach to drainage is very doubtful. When the discharge of saline drainage water from irrigated fields into the rivers is itself the major cause of the problem, how then can it be a part of the solution? Instead, disciplined minimal water use for irrigation along with biodrainage to control waterlogging and to eliminate discharge of saline drainage water into the rivers should be the most appropriate option.

(iii) The Indus Basin, Pakistan

About 14 million ha area is provided irrigation with the use of 122 bcm of canal freshwater (salts 200–250 mg/l) and 52 bcm of groundwater (salts 900–4000 mg/l). Degradation of soils has occurred due to waterlogging and salinity in 25 per cent of the area.

A surface drainage network of nearly 15000 km has been constructed and more are planned. Much of the constructed drainage network is in a state of ill-repair and results are much less than expected.

The quality of canal water used for irrigation is quite good. The total annual addition of salts in the basin with irrigation water can be matched with total annual salt removal by the harvested biomass from the basin. Therefore, biodrainage can be an effective tool in controlling waterlogging without the risk of progressive increase in soil salinity.

(iv) North Victoria (Australia), Plantation Study

A 2.4 ha site at the ISIA – Dairy Research Centre in Kyabram in the Goulburn Valley, North Victoria was planted with eucalyptus trees in 1976. The density of tree plantation was 730 trees/ha. The trees were irrigated up to the year 1982 and not thereafter. The groundwater table underneath the plantations generally fluctuated between 3 to 5 m below the ground surface. The draw-down impact of the plantations on the water table did not extend to any significant distance beyond the edge of the plantations. In the surrounding irrigated pasture lands the groundwater table was at 1 m – 2 m depth from the ground surface.

The average annual evaporation (A_{pan}) at the site was 1503 mm and rainfall 462 mm.

Accumulation of salts was observed underneath the plantation in the capillary fringe and water fluctuations zones (at 3 to 6 m depth). The average increase in soil salinity during 9-year period was estimated as 2.21 ds/m. The increase in groundwater salinity below water table was about 1.0 ds/m during the same period.

Tree water use (TWU) from groundwater was estimated as 0.4 A_{pan} in the study. Because of salt accumulation in the root zone, it was concluded that without salt removal from underneath the tree plantation, trees cannot be sustained in the long term. New plantations would need to be established at adjacent sites once every 20 years or so. The limited draw-down impact on the water table outside the plantation area was considered as a negative factor.

It is observed that while estimating the TWU from groundwater, several factors like the uptake of water directly from the groundwater table surface by tree roots by skimming off were not accounted for. On correct analysis the TWU would be found to be much more than estimated in the study. The inference that due to increasing soil salinity in the root zone, the trees would not be able to survive after 20 years or so does not take into account that the soil salinity increase is confined in the capillary fringe and water fluctuation zone. The increase in groundwater salinity below the water table surface is minimal. The tree roots should be able to uptake water directly from the surface of the groundwater table and live their full life. The tree plantation area was small (2.4 ha). An annual calculated uptake of 430 mm from the water table is equivalent to a withdrawal rate of just 1.18 m^3/hr (from plantation area of 2.4 ha) which is too small compared to the recharge rate from a large surrounding irrigated area. It is, therefore, no surprise that biopumping had only marginal impact in depressing the groundwater table beyond the boundary of the plantation area.

The reported results and findings of the study need re-evaluation.

(v) San Joaquin Valley California (USA), Plantation Study

On the west side of San Joaquin Valley, about one-third of the 9,30,000 ha of irrigated land is waterlogged and salt-affected. The groundwater table lies at approximately 2.0 m below the soil surface. A plot measuring 9.45 ha was planted with *Eucalyptus camaldulensis* trees. To isolate the site from the external lateral and salt flows, a perimeter drain was installed at a depth of 2.4 m around the plantation. Drainage water (8000 mg/1 – TDS) was used to irrigate the plantations. And 1251 mm of water was applied (irrigation and precipitation) during the period of 220 days. Tree evapotranspiration during the same period was assessed as 1150 mm.

A finite element model was used to consider the macroscopic root water approach and the effect of the osmotic potential on the soil-water movement into tree roots and through the soil matrix.

The TDS of applied and drainage waters, the salt load in and out, and the ratio of salt out to salt in over the 220 days simulation period were plotted. The cumulative salt mass entering the system was 6183 kg while cumulative salt mass discharged from the system was 2355 kg or 38 per cent of the incoming salts. The TDS of drainage water from the plantation area was 14000 mg/l.

The study shows that eucalyptus tree plantation can be successfully irrigated by saline water containing 8000 mg/l TDS (10 ds/m) with TWU of 1150 mm in 220 days. It is, however, not clear that with major quantity of salts (almost 60%) contained in irrigation water left behind in the plantation area, how was there only marginal deterioration of shallow groundwater.

(vi) CSSRI Karnal, India Study

An experiment was conducted at the Central Soil Salinity Research Institute, Karnal, India, using RCC lysimeters of 1.2 m diameter and 2.5 m depth filled with sandy loam soil with provision to maintain water table depth at 1, 1.5 and 2 m from the surface and groundwater salinity at 0.4, 3, 6, 9 and 12 ds/m. Three months old saplings of eucalyptus or bamboo were planted in the centre of the lysimeter. The applied water entered the lysimeters from below as groundwater. Another set of lysimeters with similar combinations of water table and salinity levels but without trees was used to monitor the loss of water through evaporation (E_S). The amount of water biodrained (W_b) by the tree was calculated by subtracting the evaporation (E_S) from evapotranspiration (E_t). The groundwater salinity was maintained by applying saline water containing Ca: Mg and SO_4: Cl in the ratio of 1:2 and 1:3 respectively. It had no RSC and SAR was less than 5. The amount of water biodrained by *Eucalyptus tereticornis* and bamboo (*Bambusa arundinacea*) was monitored over a period of five years by measuring the water needed for maintaining the water table daily.

The trees absorbed and transpired water throughout the year, more in summer than in the winter season. The eucalyptus plant biodrained 2880, 5499, 5518, 5148 and 5019 mm of water in the first, second, third, fourth and fifth years from non-saline groundwater and a water table depth of 1.5 m. The TWU declined as the groundwater salinity increased. It was found to be 81, 64, 63 and 53 per cent at groundwater salinity levels of 3, 6, 9 and 12 ds/m respectively.

An interesting finding was that although the amount of water biodrained was maximum in summer and minimum in winter, the ratio W_b/E_o was maximum in winter and lower in summer. The mean W_b/E_o ratio was found to be 2.59 and W_b/E_s was 13.99.

The zone of maximum salt accumulation in the lysimeters with trees was at depth between 45 and 120 cm, 45 and 150 cm and 60 and 210 cm in case of 1.0, 1.5 and 2 m depths of water tables respectively. This was apparently the capillary zone from which the tree roots uptook water leading to concentration of salts.

The study shows that:

- Tree water use (TWU) by eucalyptus, under abundant water available conditions can be very high, as much as between two to three times of potential evapotranspiration.
- TWU declined with an increase in groundwater salinity. It was about 53 per cent when soil water salinity increased to 12 ds m.

- The soil salinity in the capillary zone increased by 2–6 ds/m during the period of four years but this did not affect the TWU. This was because the tree roots had direct access to the groundwater of lower salinity.

(vii) Eucalyptus, Water and Sustainability Study, Karnataka, India

The study was funded by the British Governments' Overseas Development Administration (ODA) and was conducted jointly by the Research teams of Karnataka Forest Department, Mysore Paper Mills and the Institute of Hydrology, Willingford. The research results did not show Eucalyptus to be the villain which have often been portrayed, but neither do they show Eucalyptus to be without hydrological disadvantages.

Hydrological impact studies were carried out. The conventional methods for estimating water use are based on the assumption that evaporation is primarily determined by meteorological 'demand': a function of solar radiation, atmospheric temperature, humidity and turbulence with some regulation to the water 'supply' being imposed through stomatal controls exerted by the plants. This method is appropriate under temperate conditions but in dry zones the roles are reversed: evaporation is primarily determined by water supply rather than by meteorological demand. On the hypothesis that the volume growth of a tree is linearly related to the volume of water transpired, a water use and growth model was developed. The study concluded that the transpiration rate of a Eucalyptus tree can be estimated from the cross-sectional area of the tree trunk.

It was found that the annual water use of forest was higher than that of annual agricultural crops (about two times higher than finger millet).

(viii) Integrated System for Agricultural Drainage Management on Irrigated Farm Land, Five Points California, USA

The study was carried out at Red Rock Ranch on a 256 ha farm. Vegetable crops were grown over 73.4 per cent of the farm area using irrigation water from a canal or a well. Salt tolerant commercial crops were grown on 20.3 per cent of the farm area using drainage water from the vegetable area for irrigation. Salt tolerant grasses and trees over 2 per cent of the farm land used drainage water from salt tolerant crops for irrigation. At the last step of sequential reuse, drainage water from salt-tolerant grasses/trees was used to irrigate halophytes grown over 0.8 per cent of the farm land. Finally, the drainage water from the halophyte area was discharged into a solar evaporator (Figure 6.10).

The sequential reuse of water increased the overall efficiency of water use. High crop yields and production of high value vegetable crops grown on the non-saline (75%) area of the farm provided remunerative returns.

The total canal and drainage water used for irrigation was 148 ha m and 24 ha m respectively. About 90 per cen of the drainage water was consumed within the farm and the requirement of canal water was reduced by 22 per cent. The net remaining 10 per cent drainage water was taken care of in the solar evaporator by evaporation and manufacture of salts. There was no drainage water or salt discharge out of the farm area.

The scheme is a commendable effort for the sequential use of irrigation and drainage water in a productive manner. It facilitates better water use efficiency. Most importantly, all saline drainage water is consumed within the farm area. Salt-tolerant crops, grasses/trees and halophytes are used to consume most of the saline drainage water.

The total annual salt inflow into the farm alongwith irrigation water is 414 tonnes. The total salt actually received in the solar evaporator during 3 years (1996–1998) was 122 tonnes with an average of 41 tonnes per year, which is expected to increase to 60–70 tonnes per year after the installation problems are overcome. But this would account for only about 16 per cent of the salt input quantity. It has not been explained what happened to the balance 86 per cent quantity of salt imported in the farm land. Was it left behind in the groundwater in the farm area. If so, can the Integrated on-Farm Drainage Management (IFDM) system be sustainable in the long run?

Perhaps, most of the salt was consumed by the vegetable crops, grasses, trees and halophytes growing on the farm land. The study does not consider this aspect at all. This can be easily done by determining mineral content in all types of biomass produce and estimating the total quantity of mineral salts in the total utilisable biomass produced on the farm. In case this can explain for most of the unaccounted salt, the IFDM system can be sustainable in the long term.

But that would raise another question. If most salts, in any way, are consumed and taken care of by the biomass produce, what is the need for the elaborate system and the solar evaporator? Can biodrainage by the plantations in a part of the farm area not fully take care of the net drainage water that percolates down to the groundwater table? The whole area of the farm can be divided into two parts, one for growing commercial vegetables and the other for biodrainage plantations in compact blocks suitably dispersed in the farm area. The cost on subsurface drainage system and solar evaporator would be totally saved. There would be no drainage outflow from the farm and the water table can be maintained at safe depth.

(ix) Serial Biological Concentration of Salt, Undera SBC Site, North Victoria, Australia

A 20 ha farm was severely affected by salinity and parts of the farm were retired from productive use. The groundwater table was less than 1 m from the surface. A groundwater pump was installed on the farm to pump out groundwater to lower the water table and reclaim the farm for productive agriculture. The pumped groundwater with salinity ranging between 9–11 ds/m was used to irrigate a 3 ha tile-drained SBC area planted with *Eucalyptus camaldulensis,* a range of other tree species and a combination of salt bushes (Atriplex spp). This 3 ha tile-drained area was located outside the area of influence of the pump. The drainage water from the 3 ha plot was led into the aquaculture ponds and from there to the salt harvesting ponds (Figure 6.11). The average soil salinity under the trees and salt bush was predicted not to exceed 16 ds/m. Selection of tree species, eucalyptus and melaleuca was made to produce commercial volumes of oil. Trials were conducted to grow most suitable oyster, prawn and fish species in the agriculture ponds.

The SBC trial is an engineering approach to put saline drainage water for biological use in a profitable manner.

Details of the quantity of water pumped and used to irrigate 3 ha plot, the quantity of salt contained in this quantity of water and the salt actually received in the salt harvesting ponds are not reported. It, therefore, cannot be ascertained what part of salts contained in the pumped groundwater used for irrigation is removed via the salt harvesting ponds, how much salts are consumed by the trees and salt bush and how much salt is left behind in the soil water mass underneath the plantations.

A simpler alternative could be to set up biodrainage plantation in a plot within or adjacent to the 20 ha farm to biopump the groundwater and depress the groundwater table in the farm area. The area needed for plantation can be estimated which is likely to be smaller than the 4.5 ha area used in Undera trials. The salt balance can also be checked by considering the amount of salt removed by agriculture on the 20 ha farm, together with that removed by the biomass produce from the plantation. The cost of the subsurface drainage system and on groundwater pumping in the Undera trial would be saved. If necessary, biopumping can be supplemented by mechanical/electrical pumping of groundwater in the plantation area and using the water for irrigating the tree plantations.

(x) Study on Biological Control of Waterlogging, Bhawalnagar, Pakistan

The project area was an 18.2 ha canal irrigated flatland with major crops of cotton, rice, sugarcane and *'Kharif'* fodder. A 4 ha area was planted with eucalyptus trees with a density of 1340 plants per ha which were about six years old at the time of study. The data on water table depth was collected from observation wells both within and outside the plantation area during the period July–October 1999.

The water table in different observation wells varied during different periods and at different locations between 1.35 m to 2.58 m. The water table was higher towards cropped area and lower towards the plantation area. Due to plantation of eucalyptus trees, the groundwater was moving as a front towards the plantation area from north to south direction.

The study carried out during the period of most adverse conditions (rainy season) clearly demonstrates the effect of tree plantations in depressing the water table underneath the plantation area as well as under the surrounding areas. It would be interesting to know the conditions that prevailed before the plantations were raised and compare them with the present conditions.

The water applied for irrigating crops near the plantation was recorded but has not been reported. Estimates of recharge of the groundwater table and actual tree water use by the plantations are not available. It is not clear how the project area is reckoned to be limited to 18.2 ha when surrounding area up to long distances is cropped and irrigated.

The study should be continued and can yield very useful results.

(xi) Water Table Control by Biodrainage, Israel

The biodrainage approach was tested at five different waterlogged and salinity affected plots of 0.62 ha. each in the Yizréel valley, Israel. The groundwater tables declined constantly during the five-year monitoring span and dropped to more than 3 m below the soil surface at each of the sites in summer of 1999. The eucalyptus trees exhibited impressive growth rates with biomass yields of up to 57 tonnes per ha per year. It was concluded that, without irrigation, under current conditions, commercial forestry is a viable agricultural alternative that could bring higher returns than wheat, corn or cotton.

PRINCIPLES AND IMPORTANT FEATURES OF BIODRAINAGE PLANNING AND DESIGN

For biodrainage to be effectively adaptable, the following requirements are to be met:

- Water balance: the quantity of water removed from the groundwater annually should equal the quantity of recharge;
- Salt balance: the quantity of minerals removed annually should be nearly equal to the quantity of mineral import;
- Area under plantation: this should not be so large as to affect agriculture;
- Water for plantation: the demand on irrigation water should be minimal;
- Groundwater quality: this should not be too saline, say, not more than 12 ds/m; and
- Groundwater table depression: its impact should extend to the whole area threatened by waterlogging.

(a) Water Balance

The requirement of water balance and area needed for plantation can be estimated by the following equation (Figure 8.1)

$$P / C = \frac{R_F \times A_F \times I_R}{A_{\text{pan}}}$$

where

P: Plantation area

C: Total culturable command area

A_F: Annual intensity of irrigated agriculture

I_R: Gross irrigation supply

R_F: Recharge factor

A_{pan}: Tree water use assumed equal to surface evaporation from a standard pan.

As an illustration, if $R_F = 0.3$, $A_F = 1.0$, $I_R = 600$ mm and $A_{pan} = 1500$ mm, P/C would be 0.12 that is a 12 per cent culturable area under plantation can provide needed biodrainage.

(b) Salt Balance

The quantity of minerals imported annually with irrigation water is $m_w \times V \times 10^{-6}$ tonnes where m_w the amount of dissolved minerals in irrigation water in milligrams per litre (mg/l) and V is the volume in cubic metres of imported water used for irrigation.

The quantity of minerals removed by agriculture crops in a year is $m_c \times A$ where m_c is the percentage mineral content in the crop biomass and A is the total annual biomass produce from agriculture that is harvested and removed from the field. The quantity of minerals removed from tree plantation area P (in ha) is $m_p \times b \times P$ where m_p is the percentage mineral content is utilisable biomass produce and b is biomass produce in tonnes/ha/year.

As an illustration, if total dissolved salts in irrigation water are 150 mg/l (corresponding to mineral cation content of approximately 50 mg/l) and the gross quantity of irrigation water used annually is 600 mm then the annual mineral (cation) import with irrigation water would be 300 kg/yr per hectare of irrigated area. If the dry biomass produce (grain plus fodder) from crop is 9.5 tonnes/ha with mineral (cation) content of 3.33 per cent, the mineral removal by agriculture crop would be 313 kg. Since quantity of salt removed by agriculture would equal or exceed the quantity brought in by irrigation water there would be no increase in groundwater salinity. It can be shown that biomass produce from plantations over 12 per cent area would enable use of irrigation water with total dissolved solids (TDS) up to 200 mg/l without any increase in groundwater salinity.

With proper planning of irrigation intensity, crop selection, water management and biodrainage plantation, it can be possible to achieve salt balance using river water of normal salinity.

(c) Water for Plantation

The plantations for biodrainage are proposed to draw almost all their requirement of water from the shallow groundwater table. Fresh canal water may be needed in a limited quantity in the initial stages and for nurseries. The plantations would, therefore, not compete with the crops for the supply of irrigation water.

(d) Groundwater Quality

The groundwater quality in irrigation projects generally improves with the passage of time and it is rare that shallow groundwater would have salinity levels exceeding 12 ds/m. Poor groundwater quality may reduce the effective tree water use capability but would be no threat to practising biodrainage. Appropriate tree water use (TWU) values corresponding to the prevailing groundwater salinity levels should be used in planning the plantation area.

(e) Groundwater Table Depression

The effect of tree plantation in lowering the groundwater table can be estimated by the Donnan equation (Figure 5.8):

$$L^2 = \frac{8kY_oh}{R} + \frac{4kh^2}{R}$$

where L: is distance between plantations

Y_o: is height of water level above barrier layer underneath plantation

k: is hydraulic conductivity

h: is head difference

R: is rate of recharge

As an illustration, with a rate of recharge (R) equal to 0.5 mm/day, head difference (h) of 10 m, depth to barrier layer under plantations (Y_o) as 10 m and hydraulic conductivity value of 100 mm/day, plantations can be spaced 500 m apart. With hydraulic conductivity values of 1000 mm/day and 10 m/day, the required distance between plantations would work out to 1550 m and 155 m respectively.

Biodrainage plantations can be planned in blocks or strips and spaced so apart that the groundwater table at the farthest point from the plantation does not rise above, say 2 m below the ground surface, to eliminate the threat of waterlogging.

Criteria for Selecting Drainage Measures

The selection of most appropriate drainage measure between horizontal, vertical and bio-methods of drainage should be made by consideration of factors like efficiency and dependability, cost-benefit, expertise, requirement of water, energy requirement, flexibility of operation, environmental impact and social acceptability.

On most counts, biodrainage is the most preferable choice. A combination of two or all the three methods can also be practised.

Economy of Agroforestry Systems

The agroforestry system can provide comparable or even better returns than from agriculture. There are many studies which show attractive Net Present Worth, Benefit-Cost Ratio and Internal Rate of Return from forestry.

The biodrainage benefit of tree plantations can therefore be a zero-cost benefit or a free bonus.

Management Aspects

Innovations are rarely accepted readily. Doubts, apprehensions and apathy are more often greater stumbling blocks than lack of knowledge or resources. Biodrainage may be technically sound, economically attractive and environment-friendly but its acceptance and adaptation would depend upon the management skills with which it is propagated and practised.

Biodrainage should be compatible with the prevailing conditions of crop pattern, landholdings, land and water laws, water management and distribution practices. In case any changes are needed in these, they should be brought about carefully in consultation with people who are likely to be affected. The involvement of water users from the very beginning is desirable. Government and public agencies can contribute a good deal in taking initiative, laying down policy guidelines conducting research and providing resources but the participation of the local people in planting operations and their upkeep would be desirable.

Benefits to Users (Society)

The important benefits of biodrainage and tree plantations are summarised below.

Biodrainage would:

- Prevent waterlogging and salinisation;
- Make irrigation permanently sustainable;
- Be a no-cost remedy;
- Improve the environment;
- Provide timber, firewood and fodder;
- Act as wind breaks and prevent soil erosion; and
- Provide abode for habitation of birds and animals.

RECOMMENDATIONS OF INTERNATIONAL ORGANISATIONS

Biodrainage is a new science. Some aspects need further study and research. These may relate to irrigation and drainage engineering, soil science, hydrology, plant physiology, forestry, agriculture and agronomy, social science and the environment. It needs the attention of national governments, international organisations like the FAO, ICID, IPTRID, the World Bank and research institutions to take up and do all that is needed to develop this science, which holds great promise for human welfare and environmental protection.

The 'New Delhi Declaration' made at the Eighth ICID International Drainage Workshop (31st January to 4th February 2000), incorporates the following observations:

- An estimated one-third of the irrigated land in the arid and semi-arid zones face salinisation hazards.
- The development of drainage should increasingly be seen within the context of broader integrated water and environmental management strategies at the catchment level.
- For the removal of drainage water and for maintaining the groundwater levels, the concept of 'biodrainage', an environment-friendly measure, could hold promising possibilities. Further research is needed on some aspects of biodrainage such as salt balance.
- Due attention needs to be given to the downstream impacts of drainage interventions, including the pollution of drainage by agro-chemicals or domestic and industrial waste. The disposal of low quality drainage waters in an environmentally sustainable way, requires continued attention and an integrated management approach.

The recommendations made at the workship include:

- Subsurface drainage projects, in general, improve environments in the area serviced by them, but can also cause some undesirable impacts particularly in the downstream. Preparing a proper environmental plan and monitoring it closely are necessary.
- Biodrainage could hold promising possibilities in controlling groundwater levels. Further research on biodrainage and its impact on salt balance is needed.

The International Programme for Technology and Research in Irrigation and Drainage (IPTRID) under the Food and Agriculture Organisation (FAO) published a knowledge synthesis report (2002) on the subject of 'Biodrainage Principle, Experience and Applications'. The report concluded that biodrainage can effectively contribute to strongly reducing the problems as experienced from waterlogging in irrigated and non-irrigated agriculture. The problems associated with rise in salinity in the root zone can be effectively delayed using biodrainage systems in semi-arid and arid areas. Positive and negative aspects of biodrainage systems were described. Recommendations were made for future R&D work on the subjects of tree water use, salt uptake, salt tolerance, research methodology, tree species and sustainability.

RECENT RESEARCH IN INDIA (2004-2012)

Under the aegis of the Indian National Committee on Irrigation and Drainage (INCID) and the Ministry of Water Resources (New Delhi – India), many studies on identified subjects relating to biodrainage were carried out during the period 2004–2012 at different Agricultural Universities, Water Technology Centres, Forest Department of the State of Haryana etc. These studies have yielded very valuable data relating to different physiological and climatic conditions and the findings support the potential of biodrainage in overcoming the problems of waterlogging. Description of the studies and their results are incorporated in Chapter 7 of this book.

1

Land and Water Use for Sustainable Irrigated Agriculture

▼ INTRODUCTION

Land and water are natural resources that help sustain life on earth. Agriculture depends on the availability of water. About 260 million hectares (mha) of land in the world is provided with irrigation facilities but out of this, about 52 mha (20%) of the land has been rendered saline due to waterlogging. This land has lost much of its productive value. Moreover, drainage water from irrigated fields is polluting natural river water resources, which is highly objectionable. There is thus doubt about the long-term sustainability of 'irrigation'.

In the "Earth Summit" of Rio de Janeiro (1992), under the approved programme of action under "Agenda 21", it was stressed that economic productivity, social responsibility and environmental protection belong together as an inseparable whole, with a view to guaranteeing equitable development opportunities for all countries and preserving natural resources for the coming generations. At the World Food Summit held in Rome, November 1996, the need for the global protection of soil resources and their sustainable management was emphasised as one of the central conditions for securing food for future generations. We need a worldwide environmental partnership to achieve these goals. Transboundary eco-regional land conservation and basin-wide watershed development can be facilitated under the auspices of international conventions, action programmes and regional frameworks. "Prevention" should be the leading principle in land conservation policy, since "rehabilitation" is extremely difficult and often impossible to achieve.

All vegetation, including agricultural crops, need water for growth. In regions where water, in the form of precipitation, is available in adequate quantities every month throughout the year, trees and crops can grow without any supplemental water supply.

But in dry regions or during seasons when natural precipitation is not sufficient to meet the crop water requirement, water supply needs to be supplemented by irrigation. The source of the supplemental water supply can be surface flow in nearby streams, water stored in reservoirs behind dams, or groundwater.

All major irrigation schemes face the problems of waterlogging and soil salinity that need to be faced and tackled by proper management, failing which it may not be possible to sustain irrigation indefinitely.

EARTH'S SURFACE AREA

The total surface area of the earth is 504 million sq km, of which 147 million sq km or 29 per cent is land area and the rest, viz., 357 million sq km or 71 per cent comprises the water surface area. The region-wise distribution of land area is shown in Table 1.1.

Table 1.1 *Land area distribution*

Region	*Land area in million sq km*	*Land area as a percentage of whole*
Africa	30.0	20.5
Antarctica	13.6	9.3
Asia	47.9	32.7
Australia	7.6	5.3
Europe	5.3	3.7
North America	24.2	16.5
South America	17.6	12.0

The largest land area is in Asia (32.7%), followed by Africa (20.5%), North America (16.5%) and South America (12%) in that order.

POPULATION

The total population of the world in the year 1810 was 1 billion. It increased to 2 billion in the year 1930, i.e., an increase of 1 billion occurred during a period of 120 years. Since then, the population has been increasing at a rapid pace, with an increase from 6 to 7 billion occurring in a period of just 11 years. The past trend and future projection in world population growth up to the year 2050 are shown in Figure 1.1.

The continuous increase in population is exerting pressure on land and water resources which are not unlimited. The distribution of population and agriculture produce is quite uneven as can be seen from Table 1.2.

Table 1.2 *Comparative position of population and agricultural produce*

Region	*Population (% of total)*	*Agricultural produce (% of total)*
North America	10	31
South America	12	12
Europe (Western)	16	22
Far East (excluding China)	44	19
Near East	6	6
Africa	10	6
Oceania	2	4
Total	100	100

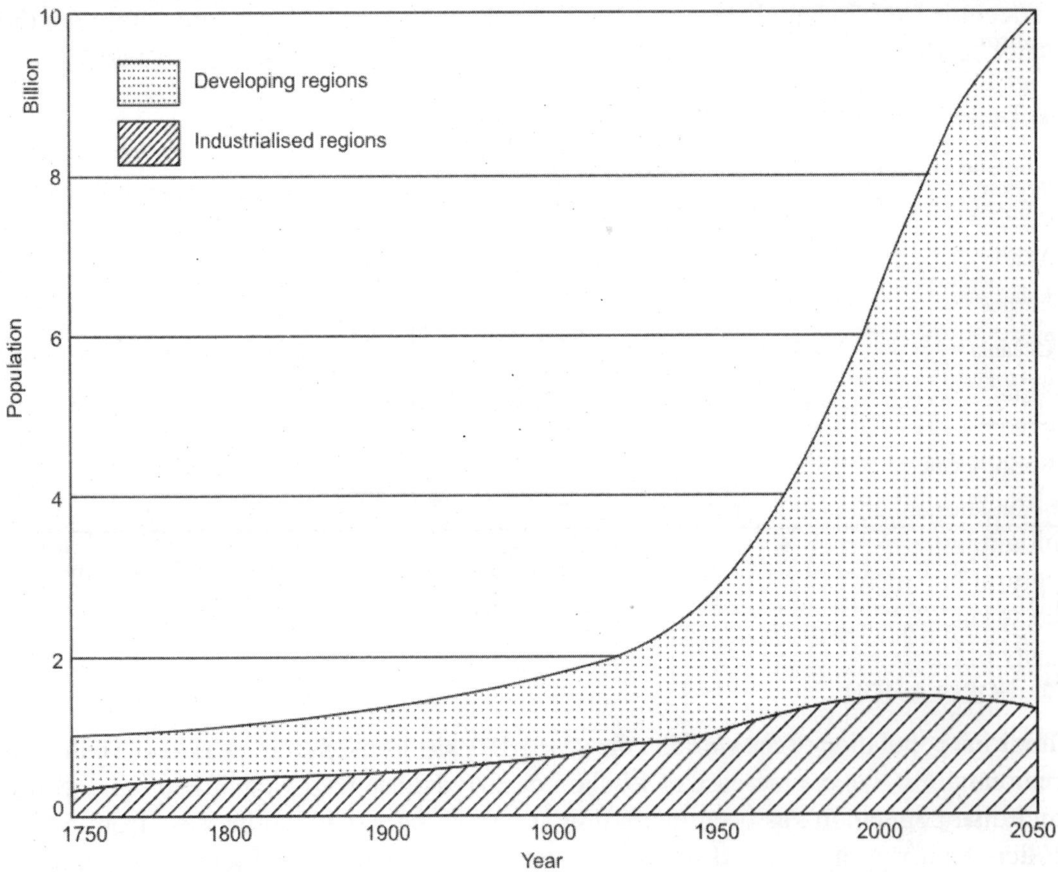

Source: World Resources 1994–95.

Figure 1.1 *Trends and projection in world population growth, 1750-2050*

The per capita agricultural produce in Asia and Africa is well below the overall world average, while it is the highest in North America.

LAND USE

About one-third (33%) of the land area in the world is under forest cover and 11 per cent is under agriculture. But there is great variation from region to region. In India, the land area under forest is 19 per cent and that under agriculture is 57 per cent. During the three decade period, 1951–80, nearly 4.3 mha of forest land in India was diverted to agriculture and other uses.

The forest area as compared to the respective geographical area in respect of world continents is given in Table 1.3.

Table 1.3 *Geographical and forest area (FAO 1982)*

Continent/Country	*Geographical area (m ha)*	*Forest area (m ha)*	*Per cent of forest area to geographical area*
Africa	3033	654	22.8
North and Central America	2242	682	30.4
South America	1782	937	52.5
Asia	2757	552	20.0
Europe	487	155	31.4
Oceania	851	150	17.6
States of the former Soviet Republic	2240	920	41.0

The agricultural productivity of land in hot deserts, cold polar regions and high mountains is the lowest, while it is the highest in tropical rain forests. The conversion of natural forests into cultivation in the tropics has led to reduction in the productivity of these lands. Data on the extent of forest and other types of land and net primary productivity of major vegetation units of the world is given in Table 1.4.

Irrigation and the ample availability of water can help convert deserts or parched lands into productive cultivable lands. This becomes feasible by storing water in dams in regions where it is available in abundance and then by transporting it via long conveyance and distribution canal systems to the dry lands. Irrigation has made vast arid areas in India, Pakistan, Africa, Middle East, China, the USA and Australia agriculturally productive. This is not to say that all desert lands in the world can be successfully irrigated. The availability of water, economic viability, physiographic topography and prioritisation of human needs are among the many aspects that have to be taken into account before implementing large irrigation projects.

Table 1.4 *Net primary productivity of major vegetation units of the world (Leith, 1976)*

Vegetation unit Area (million sq km)	Area (million sq km)	Tropical (T) Temperate (t)	Net primary productivity		
			Range tonnes/ha/yr	Approximate mean tonnes/ha/yr	Total for the area metric tonnes (10^9)
(1)	(2)	(3)	(4)	(5)	(6)
Forest					
Tropical rain forest	17.0	T	10–35	28	47.4
Raingreen forest	7.5	T	16–25	17.5	13.2
Summer green forest	7.0	t	4–25	10	7.0
Mediterranean sclerophyll forest (Chaparral)	1.5	t	2.5–15	8	1.2
Warm temperate mixed forest	5.0	t	6–25	10	5.0
Boreal Forest	12.0	t	3–12	6.5	7.8
Sub-total	50.0			6	81.6
Woodland	7.0		2–10	6	4.2
Dwarf and Open Scrub					
Tundra	8.0	t	0.6–13	1.6	1.3
Scrub	18.0	Tt	0.1–2.5	0.7	1.3
Sub-total	26.0				2.6
Grassland					
Tropical grassland	15.0	T	2–29	8	12.0
Temperate grassland	9.0	T	0.7–13	8	7.2
Sub-total	24.0				19.2
Desert (extreme)					
Dry desert	8.5	T	0–0.1	0.03	—
Ice desert	15.5	t	0–0.01		—
Sub-total	24.0				
Cultivated Land	14.0		1.0–40	6.5	9.1
Water Bodies					
Swamp and marsh	2.0	Tt	8–40	20	4.0
Lake and stream	2.0	Tt	1–15	5	1.0
Sub-total	4.0				5.0
Total for Continents	149				121.7

In some densely populated regions like India there is not much scope for extending the existing land areas under agriculture. The current environmental balance between forest and cultivated land which is in the proportion of about 3 (forest): 1 (cultivated land), as an average for the world, is already 1 (forest): 3 (cultivated land) in India.

The figures of land area and arable area for different regions of the world are shown in Table 1.5.

Table 1.5 *Land use*

All Figures in million ha

Region	*Land area*	*Arable and permanent crop area*	*Percentage of arable and permanent crop area to total land area*
Africa	2964	193	6.5
Canada	922	46	5.0
USA	916	188	20.5
South America	1753	121	6.9
Asia (total)	3085	516	16.7
India	297	170	57.2
China	933	96	10.3
Europe	473	135	28.5
Oceania (total)	849	53	6.2
Australia	768	48	6.3
USSR	2190	229	10.5
World	13048	1476	11.3

Source: FAO Production Year Book, Vol. 50, 1996.

WATER

The total amount of water on the earth is about 1400 m cu km, of which 97.3 per cent is in oceans which cover about three-fourths of the earth's surface. Of the 37.8 m cu km of fresh water, 75.2 per cent lies frozen in the polar regions, 22.6 per cent is present as groundwater and only 2.2 per cent or 0.83 m cu km is available in lakes, rivers, the atmosphere, moisture, soil and vegetation. The source of all water is precipitation which is not uniformly distributed in space or time. Annual precipitation in some parts of the world approaches 10,000 mm while in some desert regions it is less than 100 mm. In some regions precipitation occurs throughout the year, whereas in other regions it is concentrated only in a few months. Figure 1.2 depicts the global water distribution and the water cycle.

Agriculture is among the most dominant users of fresh water. Until recently, it accounted for almost 90 per cent of the total water use. The demands on water for domestic, industrial and other uses are increasing but the percentage of water use for agriculture is not expected to fall below 70 per cent.

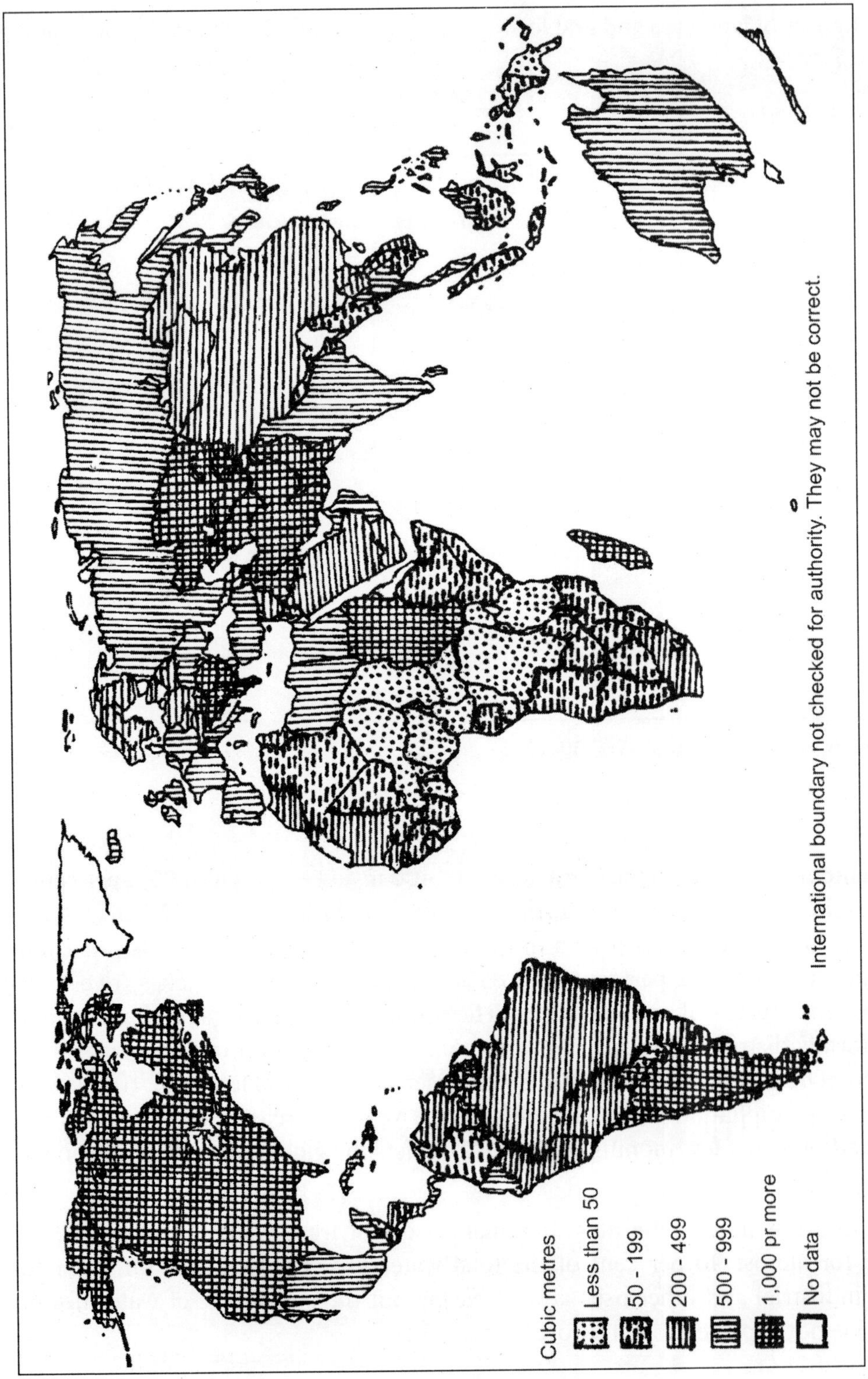

Data Source: The World Bank Atlas, 1996.

Figure 1.3 *Annual per capita water use, 1970-94*

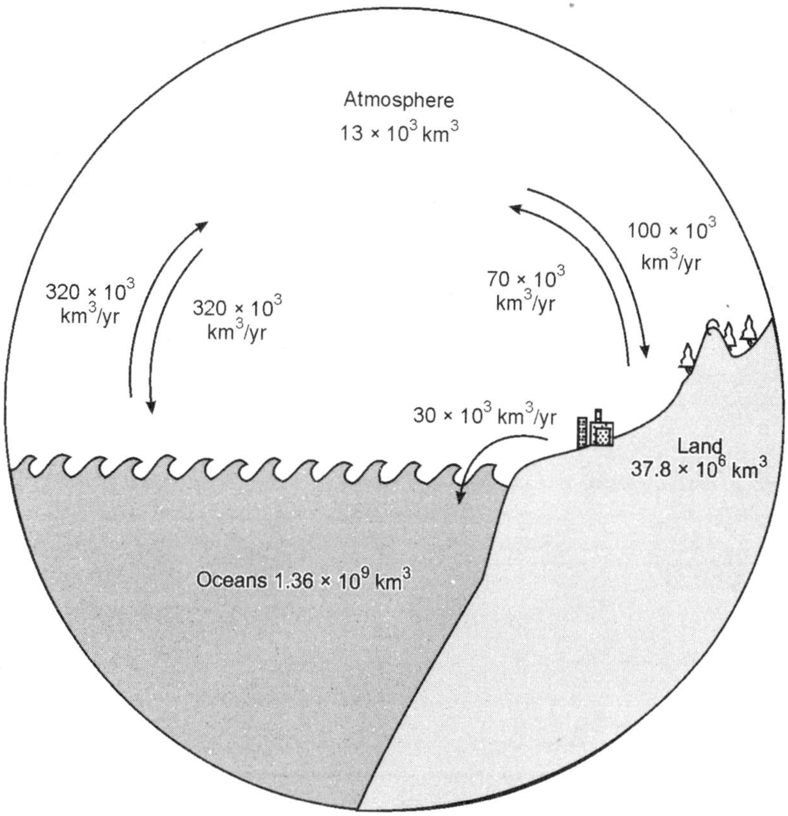

Figure 1.2 *Global water distribution and the watercycle (Todd, 1970)*

There is great variation in per capita annual water use in different regions of the world, ranging from less than 50 cu m to more than 1000 cu m as shown in Figure 1.3.

CLIMATE

Climate has a strong effect on natural vegetation and agriculture. In tropical areas, there is not much change in the climate from season to season, as it is generally hot. Away from the Equator, the climate not only turns cooler but is manifested in two distinct seasons, summer and winter. In these regions, the temperatures and their variation from season to season influence the living and eating habits of the people.

Precipitation is also an important climatic factor. In some desert regions there is little precipitation while the wettest place on earth is known to have as much as 1150 cm of annual precipitation. There are regions where precipitation occurs almost every month of the year, while in other regions, there is heavy seasonal precipitation during some part of the year and the rest of the year is practically dry.

Most of the dense forests occur in hot and wet climates. Vegetation is scarce in desert areas which receive little precipitation. The most favourable climatic conditions for agriculture are abundant sunshine and precipitation throughout the year.

The day temperatures and days of precipitation, monthwise, for some important towns in different regions and continents are shown in Table 1.6. In regions where precipitation occurs for a few days every month, there is relatively much less need of irrigation than in regions where precipitation does not occur for several months together. Plants need more water in hot dry climates as compared to cold wet conditions.

Table 1.6 *Monthly day temperatures and days of precipitation at selected locations*

Location		Jan	Feb	Mar	Apr	May	Jun	Jul	Aug	Sep	Oct	Nov	Dec
		1	2	3	4	5	6	7	8	9	10	11	12
Europe													
Helsinki	t	–3	–4	0	6	13	19	22	20	15	8	4	0
(Finland)	n	12	9	7	9	7	8	9	10	10	11	11	12
London	t	6	7	10	13	17	20	22	21	19	14	10	7
(Great Britain)	n	11	9	8	8	8	8	9	9	9	9	10	9
Moscow	t	–7	–6	0	9	17	22	24	22	16	8	0	–5
(Russia)	n	8	7	8	7	8	9	11	11	10	9	9	8
Africa													
Cairo	t	20	22	25	29	33	35	36	35	33	31	26	21
(Egypt)	n	1	1	1	0	0	0	0	0	0	0	1	1
Harare	t	26	26	26	26	23	21	21	23	26	28	27	26
(Zimbabwe)	n	15	13	8	5	2	1	0	1	1	4	9	14
Chicago	t	0	2	7	14	21	27	30	28	24	18	9	2
(USA)	n	7	7	9	10	8	9	8	6	7	6	7	7
Las Vegas	t	13	17	21	26	31	37	41	40	36	28	19	14
(USA)	n	2	2	2	1	1	1	2	2	1	1	2	1
Mexico City	t	19	21	24	25	26	24	23	23	23	21	20	19
(Mexico)	n	2	1	2	6	9	14	19	18	17	8	3	2
Montreal	t	-5	-5	2	10	18	23	26	25	19	13	5	-3
(Canada)	n	10	11	7	8	10	9	10	10	7	10	12	12
Rio de Janeiro	t	30	30	29	27	26	25	25	25	25	26	26	28
(Brazil)	n	13	11	9	9	6	5	5	4	5	11	10	12
Beijing	t	1	4	11	21	27	31	31	30	26	20	10	3
(China)	n	2	2	2	3	5	6	12	9	5	2	2	1

Cont'd...

...Cont'd

Location		Jan	Feb	Mar	Apr	May	Jun	Jul	Aug	Sep	Oct	Nov	Dec
		1	2	3	4	5	6	7	8	9	10	11	12
Colombo	t	30	31	31	31	30	30	29	29	30	29	29	30
(Sri Lanka)	n	7	6	8	14	19	18	12	11	13	19	16	10
Dhaka	t	26	28	32	34	33	32	31	31	32	31	29	26
(Bangladesh)	n	1	2	4	8	13	18	22	21	16	7	1	1
New Delhi	t	21	24	29	36	41	39	35	34	34	34	28	23
(India)	n	1	2	2	1	2	6	11	11	6	0	0	1
Jakarta	t	29	29	30	31	31	31	31	31	31	31	31	30
(Indonesia)	n	18	17	15	11	9	7	5	4	5	8	12	14
Karachi	t	24	25	28	30	31	32	31	30	30	31	30	26
(Pakistan)	n	1	2	1	1	0	1	4	2	1	0	0	1
Rangoon	t	32	34	35	36	33	30	30	30	30	31	31	31
(Burma)	n	0	0	0	2	15	27	28	27	22	11	3	1
Tel Aviv	t	18	19	21	23	26	28	30	31	31	28	24	20
(Israel)	n	10	8	8	2	0	0	0	0	0	4	6	10
Tokyo	t	9	9	12	18	22	25	29	30	27	20	16	11
(Japan)	n	6	7	10	11	12	12	11	10	13	12	8	5
Oceania													
Melbourne	t	26	26	24	20	17	14	13	15	17	19	22	24
(Australia)	n	4	3	6	6	10	10	10	10	10	9	8	9
Perth	t	29	29	27	24	21	18	17	18	19	21	24	27
(Australia)	n	2	2	3	7	10	13	13	13	7	6	5	2

t: Day temperature; n: Days of precipitation
Source: Lufthansa Climate Information, 1984.

It is obvious that in regions where temperature and precipitation are favourable, human intervention in the prevailing natural conditions is hardly required for the promotion and practice of agriculture. On the other hand, where the precipitation is inadequate or unevenly spread, there is a greater need for constructing dams and reservoirs to store water during periods when precipitation is plentiful and to carry and use it during periods of water deficiency to meet the demands of agriculture.

IRRIGATION

All forest lands depend almost entirely on natural precipitation for water supply. Agriculturally cropped area may or may not need irrigation. There are large areas where irrigation is needed but since adequate water is not available, cultivation is not possible or the yields from crops are poor. In order to meet the demand for water needed for

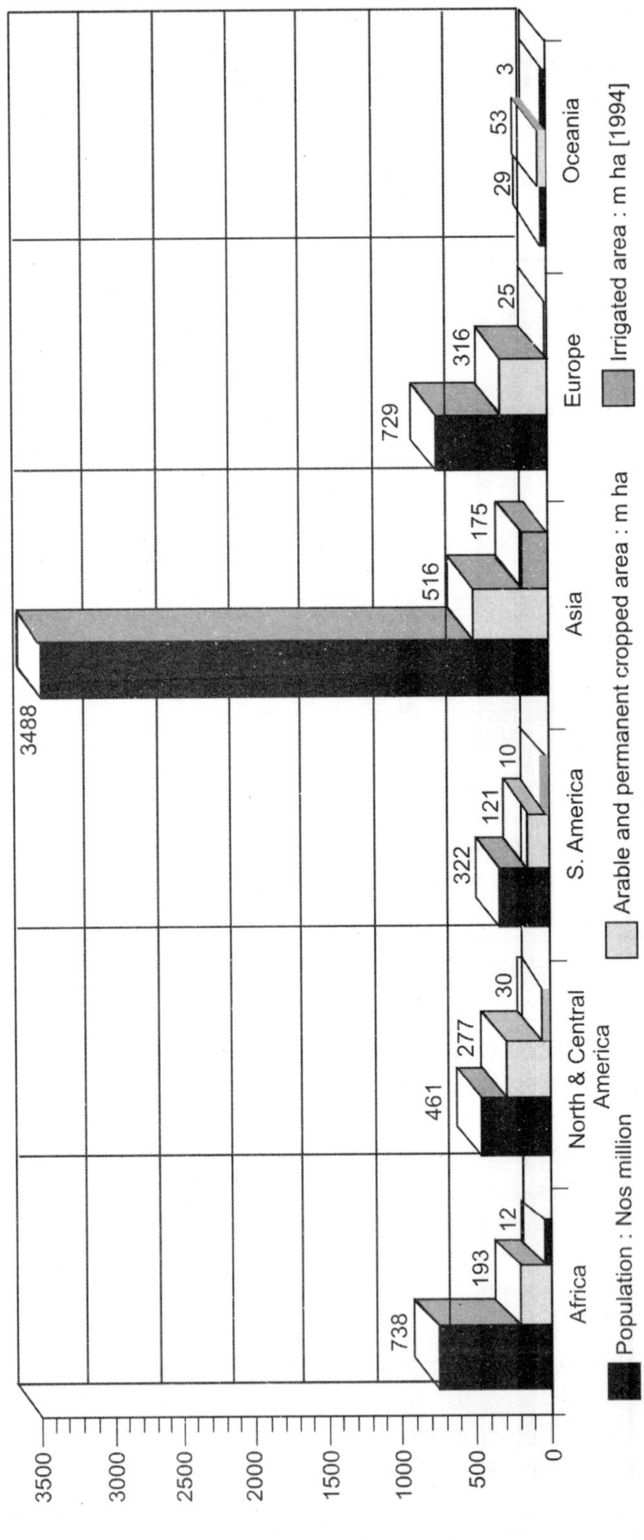

Source: FAO Production Year Book, 1996.

Figure 1.4 *Population and irrigated area*

irrigation, water resources are harnessed by constructing dams on rivers to store water, which is then conveyed to agricultural fields through elaborate canal systems. The percentage of cultivated land that is irrigated varies from 2 per cent (North-West Europe) to 50 per cent (North-East Asia), with the overall average being about 15 per cent. The position of irrigated area as a percentage of the cropped area, in different countries and regions is shown in Table 1.7.

Table 1.7 *Irrigated area and cropped area*

Country	*Cropped area (m ha)*	*Irrigated area (m ha)*	*Percentage of irrigated to cropped area*
China	90.97	44.08	40.20
India	168.99	42.10	24.90
Commonwealth of Independent States	232.57	20.48	8.80
United States of America	189.91	18.10	9.50
Pakistan	20.76	16.08	77.50
Iran	14.83	5.74	38.70
Thailand	20.05	4.00	19.90
Egypt	2.69	2.69	100.00
Australia	47.11	1.83	3.90
Argentina	35.75	1.72	4.80
South Africa	13.17	1.13	8.60
Sub total	842.80	158.70	18.80
World	1473.70	227.11	15.40

Source: Ghassemi et al., 1995: Global salinisation of land and water resources; Human causes, extent and management, Centre of Resource & Environment Studies, Australian National University.

FAO (1996) gives the position of population, cropped area and irrigated area as shown in Figure 1.4.

FAO (1999) estimates that irrigated agriculture provides 40 per cent of world food production on only 17 per cent of total cultivated land. FAO (2003) also estimates that 60 per cent of the extra food required must, in future, come from irrigated agriculture.

DRAINAGE

The existence of standing water on the ground surface or presence of excess water in the soil in the root zone, renders lands agriculturally unproductive. Water present on the surface of lands can be drained into the nearby natural streams, lakes or the sea. The subsoil water in the root zone is generally removed by pumping through wells or by a subsurface drainage system in which perforated pipes are laid underground to collect and carry away the infiltrating water. The cultivated land, which is provided with drainage facilities, varies from 2 per cent (North Africa–Magreb region) to 34 per cent

(North-West Europe). Currently, 12 per cent of the worldwide cultivated area is effectively drained, and this figure is likely to go up to 20–30 per cent.

The position of drained and irrigated land as a percentage of the total cultivated land is shown in Table 1.8.

Table 1.8 *Drained and irrigated land as a percentage of the total cultivated land*

Rank	*Region*	*% Drained*	*% Irrigated*
1	North-West Europe	34	2
2	South Europe	32	18
3	North America	24	10
4	North-East Asia	20	50
5	East Europe	20	10
6	Australia and New Zealand	10	4
7	Middle East	9	33
8	South-West Asia	9	37
9	South America	7	7
10	Central America	6	19
11	Central Asia	6	21
12	Russian Federation	5	6
13	South-East Asia	4	21
14	South & Central Africa	3	7
15	East Africa	3	7
16	North Africa (Magreb Region)	2	10
17	West & Central Africa	—	2
18	Sahel Africa	—	3

Source: Smedema et al. (1995), The State of Land Drainage in the World.

Large investments have been made the world over in expanding the areas under irrigation, which has made a significant contribution to world food production. In dry arid regions, rain-fed agriculture produces very few crops but irrigation facilitates a manyfold increase in the agricultural productivity of lands. But irrigation in the arid and dry regions very often leads to waterlogging and salinisation. There is also an apprehension that salinisation of land is inevitable as a result of which, irrigation schemes can have only a finite life and cannot be sustained indefinitely.

The Sumerian empire flourished about 4000 years ago in Mesopotamia, in the plains of the rivers Tigris and Euphrates, on the basis of a highly developed irrigation system. Later, large scale salinisation rendered the farmlands unproductive and this contributed to the collapse of the empire. In California's Imperial Valley, drainage water from irrigated lands is discharged into the Salton Sea, whose salinity is on the rise.

Similarly, the discharge of drainage water from irrigated lands in San Joaquin Valley, California, into the Kesterson reservoir has resulted in problems of toxicity and the discovery of selenium in the biota.

On the other hand, the ancient civilisation of Egypt which depended on irrigation from the river Nile, has survived for thousands of years. While the Aswan dam has brought about remarkable changes, extensive drainage systems have been constructed to overcome the problems of waterlogging and salinisation.

In the Indus basin, in India and Pakistan, extensive water storage and distribution systems have been constructed since the nineteenth century. These have made a tremendous contribution to agricultural productivity, but have also led to problems due to inadequate drainage, in some parts.

WATERLOGGING AND SALINITY

In humid and sub-humid regions, irrigation may not be needed at all or may be needed only as a stand-by or on a supplemental basis as in Europe, Canada, and large parts of the USA. In such regions precipitation is sufficient to leach out the accumulated salts. In spite of the high degree of precipitation, even exceeding 1000 mm annually, there is generally no threat of waterlogging. This is mainly due to the presence of natural vegetation on the land which absorbs most of the water precipitation and removes it from the upper layers of the soil strata through transpiration, allowing only a small fraction to infiltrate down to the groundwater reservoir. The recurring cycle of precipitation, surface run-off, transpiration through plants, infiltration, recharge to groundwater reservoir and subsurface outflow, creates such a water balance that the groundwater table stabilises well below the root zone of the plants and fluctuates within a narrow band due to seasonal variations. This condition of natural balance prevails till there is intervention by human activities. Large scale deforestation can be one such intervention, which, by reducing transpiration, can disturb the groundwater balance, resulting in the rise of groundwater table and consequent waterlogging. Kijne et al. (1998) report that land clearing and replacement of native trees with shallow rooted crops, contributed to the development of so-called dry land salinity in relatively humid countries such as the USA, Canada and Thailand.

In semi-arid or arid regions, irrigation is necessary for sustaining agriculture. Very often the source of water is surface water that is brought from outside the region. Losses caused in the conveyance and distribution systems as well as in the fields during the application of irrigation water, percolate down to the groundwater reservoir. Consequently, the groundwater table starts rising, with the rate of its rise depending upon the quantity of the water percolating down and the groundwater outflow from the region. As long as the inflow to the groundwater reservoir exceeds the outflow, the groundwater table continues to rise and finally encroaches into the root zone of the crops causing waterlogging. If unchecked, the groundwater appears on the surface of the ground in the form of large pools. Waterlogging destroys the potential of the land for agricultural use.

The groundwater table in the arid and semi-arid regions is often several metres below the land surface prior to the introduction of irrigation. In most irrigation projects, the groundwater table rises following the introduction of irrigation in the area. Development activities like building the canal network, and roads and railway lines in the irrigation project area also obstruct the natural drainage of rainwater thus causing the accumulation of rainwater for longer periods and consequently a greater recharge to the groundwater. The rise of the water table has been so dramatic in some of the irrigated areas in the semi-arid and arid regions that lands that used to be under dry farming before the introduction of irrigation, have become unfit for cultivation of any upland crop during the rainy season. Monitoring reports from several parts of India indicate that the rise of the water table in irrigated areas varies from 0.3 to 1 m per year. Experiences in different irrigation projects have shown that once irrigation is introduced in the arid and semi-arid areas, be it in the alluvial soil region, or the black soil region, the rise of the water table is inevitable. It may only be delayed by following proper water management practices. After the water table rises to shallow depths, the root zone may get waterlogged. Waterlogging refers to the condition when water is at or near the ground surface at least for some time in a year. According to the National Commission on Agriculture in India, an area is termed as waterlogged when the water table attains a level at which the soil pores in the root zone become saturated resulting in the restriction of the normal circulation of air, a decline in the level of oxygen and an increase in the level of carbon dioxide.

Effects of Waterlogging

Rao (1998) explains that a shallow water table may not be a curse, particularly when it is free from salinity hazards as is the case in most humid regions. This contributes substantially to crop evapotranspiration, depending on the depth and the capillary rise which is a function of soil texture and porosity. However, not all crop varieties can tolerate a very shallow depth of the water table owing to oxygen stress. Waterlogging also creates conditions for surface water stagnation and excess moisture after an irrigation or a rainfall. The prevailing wet conditions affect land preparation and sowing, as also the inter-cultivation practices. The ill-effects of waterlogging are manyfold in areas which have saline soils and/or saline groundwater. Waterlogging in general causes the following conditions that are detrimental to crop growth:

- Poor aeration in the crop root zone;
- Reduction in the activity of soil organisms;
- Obstruction to seed germination and seedling growth;
- Excessive weed growth;
- Hindrance to farm operations;

- Upsetting of the salt and water balance leading to the development of salinity; and
- Frost action in soils.

Many studies conducted by different agricultural universities and research institutes in India have shown that crop yields are affected by the depth of the water table. The international literature survey (Osterban, 1994) shows that crop yields vary in a linear form up to a threshold seasonal average water table depth and are independent beyond that depth. The water table affects the physical, chemical, biological and hydrological conditions of the soil, which, in turn, have an influence on crop growth and management factors. The relations between the water table depth and some important soil conditions are discussed below:

Soil Structure

The soil aeration and storage of soil water are favorably influenced by a good soil structure. Shallow water tables and wet conditions result in damage to the structure of the soil and reduced pore volumes and air content.

Soil Temperature

The reduced water content and increased air content brought about by a deeper water table help maintain a lower specific heat of soil. Waterlogged soil with about 50 per cent moisture content requires three times more heat to warm up than dry soil. The temperature of soil influences seed germination and crop growth.

Soil Workability

Working the soil at higher water contents gives rise to mechanical difficulties and destroys the soil structure, especially, in clay soils. The deteriorated soils can become very hard when dry. The soils may get compacted resulting in crust formation, which affects the hydrological properties of the soil such as infiltration, hydraulic conductivity and macro-pore space. The cloddy tilth may not be suitable for seed germination.

Nutrient Supply from the Soil

Nitrogen fixation and nitrification by micro-organisms and other activities that take place in the soil due to the presence of bacteria, fungi, and other micro- and macro-organisms are influenced by the aeration and drainage status of the soil.

Soil Salinity

The rise of the water table in semi-arid and arid areas may mobilise the salts that are present in the soil profile and groundwater. Initially the salts present in the soil profile are leached down and the soil in the root zone becomes salt-free thus enhancing the agricultural production. The rise of the water table may disturb the hydraulic gradients in the area and may induce movement of saline groundwater into the non-saline areas and into streams creating stream salinity problems. Once the groundwater table rises within 2–3 m of the soil surface, it contributes substantially to evaporation from the soil surface and the water uptake by plants. The upward flux of water due to evaporation and water uptake by plants, results in a gradual concentration of salts in the shallow depths of the soil profile. In the initial years, the crop yield may be reduced due to salinity, but as the severity of the problem increase, the lands may have to be abandoned completely as the excess salts render the soils completely unproductive.

Both water and air, which compete for the same positions in the soil in the root zone, are needed for plant growth. The soil moisture deficiency is abated through irrigation and the oxygen deficiency is done away with by providing drainage facilities in the agricultural field.

It is well known that, when an area is irrigated excessively over an extended period of time, the groundwater level rises. When the water table reaches a particular height, which is within the capillary lift of the soil, the soil moisture is brought to the surface where it evaporates. Salts, which were originally present in the irrigation water or which were dissolved in the rising groundwater, get concentrated on the land surface by the so-called 'tea kettle effect'. This causes soil salinity and some times alkalinity which are harmful to plant growth. When a saline water table rises and remains in the root zone for longer than about 48 hours, resulting in an abnormally high saline moisture condition, agricultural production is usually seriously affected.

The Drainage Manual NIH (1995) explains that plants require oxygen as well as water for their growth. They obtain their oxygen requirement from two sources: the soil air and the open atmosphere. Oxygen supply through the leaves and from there through the plant to the terminal oxidases in the roots, is sufficient to maintain the growth in plants adapted in aquatic condition (i.e., rice), and to support at least the upper 2 cm of roots in many cereal seedlings (Greenwood and Goodman 1971; vide Briggs and Courtney, 1985). Nevertheless it is rarely adequate to satisfy the requirements in more active and mature arable crops. Consequently, conditions affecting the supply of oxygen from the soil air are critical.

The movement of oxygen through the pore system of the size of the plant roots is only a indirect function of the size of the pores. In air-filled pores, oxygen diffusion is rapid and oxygen deficiencies are rare. In saturated pores, the effective coefficient of oxygen diffusion is much lower, possibly only 1/1000th or less of the rate in free air (Briggs and Courtney, 1985). During waterlogged conditions, oxygen diffusion is unable to sustain root or microbial requirements for any length of time. In the absence of sufficient oxygen, substances such as alcohol and cyanide may be formed in the plant tissues and plant growth may be severely curtailed. It is, however, not only a reduced supply of oxygen which inhibits plant growth in waterlogged soils. Under certain circumstances, toxic compounds may also build up as a result of oxygen deficiencies.

Saline soils occur for the most part in arid or semi-arid regions. Salts in soils formed by the weathering of minerals are not dissolved and leached down due to scanty rainfall. High evaporation rates cause the concentration of salts in surface soils and water. The drainage of salt-bearing waters from highlands to lower impervious basins that have no outlet to permanent streams, can result in the formation of saline soil or permanent salty lakes. Sea water intrusion into land and transport and the deposit of salt by winds are among the other cause of soil salinity. Ghassemi et al. (1995) report the global extent of salinisation as shown in Table 1.9.

Table 1.9 *Global extent of salinisation* (figures in million ha)

Country	*Light*	*Moderate*	*Strong*	*Extreme*	*Total*
Africa	4.7	7.7	2.4	—	14.8
Asia	26.8	8.5	17.0	0.4	52.7
South America	1.8	0.3	—	—	2.1
North & Central America	0.3	1.5	0.5	—	2.3
Europe	1.0	2.3	0.5	—	3.8
Australasia	—	0.5	—	0.4	0.9
Total	34.6	20.8	20.4	0.8	76.6

Source: Ghassemi et al. 1995 quoting Oldeman et al. in a 1991 publication of International Soil Reference and Information Centre.

A distinction needs to be made between soils rendered saline due to natural causes and due to human interventions. Irrigation, falling under the latter category, is a major cause of induced salinity, often referred to as secondary salinisation. Ghassemi et al. (1995) give a global estimate of secondary salinisation in irrigated lands as shown in Table 1.10.

Kijne et al. (1998) report that one-third of the irrigated land in the major irrigation countries is already badly affected by salinity or is expected to be in the near future. Present estimates for India range from 27 to 60 per cent of the irrigated land, 14 per cent for Pakistan, 13 per cent for Israel, 20 per cent for Australia, 15 per cent for China, 50 per cent for Iraq and 30 per cent for Egypt. There is a significant variation in the reported figures perhaps because there is no standard definition of the term "salinisation". But it is obvious that the problem of salinity is widespread in irrigated lands all over the world.

Non-saline soils can be rendered saline as a result of irrigation. When new lands are brought under irrigation, water is frequently plentiful during the early stages of development and is used excessively. The percolating water hastens the rise of the groundwater table. The irrigation water contains some soluble salts and the groundwater contains much more. When the water table rises to within 2 m of the ground surface, groundwater moves upward into the root zone and to the ground surface causing salinisation of the soil.

Table 1.10 *Global estimate of secondary salinisation in irrigated lands*

Country	*Irrigated area m ha*	*Salt affected land in irrigated area m ha*	*Percentage of salt affected land to irrigated land*
China	44.08	4.40	10.0
India	42.10	7.00	16.6
Commonwealth of Independent States	20.48	3.70	18.1
United States	18.10	4.16	23.1
Pakistan	16.08	4.22	26.2
Iran	5.74	1.72	30.0
Thailand	4.00	0.40	10.0
Egypt	2.69	0.88	33.0
Australia	1.83	0.16	8.7
Argentina	1.72	0.58	33.7
South Africa	1.13	0.10	8.9
Sub total	158.70	29.62	20.0
World	227.11	45.40	20.00

Source: Ghassemi et al. 1995: Global salinisation of land and water resources: Human Causes, Extent and Management, Centre of Resource and Environment Studies, Australian National University.

The economic impact of salinity has not been calculated with any degree of precision. Joshi and Jha (1991) as cited in Umali (1993), through a study of the farm level effects of irrigation-induced salinity in one irrigated area of India, found that the yields of rice and wheat on the degraded land were roughly half of those on the unaffected land. Correspondingly, the net incomes of the farmers in the salt-affected soils were only 10 to 20 per cent of those on the unaffected land.

Ghasemi et al. (1995) quote some estimates of damage to the economy of a few countries with irrigation induced salinity. For Pakistan, they give the cost of the damage as US$300 million per year for Punjab and the North West Frontier Provinces alone, based on estimates of Pakistan's Water and Power Development Authority. In Australia, it has been estimated that annual agricultural losses from salinisation in the Murray-Darling Basin amount to more than US$ 200 million and for the Colorado basins in California, USA, the estimate is as high as US$ 750 million per year.

Anti-Waterlogging Measures

The anti-waterlogging measures that are currently being implemented have met with limited success. Most often they are adopted to rehabilitable the damaged lands rather than to prevent the damage. This makes them both difficult and expensive to implement. They do not provide the desired environmental protection and many times, cause degradation of natural land and water resources. Such measures, therefore, ignore social responsibility towards the coming generations.

POLICY OPTIONS

The United Nations Conference on Environment and Development, Rio de Janeiro, Brazil(1992) in its Agenda 21, considered the "Protection of the Quality and Supply of Fresh Water Resources: Application of Integrated Approaches to the Development, Management and Use of Water Resources". Some of the important observations and recommendations relating to water use, management and development of water resources, that were made during the conference are as follows:

1. Water is needed in all aspects of life. The general objective is to ensure that adequate supplies of good quality water are maintained for the entire population of this planet, while preserving the hydrological, biological and chemical functions of ecosystems, adapting human activities within the capacity limits of nature and combating vectors of water-related diseases. Innovative technologies, including the improvement of indigenous technologies, are needed to fully utilise limited water resources and to safeguard these resources against pollution.
2. The widespread scarcity, gradual destruction and aggravated pollution of fresh water resources in many world regions, alongwith the progressive encroachment of incompatible activities, demand integrated water resources planning and management. Such integration must cover all types of interrelated fresh water bodies, including both surface water and groundwater, and must duly consider both the water quantity and quality aspects.
3. The productivity response and sustainability of irrigation systems have been constrained by the problems of waterlogging and salinisation. Financial and market constraints are also a common problem. Soil erosion, mismanagement and over-exploitation of natural resource and acute competition for water have all influenced the extent of poverty, hunger and famine in the developing countries. Soil erosion caused by over-grazing of livestock is also often responsible for the siltation of lakes. Most often, the development of irrigation schemes is not supported either by environmental impact assessments identifying hydrological consequences within the watersheds of inter-basin transfers, or by the assessment of the social impact on people living in river valleys.

4. Water resources management must be developed within a comprehensive set of policies for: (i) human health; (ii) food production, preservation and distribution; (iii) disaster mitigation plans; and (iv) environmental protection and conservation of the natural resource base.
5. An International Action Programme on Water and Sustainable Agricultural Development (IAP-WASAD) has been initiated by the FAO in cooperation with other international organisations. The main objective of the Action Programme is to assist the developing countries in planning, developing and managing water resources on an integrated basis to meet the present and future needs for agricultural production taking into account environmental considerations.
6. The Action Programme has developed a framework for sustainable water use in the agricultural sector and identified priority areas for action at the national, regional and global levels. Quantitative targets for new irrigation development, improvement of the existing irrigation schemes and reclamation of waterlogged and salinised lands through drainage for 130 developing countries, are estimated on the basis of food requirements, agroclimatic zones and availability of water and land.
7. FAO global projections for irrigation, drainage and small-scale water programmes by the end of the year 2000 for 130 developing countries are as follows: (a) 15.2 million hectares of new irrigation development; (b) 12 million hectares of improvement/modernisation of existing schemes; (c) 7 million hectares installed with drainage and water control facilities; and (d) 10 million hectares of small-scale water programmes and conservation.
8. The development of new irrigation areas at the above-mentioned level may give rise to environmental concerns in so far as it implies the destruction of wetlands, water pollution, increased sedimentation and a reduction in biodiversity. Therefore, new irrigation schemes should be accompanied by an environmental impact assessment, depending upon the scale of the scheme, in case significant negative environmental impacts are expected. When considering proposals for new irrigation schemes, consideration should also be given to a more rational exploitation, and an increase in the efficiency or productivity, of any existing schemes capable of serving the same localities. Technologies for new irrigation schemes should be thoroughly evaluated, including their potential conflicts with other land uses. The active involvement of water-users groups is a supporting objective. The following activities are recommended:
    - Introduce surface drainage in rain-fed agriculture to prevent temporary waterlogging and flooding of lowlands;
    - Introduce artificial drainage in irrigated and rain-fed agriculture;
    - Encourage conjunctive use of surface and groundwater, including monitoring and water-balance studies;
    - Practise drainage in irrigated areas of arid and semi-arid regions;

- Prevent adverse effects of agricultural activities on water quality for other social and economic activities and on wetlands, *inter alia*, through the optimal use of on-farm input and the minimisation of the use of external input in agricultural activities;
- Establish biological, physical and chemical water quality criteria for agricultural water-users and for marine and riverine ecosystems;
- Promote a farming approach for land and water management that takes into account the level of education, the capacity to mobilise local communities and the ecosystem requirements of arid and semi-arid regions;
- Plan and develop multi-purpose hydroelectric power schemes, making sure that environmental concerns are duly taken into account; and
- Develop long-term strategies and practical implementation programmes for agricultural water use under scarcity conditions with the competing demands for water.

Kijne et al. (1998) describe policy options to face the problem of land degradation due to irrigation induced salinity as follows:

1. The development of the world's land and water resources for irrigation has necessitated huge investments in the past and continues to require large sums for annual maintenance and rehabilitation. Undoubtedly, reclamation of salt-affected land through the installation of drainage facilities, and the prevention or mitigation of salinity damage through improved management at farm and system levels, are expensive. The costs of salinity and possible reclamation and preventive measures need to be compared with the expected production benefits resulting from the reclamation of affected lands and the prevention of land degradation, to ascertain the level of worldwide investments in land reclamation and prevention of salinity that are economically justified. The challenge for policy makers is to weigh the costs of addressing short-term needs, which indeed are often very pressing in developing countries, against the long-term benefits that are likely to accrue from investments in reversing the present trend in land degradation due to irrigation-induced salinity.
2. The irrigation community can no longer depend solely on the enhancement of the scientific insight that is generated in the developed countries, into the physical and chemical processes involved in the salinisation of soil and water resources. Sophisticated models that were developed have only a limited application to salinity and sodicity problems in the developing countries.
3. Research programmes for developing countries should include applied, site-specific studies to identify actions for the prevention, mitigation and reversal of resource degradation in irrigated agriculture.

The choice between different drainage methods should depend upon the feasibility, sustainability and economic viability of the proposed drainage measures. If alternative methods are possible with the above considerations, then the least cost method would be the most appropriate. Sometimes a combination of two or more methods may give the best results.

Environmental considerations should also receive due attention. In regions, denuded of forests, tree plantation is very desirable and biodrainage, if and to the extent feasible, should be preferred. The pollution of surface water by drainage water should be avoided.

Social factors, cultural practices and the needs and preferences of the local population should also receive attention while planning drainage measures.

CONCLUSIONS

The following inferences may be drawn:

1. The growing world population is putting pressure on land and water resources, which are not unlimited. The productive use of these resources should not be at the cost of their degradation.
2. Agriculture is the dominant user of good quality water. Irrigation greatly helps in increasing the productivity of lands, but it also causes waterlogging and salinisation. The long-term sustainability of irrigation is therefore, in doubt.
3. The remedial measures, besides providing effective drainage, should also be environment-friendly.

REFERENCES

Briggs, D. and F. Courtney (1985). Agriculture and Environment, 1st Edition, Longman, New York, p. 442.

Food and Agriculture Organisation (FAO). *Production Year Book, Vol. 50–1996, Source: ICID Publication WATSAVE SCENARIO.*

— (1999). IPTRID Issue Paper No. 1, p. 1.

— (2003). IPTRID Issue Paper No. 4, p. 1.

Ghassemi, F., A.J. Jakeman and H.A. Nix (1995). *Salinisation of Land and Water Resources: Human Causes, Extent, Management and Case Studies,* Centre for Resource and Environmental Studies, Australian National University, Australia.

Greenwood and Goodman (1971). 'Studies in Soil Physics 1, Flow of Air and Water Through Soil', *Journal of Agricultural Sciences,* 4: 1–24.

International Commission on Irrigation and Drainage (ICID) (1997). *World Resources (1994–95),* Source: ICID Publication WATSAVE SCENARIO.

Joshi and Jha (1991). Citing Umali, in J.W. Kijne, S.A. Prathapar, M.C.S. Wopereis and K.L. Sahrawat (1998). 'How to Manage Salinity in Irrigated Lands: A Selective Review with Particular Reference to Irrigation in Developing Countries', SWIM Paper 2, International Irrigation Management Institute, Colombo, Sri Lanka.

Kijne J.W., Prathapen S.A., Woperies McS, Saharawat K.L. (1998): SWIM paper 2 (IIMI), How to manage salinity in irrigated lands. A selective review with particular reference to irrigation in developing countries.

Leith (1976). Reproduced from A.P. Dwivedi (1992). *Principles and Practices of Indian Silviculture*, Surya Publications, Dehradun.

Lufthansa (1984). *Climate Information (Airline Pamphlet)*, National Institute of Hydrology (NIH), Roorkee, India (1995), *Drainage Manual.*

National Institute of Hydrology (NIH), Roorkee, India (1995), Drainage Manual.

Oosterban, R.J. (1994). 'Agriculture Drainage Criteria', Chapter 17 in H.P. Ritzema (ed.), *Drainage Principles and Applications*, 2nd Edition, ILRI Publication 16, Wageningen, The Netherlands.

Rao, K.V.G.K. (1998). 'Waterlogging in Irrigated Agriculture and Its Management', Proceedings of the National Workshop on 'Strategy for Prevention and Reclamation of Waterlogged Areas in Irrigation Commands', UN Conference on Environment and Development (1992), Rio de Janerio, Brazil, June 3–14: Agenda 21, Chapter 18 'Protection of the Quality and Supply of Fresh Water Resources: Application of Integrated Approaches to the Development, Management and Use of Water Resources'.

Smedema, Lambret K. and Walter J. Ochs (1995). 'The State of Land Drainage in the World', Presented at National Seminar Jaipur, India.

Todd, D.E. (1970). *The Water Encyclopedia*, Water Information Centre, Port Washington, New York, p. 559.

U.N. Conference on Environment and Development, (1992) Rio de Janerio, Brazil June 3–14, Agenda 21 chapter 18 'Protection of the quality and supply of fresh water resources: Application of irrigated approaches to the development management and use of water resources'.

World Bank Atlas (1996). Source: ICID Publication WATSAVE SCENARIO.

2

Salts in Soils and Water

SALTS IN SOILS

Weathering

The formation of naturally occurring salt-affected soils is a part of the continual geochemical processes that have existed since ancient geological time. The ultimate source of soluble salts in soil comes from the weathering of primary minerals in the rocks forming the earth's crust. Wind and rainfall transport the weathering products, the latter moves soluble salts to the groundwater system, into streams and eventually into oceans. Rainfall in humid climates is usually sufficient to transport soluble salts out of a region as fast as they are formed. In arid regions where rainfall is not sufficient to provide leaching and the transport of salts from the area, salt accumulation may occur in soils in localised pockets. Excess salts also accumulate in areas that receive surface or subsurface drainage from surrounding lands, with subsequent evaporation leaving the transported salts in the soil.

Secondary Deposits

Throughout geological history, large portions of the earth have been covered by saline areas. The marine sediments that were laid down during extended periods of inundation have since been uplifted and have served as a parent material for large areas of soils in the arid regions. These secondary deposits include shales, sandstones, mudstones and conglomerates, and can serve as substantial sources of salt. The original source of these secondary materials was, however, the weathering of continental rock.

SOURCE OF SALTS

The original and, to some extent, the direct source of all salts in the soil is the primary minerals found in soils and in the exposed rocks of the earth's crust. Sodium, calcium and magnesium occur to an extent of 2 to 3 per cent in the earth's crust. During the process of chemical weathering, which involves hydrolysis, hydration, solution, oxidation and carbonation, these constituents are gradually released and made soluble.

Although the weathering of primary minerals is the indirect source of nearly all soluble salts, sufficient salts rarely accumulate from this source alone to form saline soil. Saline soils usually occur in areas that receive salts from other locations, and water is the primary carrier. The most common direct source of salts is the surface or groundwater, which contain dissolved salts, with the concentration depending upon the salt content of the soil and geological materials with which the water has been in contact. Oceans can be the source of salts in low lying soils along the sea coasts and sometimes salts can also be transported by strong winds.

Dhir (1998) describes the origin and accumulation of salts in soils and groundwater. The presence of an excessive amount of dissolved salts in the soils and groundwaters is a characteristic feature in many parts of the world. The salts have sodium, magnesium and calcium as the main cations, and chloride, carbonate, bicarbonate and sulphate as the associated anions. Potassium, nitrate and borate are also present but normally in minor concentrations only. Since the quality of land and potentiality of water for potability and irrigation purposes depend on the concentration and composition of salts, the salinity problem has received considerable attention. The appearance of waterlogging and salinity in soils of irrigation commands has accentuated the gravity and urgency of the problem. However, in all the scientific endeavours, the emphasis has been on the assessment and characterisation of the problem, identification of causative factors and the development of a control or management approach. The issue, like the origin and mechanism of the initial concentration of salts in soil sub-strata or groundwater in the first place, has received scanty attention in these studies. This fascinating subject of the geochemistry of the landscape, regional salt balance and dynamics has, however, drawn some attention in the western world but elsewhere it is in a state of infancy. There are several reasons for this. The salt-enrichment process is spread over thousands to millions of years during which the determining parameters, including rainfall and hydrology, have not remained constant. The salts, being very mobile, often get concentrated far from the source and the latter is often diffused and spread over a large area, e.g., silicate mineral weathering. Rainfall, wind and stream waters are important original sources and also vehicles for the redistribution of the already existing salts. The current period data on their relative contributions are inadequate, but the topic is very important for the long-term management of salinity problem. Even the durability of reclamation projects that are now being undertaken at an immense cost, will depend on a proper understanding of the landscape hydrology and geochemistry, salt-input through mineral weathering, rainfall, irrigation water and various redistribution and sink mechanisms.

Oceans constitute the largest storehouse of salts on the earth's surface. The sea water has 42×10^{15} tonnes of dissolved salts, of which 85.65 per cent is sodium chloride. Although the early seas had a very small volume and the water was charged mainly with carbonates of alkali and alkaline earths, the release of huge amounts of water vapours, chlorine and other volatiles from the degassing of the earth's interior and mantle led to its enrichment with sodium chloride. This happened 100 to 600 million years ago and since then, the sea water is supposed to have maintained a composition similar to that prevailing today. The rivers on the earth annually deliver 3.85×10^{12} kg of soluble salts to the oceans. Of this, the entire sodium chloride content is believed to be cyclic, i.e., it got onto land via sea sprays and atmosphere from the oceans but the excess sodium and rest of the salts owe their origin to the continents. However, while the sea is receiving a continuous load of salts from the land, it is also losing at the same rate through precipitation and other processes, thus giving sea water a more or less constant composition. The potentiality of the ocean and land as a source of continental salts is discussed below.

Silicate Mineral Weathering

The rocks and the constituent silicate minerals form the original source of salts on the continents and the oceans. Although 75 to 85 per cent of the igneous rocks are made up of the oxides of silicon, aluminum and iron, the rest is mainly made up of oxides of calcium, magnesium, sodium and potassium. Each of these alkali and alkaline earth elements is present in the range of 2–3 per cent, though basalts have greater abundance of calcium and magnesium and synites that of sodium (Table 2.1). The elements are concentrated in feldspars and ferromagnessian minerals, e.g., soda feldspar contains 11.8 per cent Na_2O, calcic feldspar 20.2 per cent CaO, and microcline 16.9 per cent K_2O. Pyroxines, amphiboles and micas contain all the three in different proportions. The sulphur content is very low (mostly 0.02–0.03%) and chlorine content even less (mostly around 0.001–0.02%).

Table 2.1 *Elemental composition of some common rocks and mineral soils*

Element	*Ultra-mafic*	*Basalts*	*Granites*		*Syenite*	*Shales*	*Sand-stone*	*Car-bon rocks*	*Soils*
			High Ca	*Low Ca*					
Ca ($ppmx10^4$)	2.5	7.6	53	0.51	1.8	2.5	3.91	30.2	1.0
Mg ($ppmx10^4$)	20.4	4.6	0.94	0.16	0.58	1.5	0.7	4.7	0.8
Na ($ppmx10^3$)	4.2	18	28.4	25.8	40.4	0.6	3.3	0.4	6.0
K ($ppmx10^4$)	0.004	0.83	2.52	4.2	4.8	2.66	1.07	0.27	0.6
Cl (ppm)	85	60	130	200	520	180	10	150	Traces
S ($ppmx10^2$)	3.0	2.53	3.0	3.0	3.0	24	2.4	24	2.0

Source: Truckian and Wedephol, 1961; *Jackson, 1968.

Upon reaction with carbonated water, the minerals undergo hydrolysis, with a consequent release of cations and often also silica. Each molecular unit of soda feldspar produces two molecules of SiO_2 and one of $NaHCO_3$. Calcic feldspar produces one molecule of $Ca(HO_3)_2$. Garrels (1967) studied waters from a variety of igneous rocks and found that cation composition of dissolved salt was close to that predictable from mineral composition. In rhyolite the sodium element came from soda feldspar, magnesium from hornblende, and calcium both from hornblende and albite. The potassium content was low because of low weatherability of orthoclase. In basalts, the biotite, pyroxene and albite were the major sources of cations in solution. The rhyolite had much higher concentration of dissolved silica and sodium. Bicarbonate was by far the dominant anion. The presence of chloride was attributed to atmospheric contamination and fluid inclusions in rocks and sulphate from associated sulphidic minerals. Although some of the released cations, particularly K and Mg, do precipitate in neo-synthesis, there is an overall surplus. The fact that, there is a substantial net loss of alkali and alkaline earth elements and also of chloride during the weathering of rocks, and the process of soil formation is evident from the amounts in which various elements are present in normal soils.

Gunn and Richardson (1979) studied soluble chloride and sodium from weathered and unweathered widespread sedimentary and igneous rocks in east Australia and concluded that their salt content was sufficient to explain the salt content in the soil and the landscape, given sufficient time for rock weathering. Many sedimentary rocks on the continents, from the oceanic shelves and basins or when formed on land, get submerged with oceanic water and are thus charged with sodium chloride. But, in Australia, even the fresh water sedimentary rocks were found to contain considerable quantity of this salt. But geochemists believe that continental sedimentary and igneous rocks are a rather poor source of chloride. Rankama and Sahama (1950) presented data to show that the ocean contains more than 100 times as much chloride that could be obtained from igneous rocks on the crust of the earth.

Atmospheric Accession of Salts

Rainfall and dry particulate dust-fall are the two main sources of atmospheric accession of dissolved salts. Rainwater is never pure. An analysis from northern Europe (62 stations, 30 month period) showed an average composition (ppm): Na 2.05, K 0.35, Ca 1.42, Mg 0.39, Cl 3.47, So_4 2.19, NO_3 0.27 and NH_4 0.41 (Caroll, 1972). Junge and Werly (1958) reported rainwater analysis from several stations in the USA. The samples collected from stations located 100 km or more from the coast had sodium mostly in the range of 0.3 to 0.4 ppm and chloride of 0.28 to 0.35 ppm. Calcium was far greater than sodium, and sulphate far more than chloride in all the samples. However, the sodium chloride ratio was close to that in ocean water. Despite this, these results as also those from Australia (Hungston and Gallitis et al., 1977) and other parts of the world clearly bring out that the composition of salts in inland rainwater is far different from that of the salts in ocean water and the rainwater shows a high relative enrichment in potassium, calcium and sulphate. The source is obviously continental. Therefore, considering the composition, a major part of the dissolved salts in rainwater is of continental origin though a considerable part of sodium chloride could still have come from the oceans. In coastal situations the rainwater contains much more chloride than in inland situations, e.g., in south-western Australia (Table 2.2). Taking the amount of rainfall into account this could indicate an annual accretion of 12 to 200 kg ha^{-1} of chloride alone into the soil. However, the chloride content falls rapidly with distance and reaches a value of 1–2 ppm that is typical of inland rainwater but even this value makes rainwater an important source for inland sodium chloride. Hem (1965) showed that assuming an average value of 1 ppm chloride in rainwater, the weight of chloride accretion on land through rainfall would be of the order of 99×10^9 kg per year. However, by the same yardstick, the additions of Ca, SO_4 and Mg should be much greater, since inland rainwater has far higher concentrations of these. Of course, calcium can precipitate as carbonate, but magnesium and sulphate should still dominate. Gunn and Richardson (1979) also discounted a major role of rainwater. In 0.5–1 m deep black earth soils on basalts, which are in place for at least a million years and are located only 250 km away from the sea, there was no significant accumulation of salts. But in deeply weathered basalts in the same area, appreciable sodium chloride was present. They attributed salts to silicate weathering rather than to a cyclic source, i.e., rainwater.

Table 2.2 *Variation in composition of rainwater with distance from coast*

Distance from coast (km)	*Rainfall (mm annum^{-1})*	*Na (ppm)*	*Na: Cl*	*Na: Ca*
1.6	800	13.8	0.9	6.8
32	828	5.5	1.0	7.8
104	617	2.5	1.0	3.9
192	447	0.9	1.5	1.4
256	335	1.1	1.5	0.9
320	264	2.3	1.5	0.6

Source: Hutton and Leslie (1958).

Aeolian Accession

Breaking ocean waves generate small droplets of ocean water, which soon dry up to form aerosol. Although the diameter of these particles is generally so large that they can be carried only for a short distance inland, the finer fraction can be carried several hundred kilometres inland. Of course, much of it comes down with rainwater, but in arid areas it could fall as dry dust (Holland and Christi, 1909; Hem, 1965). Blackburn (1974) showed that the aeolian transport of salt is significant in those parts of Australia where the salty surface lies exposed during hot windy periods. Hubble and Isbell (1958) attributed the high salt content of grey clays of the sub-humid region to accession from the arid interior by aeolin transport. However, this amounts to only a regional redistribution of salt and not to a net continental accumulation. But sufficient data are not available globally to assess the role of dry particulate salt deposition from oceans in inland situations.

Marine Rocks and Evaporites on Continents

Many parts of the continents have been under the sea during their geological history. The upliftment of these areas and the receding of the sea have left behind sizeable masses of marine rocks and evaporites of land. According to Livingstone (1972), there are 3×10^{15} tonnes sodium in these sediments, of which 0.4×10^{15} tonnes is in the form of rock salt. Evaporites, mostly of marine origin, occur on all continents. The Pearmain evaporites of west Texas and New Mexico extend over tens of thousands square kilometres and have a combined thickness of 1,356 m. The saline formation extends from Michigan to New York, in the USA, over 0.4 million km^2. Likewise, much of western Europe has extensive evaporites. These evaporites are mainly made up of halite and anhydrite, with an occasional presence of potash beds. The leaching of salts from these saline formations forms an immediate source for the adjoining areas. In many arid areas, the salinity of soils and groundwater can be explained by the occurrence of evaporites in the region.

Salty Springs

Although igneous rocks are low in chloride, volcanic gases and waters from the interior of the earth contain large quantities of chloride. The salt springs in the Gila river system bring in 450 tonnes of sodium chloride everyday. Thus, such sources could be important locally.

Salt Movement through Soils

Saline soils do not occur in humid regions because soluble salts are carried downward into the groundwater and are ultimately transported by streams to the oceans. In arid regions, the leaching and transportation of soluble salts is incomplete. The drainage of salt-bearing waters from the higher lands of the basin may raise the groundwater level to the soil surface on the lower lands, causing the formation of permanent salty lakes. The upward movement of saline groundwater or the evaporation of surface water results in the formation of saline soil.

The salinity problem generally arises when previously non-saline soil turns saline as a result of irrigation. During the early development of irrigation projects, water is frequently plentiful and there is a tendency to use it in excess. This hastens the rise of the water table.

Salts in Irrigation Water

A major source of salt input in an irrigated region, is the salt content in irrigation water imported from outside the region. As long as the water table is deep, the salts are washed down by the percolating water to the groundwater. But when the groundwater table rises and comes up near the ground surface, the salts contained in the groundwater as well as in the irrigation water contribute to the salinisation of the soil.

While considering the salt balance, on the introduction of irrigation, the impact of weathering, oceans and winds, except in special situations and conditions, can be ignored. The salt content in groundwater does not affect the salinity levels of soils for agriculture as long as it is deep. The net import of salts with irrigation water can be estimated with a knowledge of the salt content and volume of irrigation water.

Salts with Fertilisers

Some salts are also added to soils in the form of fertilisers while practising irrigated agriculture. These are intended to provide nitrogen, phosphorus and potash, among other substances to plants to promote their growth. The use of fertilisers, in excess of the uptake by plants, can result in the accretion of salts in soils. But this is controllable, and there is no need to add excessive fertilisers that would increase soil salinity.

Removal of Salts

The removal of salts from an irrigated area can be the result of weathering, winds, surface and groundwater outflow. Their impact can be ignored just as their import effect was neglected. Some salts are removed from the soils by agricultural crops, grasses and trees. In the final analysis, whether the concentration of salts in an irrigated area would increase with time or not, would depend upon the balance in the quantity of salts imported with irrigation water and the quantity of salts removed from the soil by the biomass produce. In case the balance is favourable, in the sense that the rate of removal of salts is equal or exceeds the rate of import, irrigation should be indefinitely sustainable. If the balance is unfavourable and the rate of accretion of salt exceeds the rate of removal, irrigated agriculture can still be indefinitely sustainable so long as the rise of the groundwater table can be kept in check, though the quality of groundwater may progressively deteriorate. But under conditions of adverse salt balance and inadequate check on the rise of the groundwater table, irrigated agriculture can have only a limited period of productive life on such soils.

SALT-AFFECTED SOILS

According to Sharma (1998), salt-affected soils differ from arable soils with respect to two important properties, viz., the soluble salts and the soil reaction. A build-up of soluble salts in the soil may influence its behaviour for crop production through changes in the proportions of exchangeable cations, soil reaction, the physical properties and the effects of osmotic and specific ion toxicity. Salt-related properties of soils are subject to rapid change. Therefore, to facilitate a discussion on soil management and the influence of the two common kinds of salts (neutrals and alkali salts) on soil properties and plant growth, salt-affected soils are broadly grouped as either saline or alkali soils.

Saline Soils

These soils, with white salt encrustation on the surface, have predominantly chlorides and sulphates of Na, Ca and Mg. The soils with neutral soluble salts have saturation paste pH < 8.2. The electrical conductivity of saturation extract, ECe of saline soils, is generally more than 4 dSm^{-1} at 25°C. Such soils invariably have a sodium adsorption ratio (SAR) of the soil solution > 15. In saline soils, the excess of neutral salts restricts the normal plant growth.

Alkali Soils

These soils contain excess salts capable of alkaline hydrolysis such as sodium carbonate, sodium bicarbonate and sodium silicate; and sufficient exchangeable sodium to impart poor physical conditions to soils, thus affecting the growth of most plants. These soils have saturated paste pH > 8.2; exchangeable sodium percentage (ESP) > 15 and different levels of ECe. The pH of the calcareous alkali soils is highly related to their ESP.

In saline soils when the chlorides and sulphates of Ca and Mg are the predominant salts, the SAR usually remains less than 15. Such soils are termed saline-sodic. Many saline-sodic soils contain, in addition to soluble carbonates, an excess of neutral salts. Such soils manifest alkaline properties. From the management point of view, the saline-sodic soils that do not contain soluble carbonates are grouped with saline soils and the others with alkali soils.

CHARACTERISTICS OF SALT-AFFECTED SOILS

Alkali Soils

Soluble salts in alkali soils comprise measurable amounts of carbonate and bicarbonate of sodium. The soil exchange complex is distinguished by a conspicuous presence of sodium ions. Hence, the exchangeable sodium percentage (ESP) of these soils often exceeds 15. In barren alkali soils, the exchange complex may be completely saturated with sodium. A large presence of exchangeable Na deflocculates colloidal particles resulting in poor physical conditions of the soil. The infiltration rate is reduced to such an extent that most of the rainfall received is either lost as run-off or the water stagnates on the soil surface for prolonged periods. The presence of sodium carbonate and hydrolysis of exchangeable Na increase the soil pH, which is always more than 8.2. In highly deteriorated soils, the pH may be as high as 10.5. Alkali soils are also referred to as sodic soils by some workers. The factors inhibiting tree growth in alkali soils are:

(i) High pH throughout the profile causes problems in nutrient availability.
(ii) High ESP reduces the availability of Ca, Mg and K and causes soil dispersion. Alkali soils exhibit highly deteriorated soil structure. Poor water transmission characteristics lead to water stagnation and reduced aeration of roots.
(iii) The presence of a hard $CaCO_3$ layer at a shallow depth in the profile acts as a physical barrier for the vertical penetration of roots. The location of this layer in the profile and its thickness vary in different soils. Often compact subsurface horizons also restrict root penetration in alkali soils.

Saline Soils

These soils contain an excess of neutral soluble salts, generally chlorides and sulphates of sodium, calcium and magnesium. Saline soils, rich in divalent cations, have low ESP and pH, and are in good physical condition. Owing to the flocculating effect of neutral salts, saline soils are very permeable and can be reclaimed by leaching with good quality water, provided the groundwater table is deep. The factors inhibiting plant growth in saline soils include:

(i) Salinity-induced high osmotic pressure of soil water;
(ii) Toxic effects of specific ions;
(iii) Nutritional disorders due to competitive uptake of ions;
(iv) High water table/waterlogging associated with such soils; and
(v) Groundwaters met in saline areas, which are often of poor quality while fresh water is scarce.

Irrigation Water as a Source of Salt

Salt-affected soils that occur naturally, have first to be reclaimed to reduce the amount of salt existing in the root zone and/or exchangeable sodium percentage to make them suitable for agriculture.

More important, both historically and economically, are the man-induced salt affected soils, which result from the application of irrigation water. All irrigation waters contain some dissolved salts. The salt that is introduced to the soil by irrigation water is added to any salt that occurs naturally in the profiles. For every 100 mg/l salt contained in irrigation water, 1 tonne of salt is added to the soil per hectare metre (ha m) of applied water.

Normally, in surface irrigation practice, the quantity of water in excess of the evapotranspiration requirement, percolates down into the ground and its quantity is adequate to carry down most salts brought by irrigation water to the groundwater table. The soil in the root zone is rendered relatively free of salts and agriculture is not affected.

The composition of average river waters of the world is shown in Table 2.3. Most of natural water in rivers is of very good quality for use in irrigation. Its mineral content and electrical conductivity values are quite low. But the natural quality of river water can be greatly affected by a reduction in the quantity of normal flow, discharge of pollutants, industrial affluents, irrigation saline drainage water into the river stream and other human activities.

Table 2.3 *Composition of average river waters of the world*[a]

Region	Ec_e^b (m mhos/ cm at 25°C)	Total concentration		B (Mg/l)	Ca	Mg	Na	K	Alkalinity[c]	SO_4	Cl	NO_3	SAR
		mg/l	meq/l				(meq/l)						
North America	220	142	1.89	—	1.05	0.41	0.39	0.04	1.11	0.42	0.23	0.02	0.5
Europe	270	182	2.28	—	1.55	0.46	0.23	0.04	1.56	0.50	0.19	0.06	0.2
Australia	95	59	0.58	—	0.19	0.22	0.13	0.04	0.52	0.50	0.28	trace	0.3
World	190	120	1.42	—	0.75	0.34	0.27	0.06	0.96	0.23	0.22	0.02	0.4

a Adopted from Rhodes and Bernstein, 1971.
b Electrical conductivity
c Alkalinity implies titrable bases made up mostly of HCO_3^-, with small amounts of CO_3^{-2} and OH.
Source: James David W. *Modern Irrigated Soils* (1982).

Total dissolved solids (TDS) in the canal water in IGNP, India whose source of water supply are reservoirs on the rivers, are 125 milligram per litre (mg/l) on average, more than three decades after the reservoirs were commissioned.

Masood Ahmed and Gary P. Kutcher in a case study of Pakistan's Indus basin report that 66.3 billion cubic metres (bcm) water containing 200 mg/l salts is used by canal diversions in the Punjab Province and 55.6 bcm containing 250 mg/l salts in the Sind province. The groundwater salinity concentration in fresh areas ranges from 900 to 1500 ppm and in saline areas from 3000 to 4000 ppm. The data shows that very large volumes of river water used for irrigation contain salts less than 250 mg/l and the groundwater salinity under most irrigated areas extending over 16 million ha area does not exceed 4 dS/m. The river waters of Narmada and Chambal (India), which are used to irrigate very large areas contain no more than 150 mg/l salts.

In the Murray basin, Australia, Ghassemi et al. (1995) report an increasing trend of salinity in the lower reaches of the river during the period 1940–1985. The measured salinity of river water at Morgan was 0.61 dS/m. The contribution to the salinity of the river water by groundwater inflows was to the extent of 51.6 per cent, the rest is contributed by the surface waters from the tributary rivers. The mean pre-European settlement flow of the river was 13.2 billion m^3/yr at a salinity of 0.61 dS/m as a result of upstream diversions for irrigation water supply (Evans, 1989).

The presented data show that salt content in the river water at Morgan during the period 1945–85 increased from about 200 mg/l to about 400 mg/l. About one-half of the salt is presently contributed by the groundwater inflows.

The USSLS (1954) has presented data on chemical composition of twenty-five river waters used for irrigation in the western United States. The range of dissolved solids was found to be less then 150 mg/l in four cases, between 150–500 in eight cases, between 500–1100 in 11 cases, between 1000–2000 in one case and above 2000 in one case. The range is quite wide. It is explained that the development of irrigation and management practices have caused serious problems of water quality. There is a tendency to divert all the available water for irrigation. Over a period of years the downstream diversions have changed from uncontaminated river water to a substantive proportion of drainage return flow of poor quality.

The salt load in the river waters in many regions of the world, is found to increase in the lower reaches. This is due to pollution caused by human activities. The discharge of drainage water from agriculture fields is a major source of contamination. Dayem (2000) reports that in Pakistan about 9 million tonnes of salts are discharged annually with drainage water into the river Indus. The Amu Darya river in Uzbekistan receives an average of 6.5 billion m^3 of drainage water annually causing an increase of 1000 mg/l salt in downstream river water during normal flows and 2000 mg/l salt during the low flow season. Kitamura et al. (2000) reports that in the lower river basin of the Syr Darya, drainage water from the irrigated agriculture of rice-based cropping system has increased the salinity in river water from 400–600 mg/l to 1300–2000 mg/l during the last three decades. Xie et al. (1998) report that 1.1 billion m^3 of drainage water containing 1.2 million tonnes of salt is discharged annually from the Yinebic irrigated area, in Yellow river basin, China.

A distinction must, therefore, be made between irrigation systems using natural river water and those using polluted water. In the former case, irrigation should never result in salinisation of lands. In the latter case, the cause of the salinisation problem can be traced to the discharge of saline drainage water into the rivers by the upstream users. The subsurface drainage measures adopted in the upper reaches do not resolve the salt problem; they merely transfer the salt problem in an aggravated form to the downstream regions.

QUALITY OF IRRIGATION WATER

The quality of the water used for irrigation should receive adequate attention, as it affects the soil properties and crop growth. Regardless of its source, irrigation water always contains some soluble salts. Apart from the total salt concentration of the dissolved salts, some of the individual ions, though in a small amount, equally affect the crop growth. The major dissolved constituents in natural waters are the salts of sodium, calcium and magnesium as their chloride, sulphate, carbonate and bicarbonates. Sometimes, potassium and nitrate ions are also present in appreciable quantities. However, some other elements such as boron, lithium, silicon, iron, manganese, fluorine, aluminum, lead, zinc, vanadium and chromium are also present as ions in minor quantities besides suspended organic matter.

Attempts have been made from time to time to classify irrigation waters for their suitability to irrigate different crops on the basis of their chemical characteristics but no single classification holds true under all conditions. It is because of a number of water-soil-plant and climatic factors, viz., the ionic composition of irrigation water, mean depth and fluctuations in water table, soil texture, structure, pH, nature and amount of clay in soil, initial soil salinity and sodicity, fertility status, presence of hard pan or indurated layer, content of lime and gypsum, if present, weatherability of soil minerals, dissolution or precipitation of lime, crops and their variety, irrigation management, fertiliser usage and other agronomic practices, climatic parameters as evapotranspiration, rainfall and its distribution affect the suitability of irrigation water. However, in view of these factors and their interactions, it is possible to develop a water quality classification for a particular area.

SALINITY

The most important criterion for determining the quality of water for irrigation is the total salt content in the water. The quantity of dissolved salts in water is generally expressed in terms of electric conductivity EC as mmhos/cm or as ds/m. For water solutions of very low electrolyte concentration, the EC is sometimes expressed in unit of μmhos/cm. The total dissolved salts TDS (mg/l) in water are approximately equal to 649 per unit of EC. The waters can be categorised on the basis of their salt concentration into the following classes:

Salt content (mg/l)	***Water classification***
Less than 500 mg/l	Fresh
500–1500	Marginal
1500–5000	Brackish
Greater than 5000	Saline
35000	Brine
Greater than 35000	Bitter

Cations like Na^{1+}, Ca^{2+}, Mg^{2+} and anions like Cl^{1-}, SO_4^{2-}, HCO_3^{1-} and CO_3^{2-} are the major salt constituents contained in saline water. Excessive salts in soil water increase osmotic pressure which reduces the availability of water and adversely affects plant growth. The excessive concentration and absorption of some ions like Na, Cl, B etc. can cause harmful toxicity in plants. Waters with an EC value of less than 4.0 ds/m are suitable for irrigation.

Sodicity

A few specific cations and anions, when present in irrigation water in sufficient concentration, can cause alkalinity and sodium hazards. The Sodium Adsorption Ratio (SAR) is useful in determining the exchangeable sodium to exchangeable calcium plus magnesium ratio in the soil. The relation is

$$SAR = \frac{(Na^+)}{\sqrt{(Ca + Mg) \div 2}}$$

when ionic constituents are in milliequivalents per litre.

The SAR value of highly saline and alkaline irrigation waters is more than 10.0. Waters with an SAR value of less than 10 are suitable for irrigation, the less the better.

Residual Sodium Carbonate

The presence of carbonate and bicarbonate ions in irrigation water results in the precipitation of calcium and magnesium of the soil and increases the sodicity hazard of waters. The RSC is expressed as

$$RSC = (CO_3^{2-} + HCO_3^{1-}) - (Ca^{2+} + Mg^{2+})$$

where all concentrations are expressed in me/l.

The RSC value of good and saline waters is less than 2.5 while that of alkali waters is more than 2.5.

Salinity and Sodium Hazard

The USSLS suggests an assessment of salinity and sodium (alkali) hazards of irrigation water as shown in Figure 2.1.

Classification of Irrigation Waters

The CSSRI suggests a classification of irrigation water as shown in Table 2.4 and salt balance rating of crops as shown in Figure 2.2.

▼ WATER QUALITY GUIDELINES FOR IRRIGATION

The quality of water that can be used for irrigation depends upon the total dissolved salts in the water, soil, climate, crops, irrigation and management practices. In India, guidelines for using saline waters are recommended as shown in Table 2.5.

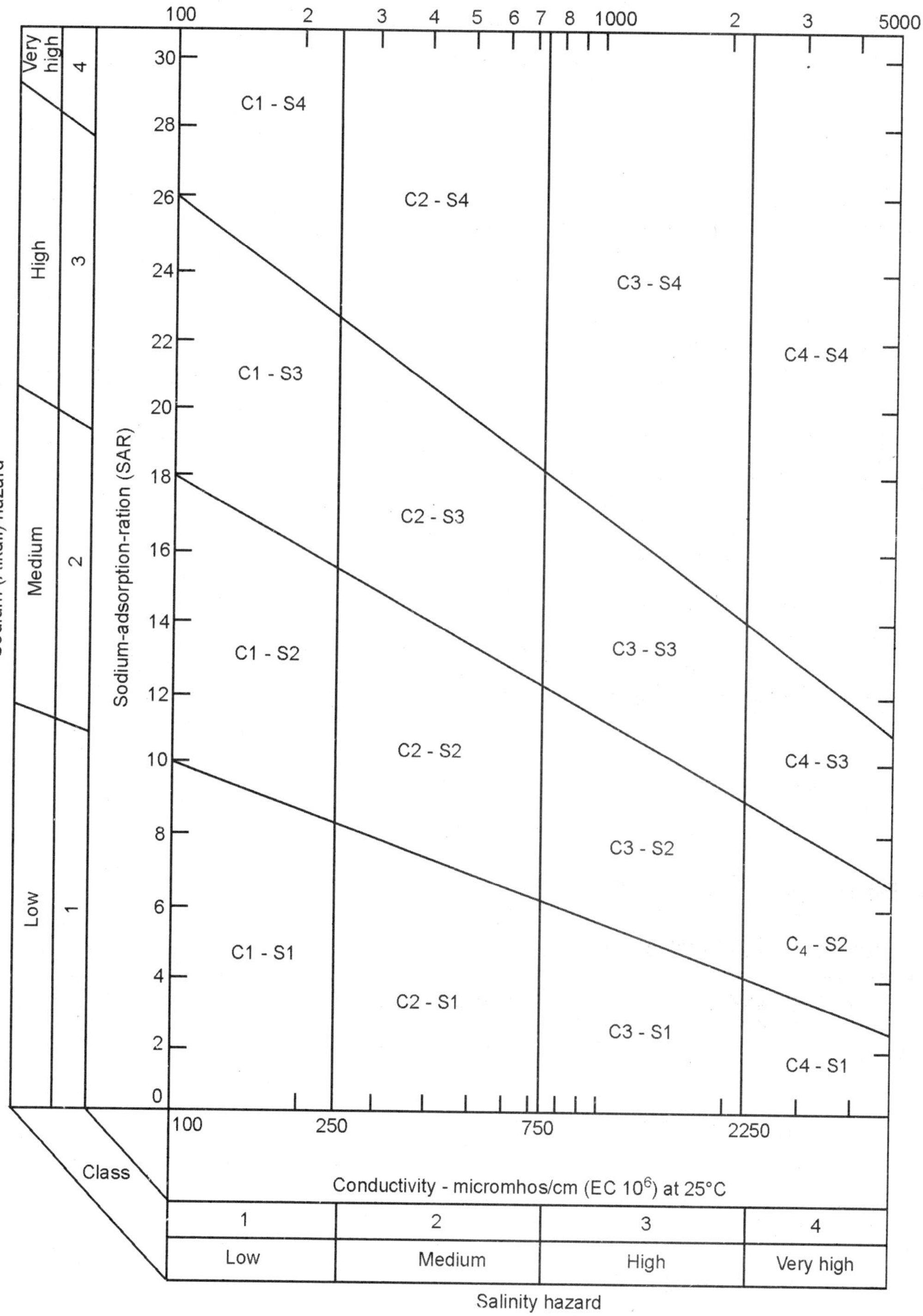

Figure 2.1 *Diagram for the classification of irrigation waters*

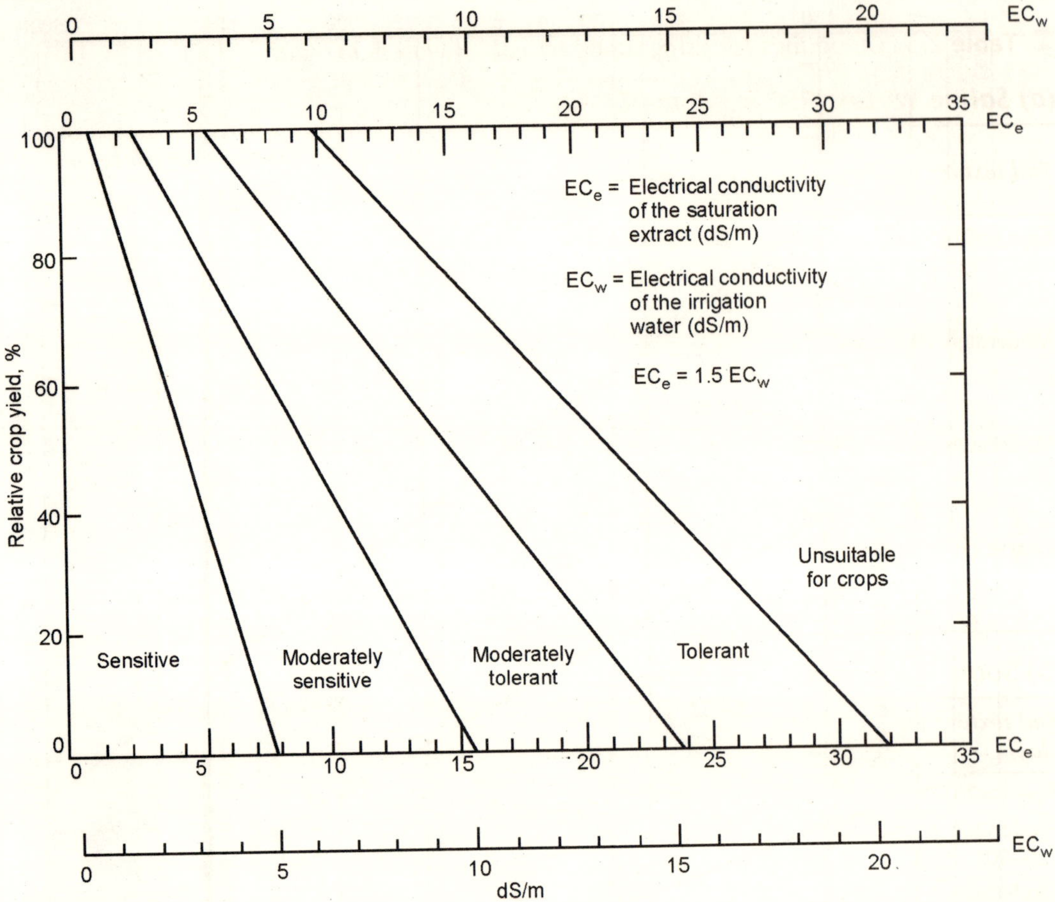

➲ **Figure 2.2** *Salt-tolerance rating of crops*

➲ **Table 2.4** *Classification of waters for irrigation*

Water quality			*ECw (ds/m)*	*SAR (mmol/l)*	*RSC (me/l)*
A.		Good	<2	<10	<2.5
B.		Saline			
	(i)	Marginally saline	2–4	<10	<2.5
	(ii)	Saline	>4	<10	<2.5
	(iii)	High-SAR saline	>4	>10	<2.5
C.		Alkali waters			
	(i)	Marginally alkali	<4	<10	2.5–4.0
	(ii)	Alkali	<4	<10	>4.0
	(iii)	Highly alkali	Variable	>10	>4.0

Table 2.5 *Guidelines for using saline irrigation waters in India*

(a) Saline Water (RSC < 2.5 me/L)

Soil texture (% clay)	*Crop tolerance*	*ECw (dS/m) limit for rainfall region*		
		<350	*350–550*	*>550 mm*
Fine (>30)	Sensitive	1.0	1.0	1.5
	Semi-tolerant	1.5	2.0	3.0
	Tolerant	2.0	3.0	4.5
Moderately fine (20–30)	Sensitive	1.5	2.0	2.5
	Semi-tolerant	2.0	3.0	4.5
	Tolerant	4.0	6.0	8.0
Moderately coarse (10–20)	Sensitive	2.0	2.5	3.0
	Semi-tolerant	4.0	6.0	8.0
	Tolerant	6.0	8.0	10.0
Coarse (<10)	Sensitive	—	3.0	3.0
	Semi-tolerant	6.0	7.5	9.0
	Tolerant	8.0	10.0	12.5

(b) Sodic Waters (RSC > 2.5 me/L, Ec_{iw} < 4.0 dS/m)

Soil texture (% clay)	*SAR (mmole/l)$^{1/2}$*	*Upper limit of RSC (me/L)*	*Remarks*
Fine (>30)	10	2.5–3.5	Limits pertain to *kharif* fallow/*rabi* crop rotation when annual rainfall is 350–550 mm. When the waters have Na<75% (Ca + Mg > 25%) or rainfall is >550 mm, the uper limit of the RSC range becomes safe. For double cropping neutralisation of RSC with gypsum is essential based on the quantity of water used during the *rabi* season. Grow low water requiring crops during *kharif*. Avoid rice.
Moderately fine (20–30)	10	3.5–5.0	
Moderately coarse (10–20)	15	5.0–7.5	
Coarse (>10)	20	7.5–10.0	

According to the Ministry of Agriculture, Government of India, the following general conclusions on the basis of their interactions with the above factors can be utilised for developing a criteria:

(i) Soil salinity increases with the salinity of irrigation water, number of irrigations during the crop period, fineness of texture and decrease in hydraulic conductivity of soil.

(ii) Soil sodicity or alkalinity increases with the increase of sodium adsorption ratio or bicarbonate ion content of irrigation water.

(iii) Even at low salinity (1 to 3 mmhos/cm), soil sodicity increases with the increase of RSC value.

(iv) The adverse effect of boron increases with the boron content in water but for most field crops it is not a serious problem in India (Table 2.7).

(v) More saline water (Na-Cl-type) can be safely used in well-drained, coarse, textured soils with low water table conditions.

(vi) The adverse effect of saline water is more on montmorillonitic than on illitic or kaolinitic soils.

(vii) More saline water can be used for more salt-tolerant crops (Table 2.8).

(viii) The adverse effect of saline irrigation water is more in arid and semi-arid than in humid regions. Generally, in India, there is no salt accumulation in regions where rainfall is more than 700 mm during the monsoon season.

(ix) High water table and restricted drainage due to presence of hard pan or lime concretions reduce the upper permissible limits of saline irrigation water.

(x) More saline water can be used in calcareous and more fertile soils particularly rich in organic matter.

(xi) Initial soil salinity and alkalinity decreases the upper permissible limit of salinity of irrigation water.

Table 2.6 shows a suggested tentative classification for Indian conditions. But more saline (7800 micromohos/cm) waters are being used in well-drained sandy soils of Haryana and Rajasthan.

In case of boron, the guidelines for upper permissible limits are shown in Table 2.7.

The sodium hazard in water is generally expressed by the residual sodium carbonate (RSC) concentration. It is safe up to 1.25, marginal up to 2.5 and unsafe above 2.5 meq/l for irrigation purposes.

Table 2.6 *Water quality-rating proposed in India (1972)**

	Nature of soil	*Crops grown*	*Upper permissible safe limit of EC of water for irrigation (micromhos/cm 25°C)*
1.	Deep black soils and alluvial soils having clay content more than 30%, soils that are fairly to moderately well-drained.	Semi-tolerant Tolerant	1500 2000
2.	Heavy textured soils having a clay content of 20–30%, soils that are well-drained internally and have a good surface drainage system.	Semi-tolerant Tolerant	2000 4000
3.	Medium textured soils having a clay content of 10–20% soils that are very well-drained internally and have a good surface drainage system.	Semi-tolerant Tolerant	4000 6000
4.	Light textured soils having a clay content of less than 10%, soils that have excellent internal and surface drainage.	Semi-tolerant Tolerant	6000 8000

*In the above proposed limit of water quality it is presumed that (i) the groundwater table is at or below 1.5 m from the surface, (ii) the soils have good drainage, (iii) the soluble sodium percentage in water is less than 70 (if more, gypsum need to be added to soil occasionally). If not, then the indicated limits need to be appropriately reduced.

Table 2.7 *Limits of boron in irrigation water for crops of different degree of boron tolerance*

Tolerant	*Semi-tolerant*	*Sensitive*
4 ppm B	2 ppm B	1 ppm B
Tamarix aphla	Sunflower	Pecon
Palm	Potato	Walnut
Date palm	Cotton	Plum
Sugarbeet	Radish	Pear
Lucern	Field pea	Grape
Onion	Barley	Orange
Turnip	Wheat	Avocado
Cabbage	Corn	Grapefruit
Lettuce	Oat	Lemon
Carrot	Line bean	

Table 2.8 *List of crops sensitive, semi-tolerant and tolerant to salts*

Sensitive	*Semi-tolerant*	*Tolerant*
Mash	Maize	Wheat
mung	Jowar	Barley
Lentil	Bajra	Oat
Gram	Rice	Sugarbeet
Toria	*Raya*	Guava
Peas	Safflower	Ber
Field beans	Sugarcane	Date palm
Ladies finger	Cotton	Coconut
Cluster beans	Musk melon	
Cowpeas	Water melon	
Citrus	Cauliflower	
Lemon	Cabbage	
Apple	Onion	
Peach	Potato	
Pear	Mango	
	Pomegranate	

Groundwater may be of good or poor quality, but its continuous rise due to recharge from irrigation can threaten to cause waterlogging and undo the benefits for which irrigation projects are planned. All groundwaters contain some salts and an unchecked rise of the groundwater table inevitably leads to waterlogging and salinisation. The groundwater in the root zone deprives the plant roots of air which is needed for their normal growth. The capillary rise of water above the groundwater table, and its evaporation or absorption by plant roots leaves behind salt that was dissolved in the groundwater. The deposit and accumulation of salts in the soil pores in the root zone increases progressively with time. When the salinity level reaches a level that the plants cannot tolerate, the lands become uncultivable due to the salinity hazard.

Salt Tolerance of Crops

There is a great variance in the tolerance capability of crops to salinity and sodicity. Crops that are most suitable to grow under given conditions of salinity or sodicity, can be selected. CSSRI, Karnal has suggested a salt tolerance rating of crops on the basis of soil salinity (ECe) and that of irrigation water (ECw) and relative crop yield as shown in Figure 2.2. The estimated yields of some important crops with irrigation waters of varying salinity under conditions prevailing in North India are shown in Table 2.9.

Table 2.9 *Estimated yields of important crops using irrigation waters of varying salinity*

Crop	*ECw (dS/m) of irrigation water for relative yield*		
	90%	*75%*	*50%*
Cereals			
Wheat	Up to 6.0	8–12	12–16
Barley	Up to 8.0	8–16	16–20
Rice	Up to 3.0	3–5	5–8
Maize	Up to 4.0	4–8	8–16
Pearl millet	Up to 8.0	8–16	16–20
Italian millet	Up to 3.0	3–5	5–9
Sorgum	Up to 3.0	5–10	10–18
Oil Seeds			
Mustard	Up to 6.0	6–12	12–15
Safflower	Up to 4.0	4–7	7–12
Sunflower	Up to 4.0	4–8	8–14
Groundnut	Up to 2.0	2–4	4–6
Soyabean	Up to 2.0	2–4	4–6
Pulses/Legumes			
Pigeon pea	Up to 2.0	2–3	3–4

Cont'd...

...Cont'd

Crop	ECw (dS/m) of irrigation water for relative yield		
	90%	75%	50%
Cluster beans	Up to 4.0	4–6	6–10
Cow pea	Up to 8.0	8–13	—
Berseem	Up to 3.0	3–4	4–5
Vegetables			
Onions	Up to 4.0	4–6	6–8
Potato	Up to 3.0	3–5	5–8
Tomato	Up to 3.0	3–5	5–8
Okra	Up to 3.0	3–6	6–10
Brinjal	Up to 3.0	3–6	6–8
Chillies	Up to 2.0	2–4	4–6

Ionic Constituents

Besides the total electrolyte content, plant response also depends upon the concentration of different ions in the soil solution. The precipitation and dissolution reactions within the soil solution govern the effective salinity influence on the crops. An excessive accumulation of certain ions in the plant tissues can cause toxic effect. In general, the tolerance to salinity decreases with increase in SAR, and pH. In the presence of Ca concentrations, crops can withstand relatively higher levels of SAR. Sodium toxicity may occur when Na concentration is high and calcium concentration is low or when the Mg/Ca ratio exceeds 3.

▼ SOIL-WATER RELATIONS

Acharya and Abrol (1998) explain that soil-water relations are expressed in terms of the ability of the soil to retain, release and transmit water within and across the soil system to the atmosphere. The processes of infiltration, profile water storage, drainage redistribution and evaporation of water from the soil are governed by its soil-water relations. The extent of run-off and erosion that erode the capacity of the soil to contain and sustain life are indirectly related to soil-water relations. Similarly, salinisation and alkalisation processes are directly or indirectly influenced by soil-water relations since the soluble salts move with water. The knowledge of these relations is important for a better understanding of soil-plant-atmosphere continuum so as to devise efficient soil-water management practices for higher and sustainable crop yields.

The soil-water relations are governed primarily by soil texture, structure, nature of clay mineral and the nature and extent of ions on the exchange complex. Salty soil having an excess of exchangeable sodium is highly dispersed in nature and shows poor soil-water relations. In non-swelling soils the lower quantity of water at low suctions and higher quantity of water at higher suctions is retained with increasing level of sodium on the exchange complex. The sealing of water conducting pores by the dispersed soil particles results in a low intake of water and limits the profile-water storage after rain or irrigation. The replacement of sodium with favourable cations like calcium improves the soil-water relations of salty soils. The hydraulic conductivity of these soils is several times less than that of a normal soil of identical textural and mineralogical make-up. The greater difference in hydraulic conductivity is at a water content of near saturation and at exchangeable sodium percentage (ESP) of >15. The water movement even in a highly alkaline soil can be maintained at a higher level by increasing the electrolyte concentration and keeping low SAR of the permeating water. Great variations in the drying pattern of salty soils, compared with normal soils, are observed owing to poor transmission characteristics. The soil-moisture content in alkali soils decreases rapidly due to evaporation at the surface, but layers below the surface continue to remain wet, as moisture loss at the surface is not replenished by the lower depths due to poor hydraulic conductivity. For the same ESP, though a lower drying rate is maintained for a longer time during winter, the higher drying rate during summer results in greater cumulative loss at any given time. Hence, the deleterious effect of high ESP on crops is likely to be accentuated in summer than in the winter season.

SALT REMOVAL BY DRAINAGE

In order to overcome the threat of waterlogging and salinisation in irrigated lands, drainage measures become necessary to prevent the rise of the groundwater table above an acceptable level. Drainage measures aim to remove as much quantity of water from the groundwater as is added in the form of recharge from the irrigation system and other sources. If this is effected by installing horizontal subsurface drainage, water and dissolved salt are both removed from the sub-strata. This can enable a check on the rise of the groundwater table into the root zone. Since salts are also removed alongwith the drainage water, there is no progressive increase in the salinity of the groundwater. The handicaps, however, are the high cost of installation of the subsurface drainage system and disposal of saline affluent water into natural streams or in evaporation ponds with the attendant pollution and environmental problems.

Many times the rise of the groundwater table is arrested by pumping out water through wells, at a rate equal to or higher than the rate at which the recharge is occurring. This is quite effective in checking the rise of groundwater table or even in depressing it. Moreover, the pumped water can be used to supplement the surface irrigation supply to increase the area under irrigation. But in this method, the salts contained in the groundwater are not removed and are merely recycled. The total salts in the soil-after mass may increase progressively.

In the biodrainage system, the plants, through their deep roots, draw groundwater and remove it by evapotranspiration. If the rate of uptake of water by plants equals the rate of net recharge to the groundwater system, a further rise of the groundwater table can be checked. When the water balance condition is achieved, the groundwater table would stabilise at a level where the equilibrium is struck.

All plant bodies contain some minerals, which vary from plant species to species and the environmental conditions in which the plants grow. About 80 per cent of the plant body is made up of water. Of the dry biomass, nearly 95 per cent is made up of carbon, oxygen, hydrogen and nitrogen, all of which except nitrogen are obtained from the atmosphere. About 5 per cent of the dry biomass is constituted by minerals, which are all obtained from subsoil water. It is sometimes assumed that plants take only water from the ground and leave behind all dissolved minerals. The fact is that while plants do not absorb and uptake all minerals dissolved in soil water, they do uptake some minerals. The quantity of minerals taken up by the plant can be estimated from the biomass produce and mineral content in the biomass. In case the irrigation water contains significant minerals, the quantity of minerals taken up and removed from the soil by crops and plants may be relatively small as compared to the quantity imported and added to the soil water mass by the irrigation water. But if the irrigation water is of good quality and contains a small quantity of minerals, it would be erroneous to ignore or neglect the quantity of minerals, removed from the system by the plants and crops. In fact, it is possible to plan and implement a biodrainage scheme so that minerals imported and added to the system by irrigation water are substantially, if not entirely, extracted and removed from the system alongwith the biomass produce. Ignoring or neglecting the quantity of minerals consumed and removed from the system, by the utilisable biomass produce can give misleading results.

SALT CONTENT IN WATER AND EC

The electric conductivity (EC) of saturation extracts from soils is generally used to measure and express the degree of salinity in soils. The US Salinity Laboratory (Hand Book No. 60) gives the relationship between EC values and per cent salt content in water as shown in Figure 2.3.

The relationship between the average electrical conductivity EC and salt concentration is as follows:

EC (ds/m)	0.5	1.0	2.0	3.0	4.0	5.0	6.0	8.0	10.0	20.0	40.0
% Salt content in water	0.03	0.06	0.13	0.20	0.26	0.35	0.44	0.60	0.74	1.60	4.0

It is seen that on an average an increase of 0.08 per cent in the salt content in water results in an increase of 1.0 ds/m in the EC value.

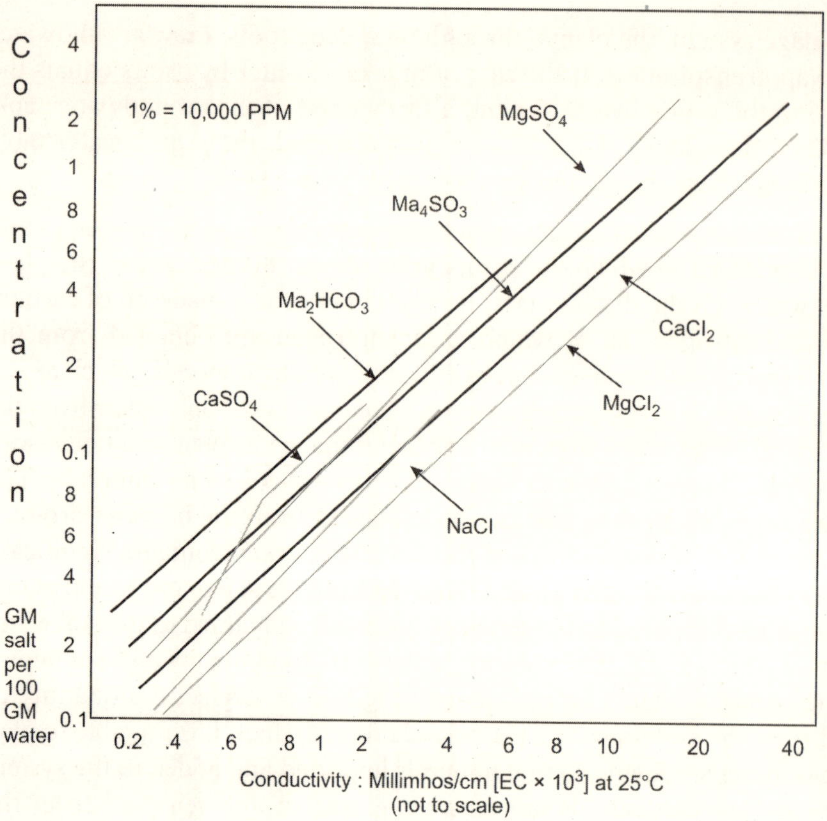

Relation between average electrical conductivity EC and salt concentration

EC (ds/m)	% Salt content in water
0.5	0.03
1.0	0.06
2.0	0.13
3.0	0.20
4.0	0.26
5.0	0.35
6.0	0.44
8.0	0.60
10.0	0.74
20.0	1.60
40.0	4.0

It is seen that on average increase of 0.08% in salt content in water results in an increase of 1.0 ds/m in the EC value

Figure 2.3 *Concentration of single salt solution in % (as related to electrical conductivity)*

INCREASE IN GROUNDWATER SALINITY

It is possible to estimate the rate at which groundwater salinity is likely to increase due to recharge from irrigation, after the groundwater table becomes stable under water balance condition. In case this water balance condition is achieved with the help of biodrainage and if it is assumed that the plants do not take any minerals whatsoever from the subsoil water mass, the rate of increase in the salinity of the groundwater would depend upon:

- Salt content in irrigation water: Si (mg/l)
- Depth of annual irrigation: Δ (m)
- Water column depth above barrier layer: D (m)
- Storage coefficient or porosity of sub-strata, in which the rise (D) of groundwater table occurs: C

The position is shown in Figure 2.4.

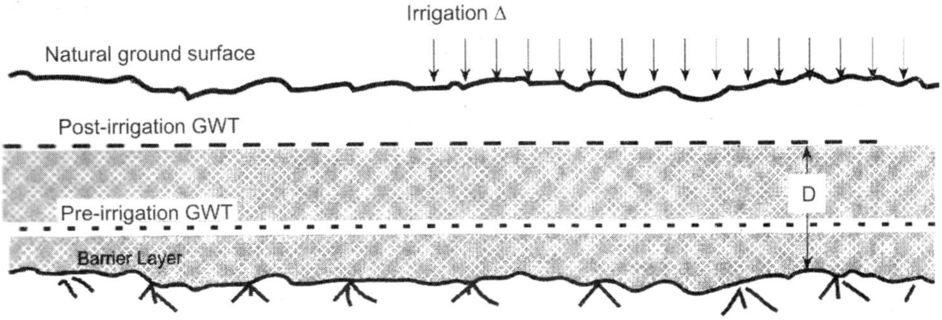

$$\%\text{increase in salt content in } GW = \frac{\Delta \times si}{D \times C} \times 10^{-4} \text{ after 1 year of irrigation}$$

$$\text{or, an increase of 'P\% ' in } GW \text{ salt content would occur in: } P \times \frac{D \times C \times 10^{4}}{\Delta \times si} \text{ years}$$

➲ **Figure 2.4** *Rate of increase in groundwater salinity (Due to irrigation)*

At the equilibrium stage when the groundwater table becomes stable, the salt balance in a column of unit area and height D can be considered as follows:

Volume of groundwater column $= 1 \times D \times C\, m^3$

Weight of this water column $= DC$ tonnes

Volume of irrigation water applied annually $= 1 \times \Delta \text{ m}^3$

Weight of total salt in annual irrigation water $= 1 \times \Delta \times Si \times 10^{-6}$ tonnes

Maximum possible percentage increase in salt content in groundwater as a result of one year of irrigation (assuming all salts in irrigation water are retained in depth 'D' of the groundwater zone) $= \dfrac{\Delta Si \times 10^{-6}}{DC} \times 100$

Or, an increase of 'p' per cent in groundwater salt content can occur in a minimum period of:

$$\frac{DC \times 10^{4}}{\Delta Si} \times p \text{ years} \qquad \ldots(1)$$

Applying the value of 0.08 for 'p' in equation (1), we get:

Minimum period in years in which EC can increase by 1.0 ds/m

$$= \frac{800\, DC}{\Delta\, Si} \text{ years} \qquad \ldots(2)$$

In dry arid regions, where the natural groundwater table is quite deep before the advent of surface irrigation, the value of D may be 30 m or more.

C, the storage coefficient or porosity of sub-strata would depend upon the nature and characteristics of the sub-strata formations. In deep alluvium soils, this may be as high as 0.30 or more. In hard rocky strata, its value may be as low as 0.15 or less.

Δ, the depth of annual irrigation water application, would depend upon the rate of crop evapotranspiration, the intensity of irrigation, crop patterns and efficiency of water usage. In dry arid regions, it may be about 500 mm. It is rarely more than 1000 mm.

Si, the salinity in irrigation water, would depend upon the source of the water supply, the distance and the strata through which the water is to be transported for irrigation. In large projects, river waters are generally stored in reservoirs and later used for irrigation through canals. Precipitation and melted snow are the dominant contributors of water supply. The salt content in such water can be as low as 100 mg/litre and rarely more than 500 mg/litre. An idea of the composition of average non-polluted river waters of the world is given in Table 2.3.

GRAPHICAL REPRESENTATION

Figures 2.5 and 2.6 indicate the likely period in years for the groundwater conductivity value EC to increase by 1.0 ds/m with the use of irrigation water containing varying levels of dissolved salt. The relationship is based on equation (2), and the range of values used are:

D = 10 m, 20 m and 30 m

C = 0.15 and 0.30

Δ = 500 mm, 750 mm and 1000 mm

Si = 100 mg/l to 500 mg/l

It is seen that with D = 30 m, the storage coefficient value of 0.30 and 200 mg/l dissolved salts in irrigation water, it would take 72 years for the groundwater EC to increase by 1.0 ds/m when the annual irrigation depth Δ is 500 mm. Similarly, with a storage coefficient = 0.15, it would take half the time, i.e., 36 years, for the groundwater EC to increase by 1.0 ds/m. In other words, if the EC value of groundwater is 4.0 ds/m when the water table stabilises under the water balance condition, it would take a period of 36 years (if C = 0.15) or 72 years (if C = 0.30) for it to rise to 5.0 ds/m with use of 500 mm irrigation water annually containing dissolved salts to an extent of 200 mg/litre. The EC value would increase to the level of 8 ds/m (from 4 ds/m) after a period of 144 to 288 years.

It is to be noted that the above exercise has been done with an assumption that plants do not take any salts/minerals from the soil. To the extent they actually do, the rate

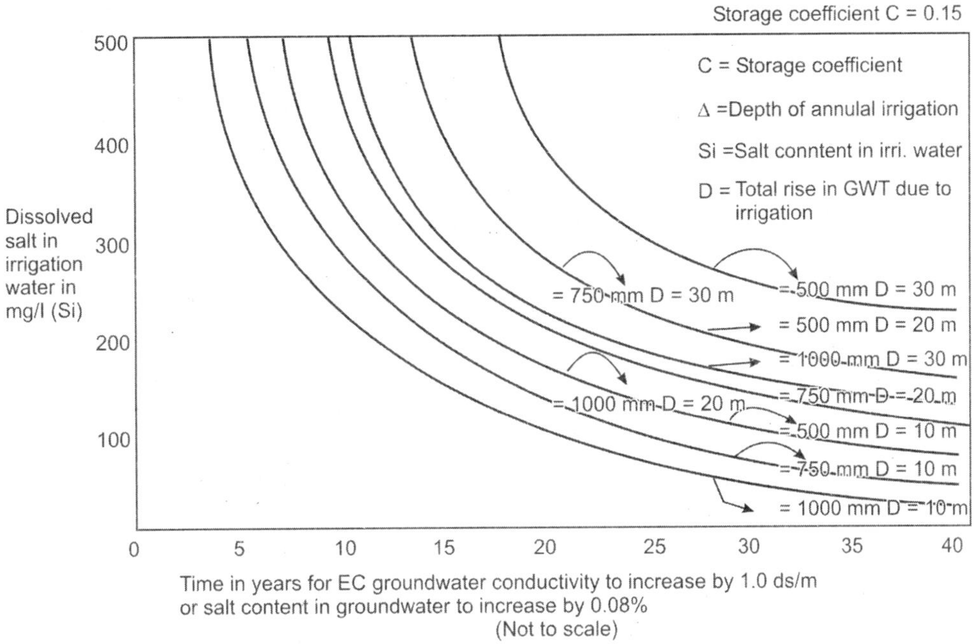

Figure 2.5 *Estimated time in years in which groundwater conductivity EC can increase by 1.0 ds/m due to irrigation*

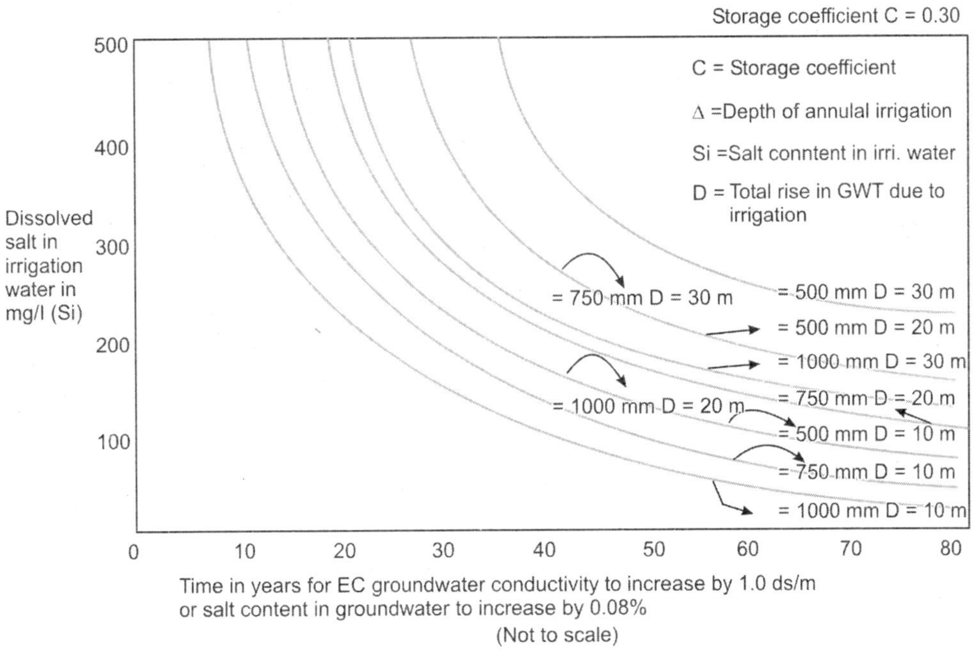

Figure 2.6 *Estimated time in years in which groundwater conductivity EC can increase by 1.0 ds/m due to irrigation*

of increase in groundwater salinity would be proportionately slowed down. Another assumption is that all salts brought in with irrigation water remain confined within the groundwater zone of depth *D*. To the extent the salts move out or go down to deeper depths, it would slow down the rate of increase in groundwater salinity. Since all crops and plants uptake some minerals and as there would always be some groundwater outflow from the irrigated lands, the rate of increase in groundwater salinity should in practice, be slower than given by equation (2).

Kijne et al. (1998) (SWIM-paper-IIMI), report that in the Shepparton region of Murray basin, Australia, the rate of groundwater degradation caused by salts in the irrigation water and rainfall under good conjunctive water management was found to be as low as 0.02 ds/m per year for a situation in which the depth from the soil surface to the base of aquifer is 10 m, and the annual infiltrated depth of the surface water applied for irrigation is 0.8 m.

In an experiment in Victoria, Australia, where the groundwater table under plantations fluctuated between 3 m to 5 m depth below the ground surface, the soil salinity was found to increase from about 1 ds/m to 5 ds/m during a period of nine years (1983–92). But at depths deeper than 6 m it increased to only about 2 ds/m during the same period. The eucalyptus plantation (2.4 ha) was not irrigated but adjoining areas were irrigated for pasture.

In IGNP, Rajasthan, India, irrigation was introduced in a dry arid region in the year 1961. The groundwater table before irrigation was quite deep and the water was brackish. The water table started rising soon after the commencement of irrigation. The annual rate of rise of the groundwater table during the period 1961–1992, averaged 0.94 m per year. The quality of groundwater improved gradually but progressively due to the dilution effect of fresh water. The change in areal extent of electrical conductivity of groundwater during the period 1983 to 1991 is shown in Table 2.10.

Table 2.10 *Change in areal extent of electrical conductivity (km^2) in IGNP Stage I*

(June 1983 to June 1991)

	June 1983	*June 1990*	*June 1991*
Fresh (EC < 2000)	460	780	860
Marginally Fresh (2000 – 5000)	1130	2180	2480
Total:	1590	2960	3340

In the year 1983, the groundwater in about 52 per cent area had an EC value of 4 ds/m or less and an EC value of more than 16 ds/m in 18 per cent area. In the year 1996, the groundwater EC was 4 ds/m or less in 70 per cent area and more than 16 ds/m in 5 per cent area. The study was carried out over a large area of 334 thousand hectares, and the position of change in groundwater salinity with passage of time is shown in Figure 2.7.

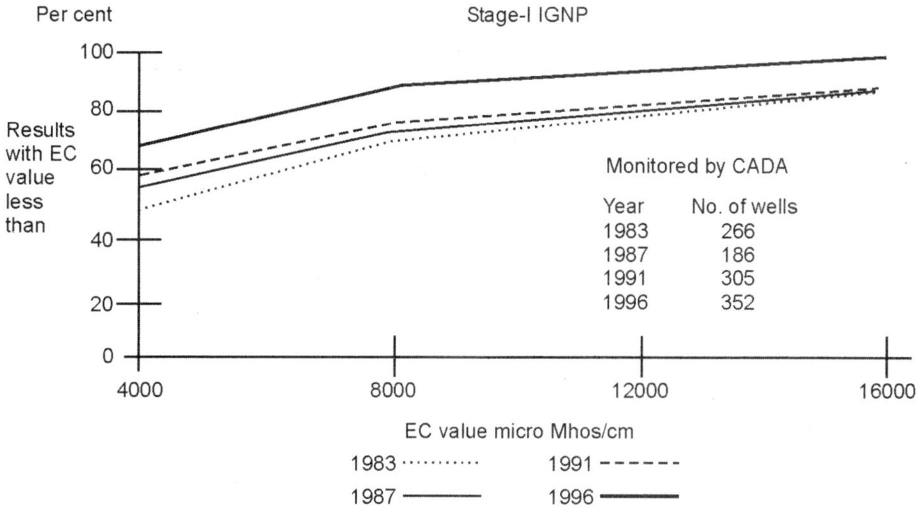

➲ **Figure 2.7** *Change in groundwater salinity with passage of time, Stage I-IGNP*

CONCLUSIONS

The following conclusions can thus be drawn:

1. The original and to some extent the direct source of all salt in the soil are the primary minerals found in soils and in the exposed rocks of the earth's crust.
2. The surface waters in nature are generally of good quality. The subsurface waters in arid regions are very often saline.
3. A recharge from irrigation generally dilutes and reduces salinity levels in groundwater, particularly in the top layers. But once the groundwater table becomes stable and if there is no draining out of salts, the salt brought in by irrigation recharge water can cause an increase in the salinity levels of the groundwater.
4. It is possible to estimate the rate at which the groundwater salinity may increase due to an irrigation recharge in case there is no draining out of salts. With the use of good quality irrigation water, the rate of increase in groundwater salinity, if any, would be too slow to be a matter of concern.

REFERENCES

Acharya, C.L. and I.P. Abrol (1998). 'Soil Water Relations', in P.S. Minhas and N.K. Tyagi (eds), *Agricultural Salinity Management in India*, Chapter 6, CSSRI, Karnal, India, p. 79 and 89.

American Society of Civil Engineers (ASCE) (1990). 'Agricultural Salinity Assessment and Management', ASCE Manuals and Reports on Engineering Practice No. 71.

Central Ground Water Board, India (1995). *Studies on Conjunctive Use of Surface Water and Ground Water Resources in IGNP Stage–I*, Rajasthan.

Central Soil Salinity Research Institute (CSSRI) (1993). 'Agroforestry in Salt-affected Soils', *CSSRI Bulletin*, Karnal, India, pp. 5–7.

Central Soil Salinity Research Institute (CSSRI) (1998). Minhas P.S., Tyagi N.K. Guidelines for Irrigation with Saline and Alkali Waters.

Minhas, P.S. and N.K. Tyagi (1998). 'Guidelines for Irrigation with Saline and Alkali Waters', *CSSRI Bulletin*, Karnal, India.

Dayem, Safwat Abdul (2000). 'Drainage Experiences in Arid and Semi-arid Regions', Paper Presented at the 8th ICID Drainage Workshop, New Delhi.

Dhir, R.P. (1998). 'Origin and Accumulation of Salts in Soils and Ground Water', in P.S. Minhas and N.K. Tyagi (eds), *Agricultural Salinity Management in India*, Chapter 4, CSSRI, Karnal, India.

Evans, R.S. (1989). 'Saline Water Disposal Options in the Murray Basin', *BMR Journal of Australian Geology and Geophysics*, 11 (2/3): 167–85.

Garrels, R.N. (1967). 'Genesis of Some Ground Waters from Igneous Rocks', in P.H. Abelson (ed.), *Researches in Geochemistry*, Vol. 2, John Wiley & Sons, New York, pp. 405–20.

Ghassemi, F., A.J. Jakeman and H.A. Nix (1995). *Salinisation of Land and Water Resources: Human Causes, Extent, Management and Case Studies*, University of New South Wales Press, Sydney, p. 526.

Gunn, R.H. and D.P. Richardson (1979). Holland and Christi (1909). 'The Nature and Possible Origins of Soluble Salts in Deeply Weathered Landscapes of Eastern Australia', *Aust. J. Soil Res.*, 17: 197–215.

Heuperman, A.F. (1995). *Salt and Water Dynamics Beneath a Tree Plantation Growing on a Shallow Water Table*, Department of Agriculture, Energy and Minerals, Victoria, Australia.

Holland and Christi (1909). The nature and possible origins of soluble salts in deeply weathered landscapes of eastern Australia. *Aust J. Soils Res.* 17: 197–215.

Hungston and Gallitis et al. (1977). Quoted from R.P. Dhir (1998). 'Origin and Accumulation of Salts in Soils and Ground Water', in P.S. Minhas and N.K. Tyagi (eds), *Agricultural Salinity Management in India*, Chapter 4, CSSRI, Karnal, India.

Hutton, J.T. and T.I. Leslie (1958). 'Accession of Non-nitrogenous Ions Dissolved in Rainwater to Soils in Victoria', *Aust. J. Agric. Res.*, 9: 492–507.

James, David W., R.J. Hanks and J.J. Jurinak (1982). *Modern Irrigated Soils*, John Wiley & Sons, New York.

Junge and Werly (1958). Quoted from R.P. Dhir (1998). 'Origin and Accumulation of Salts in Soils and Ground Water', in P.S. Minhas and N.K. Tyagi (eds), *Agricultural Salinity Management in India*, Chapter 4, CSSRI, Karnal, India.

Kapoor, A.S. (1998). 'Rate of Increase in Ground Water Salinity Due to Irrigation with and without Biodrainage' (Unpublished), Agriculture University, Bikaner.

Kapoor, A.S., Mahesh Kumar and Dalip Singh (1995). Committee Report on 'Remedial Measures for Reclamation of Water-logged Area Around Manaktheri Badopal and Bherusari Village (IGNP) Stage–I, India'.

Kijne, J.W., S.A. Prathapar, M.C.S. Wopereis and K.L. Sahrawat (1998). 'How to Manage Salinity in Irrigated Lands: A Selective Review with Particular Reference to Irrigation in Developing Countries', SWIM Paper 2, International Irrigation Management Institute, Colombo, Sri Lanka.

Kitamura, Y., T. Yano and S. Yashda (2000). 'Irrigation Induced Salinisation of Agricultural Lands and Its Remedial Measures in the Aral Sea Basin', Paper Presented at the 8th ICID Drainage Workshop, New Delhi.

Massod, A. and G.P. Kutcher (1992). 'Irrigation Planning with Environmental Considerations', World Bank Technical Paper No. 166.

Ministry of Agriculture, India (1979). 'Manual on Irrigation Water Management', Chapter VI, Government of India, Ministry of Agriculture and Irrigation Publication, pp. 111–13.

Rankama, K. and T.G. Sahama (1950). *Geochemistry*, Chicago University Press, Chicago, p. 221.

Rhodes and Bernstein (1971). Quoted by David W. James (1982). *Modern Irrigated Soils*, John Wiley & Sons, New York.

Sharma, R.C. (1998). 'Nature, Extent and Classification', in P.S. Minhas and N.K. Tyagi (eds), *Agricultural Salinity Management in India*, Chapter 3, CSSRI, Karnal, India.

Truckian, K.K. and K.H. Wedepohl (1961). 'Distribution of the Elements in Some Major Units of the Earth's Crust', *Bull. Geol. Soc. Am.*, 72: 175–92.

US Department of Agriculture (1954). *US Salinity Laboratory Diagnosis and Improvement of Aline and Alkali Soils*, Handbook No. 60, p. 69 and 77 (Table 12), p. 80 (Figure 2.5)

Xie, G., J. Breburda, A. Kollender-Szych and A. Battenfeld (1998). 'Relation between Soil Salinity and Water Quality as well as Water Balance in Yinebi Plain, PR China', in *Towards Sustainable Land Use, Furthering Cooperation between People and Institutions, Vol. I, Advances in Geolocology 31*, International Society of Social Sciences (ISSS), pp. 485–93.

3

Plant Mechanism of Absorbing and Transpiring Water

INFLUENCE OF PLANTATIONS ON HYDROLOGICAL CYCLE

Forests and plantations influence the hydrological cycle. Stebbing (1952) observed, "The entire philosophy of the role of forests is based on the ability of the forests to prolong the water cycle from its inception as falling precipitation to the final disposal of run-off into streams and oceans. The longer the water is retained on the land, the greater is the usefulness in nurturing crops and trees, in maintaining the regular supply of water in streams and in preventing soil from washing". Precipitation, which without a forest cover is rapidly disposed of as surface run-off, is (with forest cover) retained in the soil and gradually utilised for the growth of trees and crops and for maintaining a steady and sustained flow of water in streams, as depicted in Figure 3.1.

ABSORPTION OF WATER

Pandey and Sinha (1995) describe the absorption of water as follows: Life originated in an aqueous environment and in the course of evolution, it became fully dependent upon water in a number of ways. Water is now said to be the *liquid of life* or *elixir of life*. In general, water is essential for life and its importance to plants may be summarised as follows:

1. Water is the main constituent of the protoplasm comprising up to about 90 to 95 per cent of its total weight. In the absence of water, protoplasm becomes inactive and is even killed.
2. Different organic constituents of plants such as carbohydrates, proteins, nucleic acid and enzymes, etc., lose their physical and chemical properties in the absence of water.

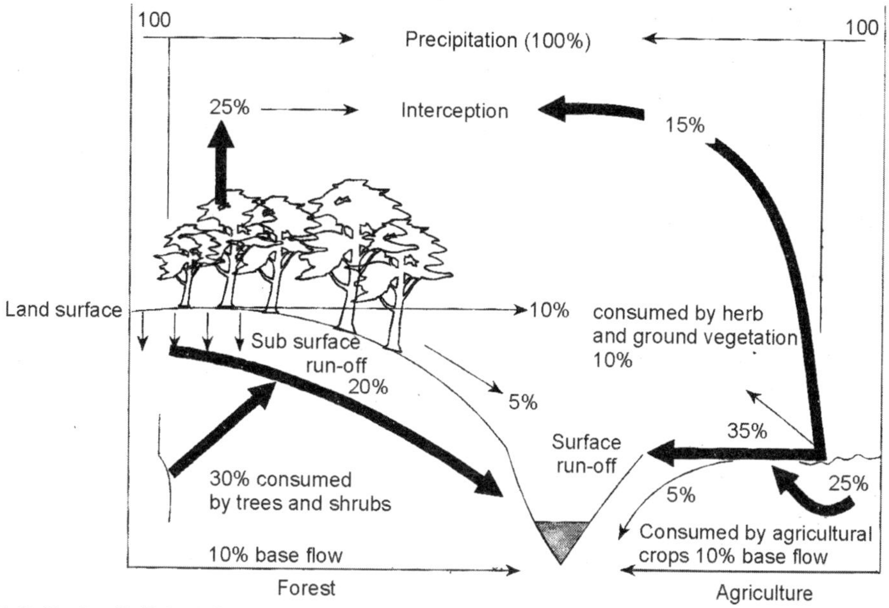

Source: A.P. Dwivedi, *Principles and Practices of Indian Silviculture* (1992).

➲ **Figure 3.1** *Role of forests in hydrology*

3. Water participates directly in many metabolic processes. The inter-conversion of carbohydrates and organic acids depends upon hydrolysis and condensation reactions.
4. Water increases the rate of respiration. Seeds respire fast in the presence of water.
5. Water is a source of the hydrogen atom for the reduction of carbon dioxide in the reactions of photosynthesis.
6. Water acts as a solvent of and carrier for many substances. It forms the medium in which several reactions take place.
7. Water which is present in the vacuoles helps in maintaining the turgidity of the cells which is a must for the proper activities of life. The turgidity of cells helps in their elongation thus resulting in growth. The difference in the amount of available water during summer and winter seasons is the reason for the formation of annual rings in higher plants. In summer, the turgidity is less and as a result, smaller cells are formed.
8. A network of thin layer of water surrounding each cell plays an important role in the entry and movement of dissolved substances.
9. Water helps in:

   (i) the translocation of solutes;

   (ii) the mobility of gametes;

(iii) the dissemination of spores, fruits and seeds; and

(iv) providing support to aquatic plants.

10. In tropical plants, water plays the very important role of thermal regulation against high temperatures. Some people think that plants lose about 95 per cent of the absorbed water just to maintain the optimum temperature.
11. Thousands of characters develop for balancing the water content of plants. Even atmospheric moisture affects plant growth.

Different plants absorb water in different ways; for instance, lower plants absorb water from their general surface, while in higher plants the roots are the organs concerned with the absorption of water. Certain orchids even absorb moisture directly from the atmosphere.

Soil Water

Plants absorb water from the soil by their roots. Water is found in different forms in the soil. For the soil, the chief source of water is rain. The total amount of water present in the soil is termed as *holard,* of which the water available to the plants is called *chesard,* while that which cannot be absorbed is termed as *echard.* After the rains, a part of the water drains away. This is known as *run-away water* and is not available to plants. A part of the water percolates downwards through large pores between the soil particles under the influence of gravity and is called *gravitational water.* It reaches the low water table and goes off beyond the plant's reach. If the gravitational water is available to the plants, it is absorbed. A good amount of the water is retained by the soil particles and is known as *field capacity moisture,* i.e., water retained due to the water-holding capacity of the soil particles. This water may be present in three forms: *capillary water, hygroscopic water* and *chemically combined water.* The water available to plants is the capillary water which fills the spaces between the non-colloidal particles or it forms a thin film around the non-colloidal smaller particles of soil. Hygroscopic water is held by the soil particles of colloidal complex due to adhesive force. This cannot be drawn by the roots.

Field capacity (the total amount of retained water) differs from soil to soil and depends upon the size of soil particles, e.g., it is 5 per cent in sandy soil, 35 per cent in loamy soil, 45 per cent in clay soil and negligible in gravel and rocks. In a fully saturated soil, the percentage of moisture held in the form of films is called water holding capacity. The percentage of water that remains in the soil when permanent wilting is attained is called wilting *co-efficient* or permanent wilting percentage.

Water Absorbing Parts of a Plant

A major portion of the water required by plants is absorbed by the roots, but the absorption of water by leaves and stems has also been found in a few plants. Hydrophytes absorb water through the general surface. Wood (1925) reported that *Kochia baosia* and *Rhagodia* are seen to absorb water in an appreciable amount from the atmosphere. In all, about 100 genera have been reported as capable of absorbing water of this category. A few of them are *Vitis, Solanum, Beta, Lycoperisicum* and *Phaseolus.* The uptake of water by leaves is influenced by the:

(i) structure and permeability of the cuticle and epidermis;
(ii) hairiness of the leaf surface;
(iii) case of the wetting surface, and
(iv) internal environment of deficiency of water in parenchymatous cells that lie close to the epidermis.

Roots play a principal role in the absorption of water. Even orchids absorbing water from the atmosphere develop modified roots for the purpose.

Roots

Roots absorb water mainly from the apical region. The apical organisation of the root shows three clear demarcations, viz., the *zone of meristematic cells,* the *zone of elongation* and the *zone of absorption or differentiation.*

The zone of differentiation consists of three different types of tissue systems, i.e., *dermal, cortical* and *stellar.* Dermal tissue includes the surface layers of cells. The epidermis in this region has an enormous number of unicellular root hairs. Cortical tissue commonly consists of parenchymatous cells but the stellar tissue system is complex and consists of pericycle, phloem and xylem among others. The important ones are described here.

Root Hair

Root hair is the special modified cell of the epidermis meant for the absorption of water. It is specialised not only in appearance but also in its internal structure. The wall of the root hair consists of cellulose and pectic substances (calcium pectate). Both are highly hydrophilic in nature. These substances have a great capacity of water absorption.

The cell wall acts as a permeable layer. Next to the cell wall is the plasma membrane enclosing the cytoplasm, nucleus and vacuole. The size of the vacuole is quite large in order to facilitate the peripheral arrangement of the cytoplasm. The role of vacuole during the absorption of water is just like that of a controller.

Endodermis

The endodermis is the bounding layer of the stele and in its cells is present a particular type of thickening substance (of suberin or cutin and sometimes of lignin). This thickening is known as 'casparian's strip'. Certain cells of the endodermis, that lie opposite the protoxylem points, have no such thickening, i.e., they are thin-walled. Such cells provide a passage for the movement of water from the cortex to the stele and are called *passage cells.*

Xylem

The xylem consists of tracheids, vessels and xylem parenchyma of which first two are thick-walled due to the deposition of lignin. In many cases, pits are present on their radial walls where there is little lignin deposition. Hence, these areas are thin-walled so as to facilitate the movement of water. Besides the pits, the thickening of the wall does not interfere in the movement of water because lignin is a hydrophilic substance and easily permits the movement of water through it. Vessels are other specially modified cells for the purpose of the movement of water. Since transverse walls of these vessels are perforated, a continuous water channel is maintained in the plant.

MECHANISM OF WATER ABSORPTION BY PLANT ROOTS

The root hair of a plant or tree maintains contact with the soil water and in nature, it acts as a sole water absorbing organ. The water diffuses into the root hair as a result of a diffusion pressure deficit gradient. The cell sap contains a more concentrated solute than the water present outside. Water enters as long as the diffusion pressure deficit of the cell sap is greater, distending the cell until the elasticity of its stretched wall is sufficient to balance the osmotic pressure of solutes.

It is not very clear how exactly water enters in the root hair and what is the precise mechanism of water absorption. According to the 'active absorption' theory, water is absorbed as a result of the activities in the root itself and is not concerned with any affair of the shoot. To explain the actual mechanism of active absorption, there are two theories viz., (a) the osmotic theories, and (b) the non-osmotic theories. According to the 'passive absorption' theory, the forces responsible for the absorption of water into the roots are governed by other cells. The governing force originates in the cells of the transpiring shoots rather than in the root itself. These forces develop due to transpiration. With the occurrence of transpiration, the diffusion pressure deficit of the leaf cells increases which results in the movement of water from the xylem cells to the adjacent mesophyll cells. Due to the presence of the continuous column of water from leaves to roots through xylem channels, the deficit is transmitted to the xylem of the roots and finally to the root hair through the endodermis and cortex.

EXTERNAL ENVIRONMENTAL FACTORS AFFECTING WATER ABSORPTION RATE

(a) Available Soil Water

The water in the soil is present in different forms such as capillary, hygroscopic, and gravitational water, of which capillary water is readily available for absorption. This is in between the field capacity and the permanent wilting percentage, where the rate of absorption of water is generally uniform and is not affected. With an increase in water beyond the field capacity, the aeration of the soil is badly affected which reduces the rate of absorption and under severe conditions, results in wilting. Similar wilting is observed in extremely dry soils and a decrease in soil water reduces the rate of absorption.

(b) Concentration of Soil Solution

A large number of elements are dissolved in soil water which is called soil solution. These elements change the concentration of the soil solution. If the soil solution is highly concentrated, it greatly increases the osmotic pressure and when it reaches higher to that of the level of the cell sap, water is not absorbed. This is one reason why the plants fail to grow in highly saline fields.

(c) Soil Temperature

The variations of temperature affect the rate of absorption. Around 20° to 30°C is the most suitable temperature for absorption. Low temperatures reduce and moderately high temperatures increase the rate of absorption. A very high temperature may kill the cell.

(d) Soil Aeration

Water is absorbed more efficiently in a well-aerated soil rather than in poorly aerated soils.

▼ TRANSPIRATION

The important external environmental factors affecting the rate of transpiration are:

(i)	Humidity of air	A higher relative humidity of the outside atmosphere lowers the rate of transpiration.
(ii)	Temperature	An increase in temperature increases the rate of transpiration.
(iii)	Wind velocity	An increase in wind velocity increases the rate of transpiration.
(iv)	Light	With an increase in the light intensity, the rate of transpiration also increases because the stomata get opened.
(v)	Atmospheric pressure	The reduction of atmospheric pressure permits more rapid diffusion of water. The plants growing on the hills thus show higher rates of transpiration.
(vi)	Water supply	The deficiency of water in the soil decreases the rate of transpiration because of a decrease in the rate of absorption.

The important internal factors that affect the rate of transpiration are:

(i) *Stomatal frequency:* This implies the number of stomata per unit area of the leaf surface. If the stomata are opened, the resultant increased stomatal frequency, increases the rate of transpiration.

(ii) *Structural peculiarities of the leaf:* A reduced size of the leaves reduces the transpiration surface. Deposits of cutin or wax-like substances on the surface of leaves (as in calotropis) check the rate of transpiration.

(iii) *The leaf area, amount of spongy tissue, inter-cellular spaces, orientation of* leaf and the *extent of the root system* are other factors that affect the rate of transpiration.

The withdrawal of groundwater by transpiration from trees depends upon:

(i) Climatic conditions;
(ii) Plant species;
(iii) Density of plantations;
(iv) Area under plantations;
(v) Water supply; and
(vi) Root system.

(i) Climatic Conditions

The climatic factors influencing the rate of transpiration from plants are temperature, humidity, wind velocity, light and atmospheric pressure. An increase in the temperature increases the rate of transpiration. Higher relative humidity of the outside atmosphere lowers the rate of transpiration. An increase in wind velocity increases the rate of transpiration. With an increase in light intensity, the rate of transpiration increases because the stomata get opened. The reduction of atmospheric pressure permits more diffusion of water. The plants growing on hills show higher rates of transpiration.

(ii) Plant Species

The important plant factors influencing the rate of transpiration are stomatal frequency, structural peculiarities of leaf, leaf area, amount of spongy tissue, inter-cellular spaces, orientation of leaf and the extent of root system. Stomatal frequency implies the number of stomata per unit area of the leaf surface. If the stomata are opened, the resultant increased stomatal frequency increases the rate of transpiration. The reduced size of leaves reduces the transpiration surface. Deposits of cutin or wax-like substances on the surface of leaves (as in calotropis) check the rate of transpiration.

(iii) Density of Plantations

The rate of transpiration from a plantation area depends upon the density of plantations, though not strictly in that proportion. Plants lying at a spacing of 3 m × 3 m give a density of about 1100 trees per hectare. More widely spaced plants would give a lower rate of transpiration per ha while plantations with closer spacing may give higher rate of transpiration per ha.

(iv) Area Under Plantation

The total volume of water withdrawn from the groundwater by plantations would depend upon the total area under plantations. In properly planned and spaced plantations, the volume of the groundwater removed by transpiration would be in direct proportion to the extent of the area under plantations.

(v) Water Supply

The availability of water supply greatly influences the rate of transpiration from plantations. The deficiency of water in the soil decreases the rate of transpiration because of a decrease in the rate of absorption. Plantations, identical in respect of plant species, density of plantations, etc., and growing in similar climatic conditions can exhibit widely different rates of transpiration. Plants are known to survive with minimal water availability conditions with low transpiration rates. The same plants transpire water at a significantly higher rate when water is available in abundance in the root zone.

(vi) Root System

The extent of the root system too has a bearing on the rate of transpiration. When groundwater is easily available to plants in the upper layers of the soil, the root system does not develop very deep in the soil. On the other hand, if water is available at deeper depths of the soil, the root system develops to greater depths. The root systems are known to go down up to a depth of 15 m and more in search of water.

TREE WATER USE

The reported results of the capacity of trees to grow and transpire water show great variance. This is not surprising because of several factors, as described above, influence the rate of transpiration. It is quite difficult to carry out experiments under controlled conditions to determine the rates of transpiration. But fairly dependable data is available on the potential rate of evapotranspiration from crops (ET_o) and rate of evaporation from free water surface (A_{pan}). Potential evapotranspiration (ET_o) is defined as the rate of evapotranspiration from an extensive surface 8 to 15 cm tall, green grass cover of uniform height, actively growing, completely shading the ground and not short of water (FAO Paper No. 24). A_{pan} is the observed rate of evaporation from the water surface in a pan of standard size and is generally 1.15 to 1.20 times that of ET_o.

The published annual tree water use values range from 0.6 A_{pan} for irrigated globules with full canopy cover in Western Australia (Marshall and Chester 1991) to 1.9 A_{pan} for *Eucalyptus camaldulensis* irrigated with seepage affluent (Morris and Wefner, 1987). Diwan quoting Greenwood et al. (1979), reports the *Eucalyptus globulus* and *camaldulensis* at the age group of 11, 16 and 24 months respectively transpired about 0.8, 6.8, 37 and 0.4, 8.5, 21 litres per day, per tree respectively of the two species.

In a study in California, USA, the total evapotranspiration from tree plantations during 220 days (April–November 1990) was estimated at 1,153 mm, which was nearly equal to the applied water.

Grattan, et al. report that the extent to which eucalyptus can reduce drainage volumes, depends on the maintenance of high rates of evapotranspiration. In non-stressed environments, the literature reports crop coefficients (*Kc*) for a full cover of Eucalyptus trees between 1.2 to 1.5 (Stribbe 1975; Sharma 1984). In case of *Eucalyptus camaldulensis* irrigated with saline drainage water (*Ec* = 10 ds/m and 12 mg/l *B*), evapotranspiration (*ET*) was estimated using two energy balance methods and *Kc* values were 0.83 (i.e., *ET* of Eucalyptus was 0.83 reference crop *ET*) (Dong et al. 1992).

In s study in a desert area of Rajasthan, India, the annual evapotranspiration from tree plantations with a density of 1,900 trees/ha was estimated at 3,446 mm which is about 1.2 A_{pan}.

Bhutta and Chaudhry (2000) report the estimated annual consumptive use of water in south-western USA for different plant species as 770–1,350 mm for Willow, 650–2,350 mm for Polar, 1,680 mm for Alder and 2,200–2,800 mm for Tamarix. The daily consumptive use of water by individual trees was estimated as 45 litres for a three-year-old Poplar plant, 60 litres for the Tamarix tree and 143 litres for a full-grown Eucalyptus tree. Investigations on the inherent water consumption by the seedlings of six trees commonly found in Pakistan were conducted by Riotzsch and Marsur. The average daily luxury water consumption by an individual tree, exhibited as a true specie characteristic, was reported as 129 litres for *Acacia arabica,* 137 litres for *Eucalyptus tereticornis,* 143 litres for *Eucalyptus camaldulensis,* 153 litres for *Moris alba,* 156 litres for *Delbergia sisoo* and 174 litres for *Tamarix articulata.* At Tando jam in the Sindh province of Pakistan, the annual water use by three to five-year-old *A. Nilotica* was 1,248 mm on the severely saline site and 2,225 mm on the mildly saline site.

Landsberg (1999) describes that under atmospheric conditions the transpiration rate depends on the surface area of leaves exposed to the atmosphere, the extent of the root system which absorbs water from the soil and the amount of water in the soil. A relatively small tree in wet soil can transpire faster than a larger tree in drier soils under the same conditions of atmospheric demand. Dye (1996) measured water uptake by E. Grandis in South Africa at two sites from depths up to 8 m. On the basis of water loss from 200 trees, the daily water use was measured as 38.5 litres/day in block plantation and 43.7 l/tree/day in scattered plantations. Landsberg found no evidence of systematic and consistent variation in the water use rates of different tree species and suggests that if primary objective is water use and the amelioration of water balance problems, the trees that grow best and biggest in a particular place will use the most water.

Hutton et al. (1999) state that although there are reports to the contrary, especially from investigations on eucalypt plantations, there is little evidence for rejecting the hypothesis that the rate of water use per unit leaf area does not vary significantly among sympatric eucalypt species in rainfall limited (soil water limited) systems.

According to Jason (1991), Morris and Wehner measured annual transpiration from irrigated redgum trees at Robinvale in north-west Victoria at over 4000 mm. Transpiration rates of four-year-old Eucalyptus Camaldulensis and Eucalyptus grandis in North Victoria have been measured respectively at over 4000 mm and 3500 mm per annum.

LYSIMETER STUDY AT CSSRI, KARNAL, INDIA

A lysimeter study was carried out at the Central Soil Salinity Research Institute, Karnal, in the Haryana State of India to determine the water use capacity of eucalyptus and bamboo trees. Chhabra and Thakur (1998) report that a series of lysimeters made from reinforced cement concrete (RCC) hume pipes with a diameter of 1.2 m and a depth of 2.5 m, were installed in an experimental field. The site has a monsoon type of climate with maximum and minimum temperatures of 45° and 0°C respectively. The potential evaporation (E_o) is highest (12 mm day^{-1}) in summer and lowest in winter (1.2 mm day^{-1}). During the four years when the experiment was conducted, the rainfall was 822, 1076, 1075 and 891 mm while the E_o was 1424, 1388, 1461 and 1307 mm respectively.

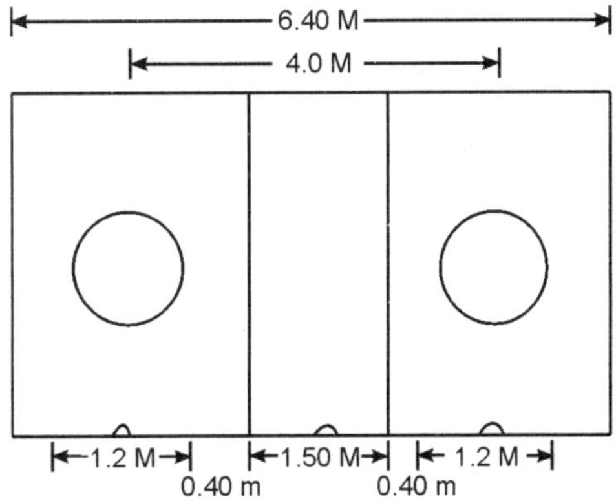

(a) Plan

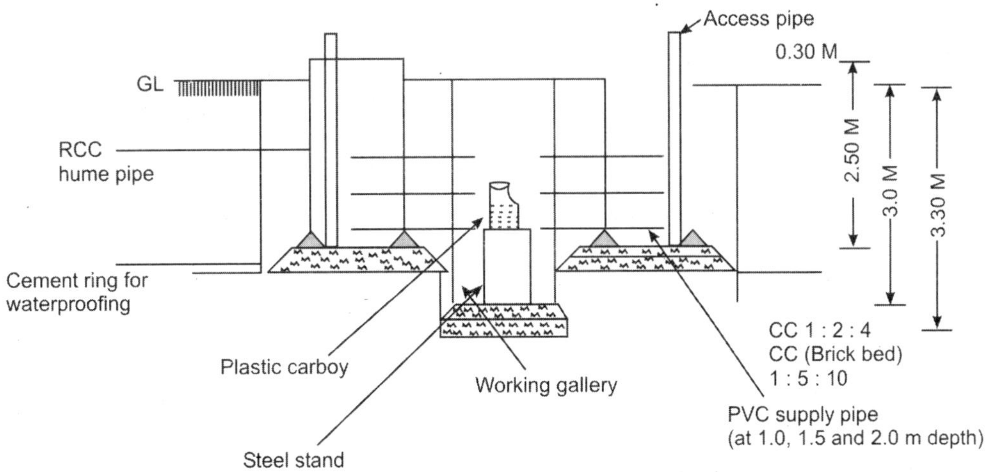

(b) Section

Figure 3.2 *Design of lysimeters with facility to maintain different water table depths*

PVC perforated supply pipes, covered with nylon gauze were inserted in the lysimeters at a depth of 1, 1.5 and 2 m, respectively to supply water of the desired salinity from a plastic reservoir. This inlet pipe was also used for drawing samples of groundwater from the lysimeter to check on its salinity level during the conduct of the experiment. Using stands, these reservoirs (20 litre capacity) were placed at the required heights, to maintain the water table in the lysimeters at the specific depths (Figure 3.2). After the construction and testing for waterproofing, each lysimeter was filled with sandy loam alluvial soil (Typic Ustochrept), with a pH of 8.2 and *EC* of 0.4 dS m^{-1}. The soil was packed to a bulk density of 1.50 g cm^{-3}. The excavated soil was filled back around the buried lysimeters to minimise temperature fluctuations in the root zone.

Each lysimeter was planted with a three-month-old sapling of eucalyptus or bamboo. These were raised earlier in 20 cm deep polyethylene bags and transplanted at the centre of the lysimeter in the month of March, 1993, following standard silvicultural practices. Except for the initial six months, when irrigation was also undertaken at the surface to help in the establishment of the trees, all the water was applied in the reservoir and thus made to enter the lysimeters from below.

The water in the reservoir was replenished everyday and taken as a measure of the water lost through evapotranspiration. The lysimeters with trees were not covered and all the rainfall was allowed to enter and accounted for while computing the water balance. In the rainy season usually there occurred a gap between the rainy and non-rainy days. As a result of that rainwater which was allowed to enter the lysimeters increased soil moisture content of the soil to saturation level and also accumulated in the reservoir, yet never came out of it. The accumulated leachate in the reservoir again entered the lysimeter during non-rainy days to maintain the water table depth, as a result of that less water was added from outside to maintain the water table depth during that period. Another set of lysimeters, with similar combinations of the water table and salinity levels but without trees, were used to monitor the loss of water through evaporation (*Es*) alone. The amount of water biodrained (*Wb*) by the tree was calculated by subtracting the evaporation (*Es*) from the evapotranspiration (*ET*) i.e., $ET - Es = Wb$. However, due considerations may be given to the fact that the Es of the soil under tree canopy will be less than that from non-vegetated soil because of the:

- mulching effect of forest litter on the soil;
- decrease in temperature;
- increase in relative humidity; and
- low moisture content of the surface soil (caused by depletion by the tree roots).

To be on the conservative side, the loss of moisture from the lysimeters with no trees but with a similar depth of the water table and groundwater salinity, was subtracted from the observed *ET* to arrive at the net transpiration by the trees (*Wb*). Hence, the *Wb* calculated through this method may be relatively underestimated. The potential evaporation (*Eo*) was measured everyday from the open pan evaporometer installed at the Institute Meteorology Observatory and compiled for each month and year.

The moisture content of the soil was monitored *in situ* through the neutron probe gauge using a calibration curve. For that an aluminum access pipe was installed in each lysimeter to a depth of 2.5 m.

Effect of Water Table Depth

The biodrainage capacity of the Eucalyptus was significantly affected by the water table depth (Table 3.1). In the first year, the maximum volume of biodrained water was 2420 mm at 1.5 m followed by 2284 and 1799 mm at 1.0 and 2.0 m depth, respectively. In spite of the higher soil moisture content at 1 m, the trees grown at 1.5 m biodrained the most. An improved performance at water table depth of 1.5 m is believed to be due to a better root environment. The biodrainage capacity of the eucalyptus decreased significantly as the water table depth became deeper (2 m) which was apparently due to the fact that in the first year, the roots had not reached to that depth. Greenwood and Beresford (1979) also found that the transpiration rate of Eucalyptus increased significantly when the roots reached the zone of higher water content. Similarly, Calder (1986) reported that Eucalyptus species can use very high quantities of water if planted in areas where the water table is shallow because there is no stomatal control when there is abundant amount of water supply in the root zone. Heinrich and Sands (1990) reported that *ET* by the Eucalyptus plant was a function of the soil moisture content as the leaf water potential controlling *ET* was low when grown in wet as compared to that in dry soil.

In the second year, the amount of water biodrained by the Eucalyptus plant was 3572, 3698, 2902 mm at water table depths of 1, 1.5 and 2 m respectively. This increased to 3873, 3995 and 3190 mm, respectively in the third year. The maximum amount (3698 and 3955 mm) of water biodrained by Eucalyptus was again observed at a water table depth of 1.5 m. On comparing the biodrainage capacity of the Eucalyptus tree in the first year with that in the second and third years, it was found that it had increased by almost one and a half times and seems to be well correlated with plant growth. But in the fourth year, the biodrainage capacity at each depth of the water table was slightly lower (3596, 3730 and 2820 mm). The growth (27.8 cm girth at 1.35 m height) of the Eucalyptus trees within the lysimeter was less than those trees planted outside the lysimeters in the non-experimental area (55 cm girth at 1.35 m height). In the third year, the girth was 26.7 cm and increased only marginally to 27.8 cm in fourth year. The height also was almost similar in the third and fourth years, which appears to be due to the stagnation of the root growth in relatively limited space in the lysimeter. Sharma (1984) had also described the pattern of root distribution as the major factor affecting the evapotranspiration of Eucalyptus trees.

Table 3.1 *Biodrainage capacity (Wb), mm, of eucalyptus for different water table depths and groundwater salinity*

	1st Year				2nd Year			
Ground-water salinity dSm^{-1}	Water table depth, m							
	1.0	1.5	2.0	Mean	1.0	1.5	2.0	Mean
0.4	2449	2880	2133	2487	4342	5499	3466	4435
3.0	2388	2305	1872	2188	3902	3641	3099	3547
6.0	2256	2399	1683	2112	3502	3277	2764	3181
9.0	2097	2334	1652	2028	3124	3114	2638	2958
12.0	2231	2185	1653	2022	2988	2959	2543	2830
Mean	2284	2420	1799	—	3572	3698	2902	—
	3rd Year				4th Year			
Ground-water salinity dSm^{-1}	Water table depth, m							
	1.0	1.5	2.0	Mean	1.0	1.5	2.0	Mean
0.4	5098	5518	3746	4787	4958	5148	3787	4631
3.0	4014	3985	3398	3799	3885	4189	3183	3752
6.0	3874	3675	3125	3558	3447	3502	2629	3193
9.0	3319	3369	2908	3199	3078	3139	2357	2858
12.0	3061	3230	2774	3021	2612	2673	2140	2475
Mean	3873	3955	3190	—	3596	3730	2820	—

Effect of Groundwater Salinity

The volume of water biodrained in the first year was 2487, 2188, 2112, 2028 and 2022 mm at salinity levels of control, 3, 6, 9 and 12 dS m^{-1}, respectively. The results showed that the biodrainage was highest when the groundwater salinity was lowest. Its magnitude decreased with an increase in the salinity of the groundwater. There was no significant difference between the volume of water biodrained at salinity of 9 and 12 dS m^{-1}.

In the second, third and fourth years, the effect of groundwater salinity was more pronounced at every level irrespective of the depth of the water table (Table 3.1). This was due to an increase in the salinity of the lower soil layers as a result of the extraction of water by the roots from those zones. These observations are in contrast to those reported by Handa and EL-Enany (1994), who observed a reduction in the transpiration rate only for increasing salinity of the shallow groundwater. The Eucalyptus tree biodrained 2022, 2830, 3021 and 2475 mm of water respectively in the first, second, third and fourth years at a groundwater salinity of 12 dS m^{-1}, which was respectively 81, 64, 63 and 53 per cent of that observed under non-saline conditions. Tanji and Karajeh (1993) also observed a decrease in the efficiency of Eucalyptus to extract soil water when the salinity in the soil built up, which in their case was due to the continuous use of saline drainage effluent as a source of irrigation. These results are in agreement with those of Sumayao et al. (1977), Babla and Soliman (1978) and Roberts (1983) all of whom reported that evapotranspiration decreases with an increase in salinity level in the growth medium.

Eucalyptus plants have an open canopy and the leaves do not overlap with each other. Due to this, it provides an ideal condition for transpiration of the extracted water at relatively high rates even from the lower storey leaves. This capacity of Eucalyptus plants to extract high amounts of soil water and to act as biopumps has previously been highlighted (Chhabra and Baddesha, 1985; Rawat et al., 1985; Calder, 1986). Alvares (1982) contended that for this reason Eucalyptus may even cause desert-like conditions in areas with low water availability.

Effect of Biodrainage by Trees on Water Table Depth

The amount of water lost through evaporation from the non-vegetative soil (E_s) was maximum from shallow (1 m) and minimum from deep (2 m) water table depth. The mean E_o/E_s increased from 3.75 to 7.91 with an increase in the water table depth from 1 to 2 m (Table 3.2). Hence, the water lost from the barren soil will be much less and thus, the natural drying of the soil will not be able to control the rise in the water table. Further, with an increase in groundwater salinity from 0.4 (Control) to 12 ds m^{-1}, the E_o/E_s increased from 5.15 to 6.61 (mean of the three water table depths). The loss of soil water was minimum from deeper water table with the highest groundwater salinity level resulting in highest E_o/E_s of 10.0. These results indicate that for a given depth of the water table, an increase in groundwater salinity will further limit the loss of soil water through evaporation.

Table 3.2 *Ratio of W_b/E_O, W_b/E_s and E_o/E_s as affected by depth of water table and its salinity under eucalyptus plantation in the fourth year*

Salinity levels	W_b/E_o				W_b/E_s				E_o/E_s			
	Water table depth, m											
dS m^{-1}	1.0	1.5	2.0	Mean	1.0	1.5	2.0	Mean	1.0	1.5	2.0	Mean
0.4	3.79	3.94	2.90	3.54	13.85	17.75	21.16	17.59	3.65	4.50	7.30	5.15
3.0	2.97	3.21	2.44	2.87	10.76	15.99	15.68	14.14	3.62	4.98	6.45	5.02
6.0	2.64	2.68	2.01	2.44	9.63	15.70	12.95	12.76	3.65	5.85	6.45	5.32
9.0	2.36	2.40	1.80	2.19	9.65	13.13	16.84	13.21	4.10	5.46	9.35	6.30
12.0	2.00	2.05	1.64	1.89	7.46	12.43	16.34	12.08	3.73	6.09	10.0	6.61
Mean	2.75	2.85	2.16	2.59	10.28	15.01	16.67	13.96	3.75	5.38	7.91	5.68

The experiment showed that the amount of water biodrained by Eucalyptus trees

(Wb) was 1.64 (groundwater salinity 12 dS m^{-1}) to 3.94 (groundwater salinity 0.4 dS m^{-1}) times the potential evaporation from a Standard Pan (E_o or A_{pan}). The relationship is shown in Figure 3.3.

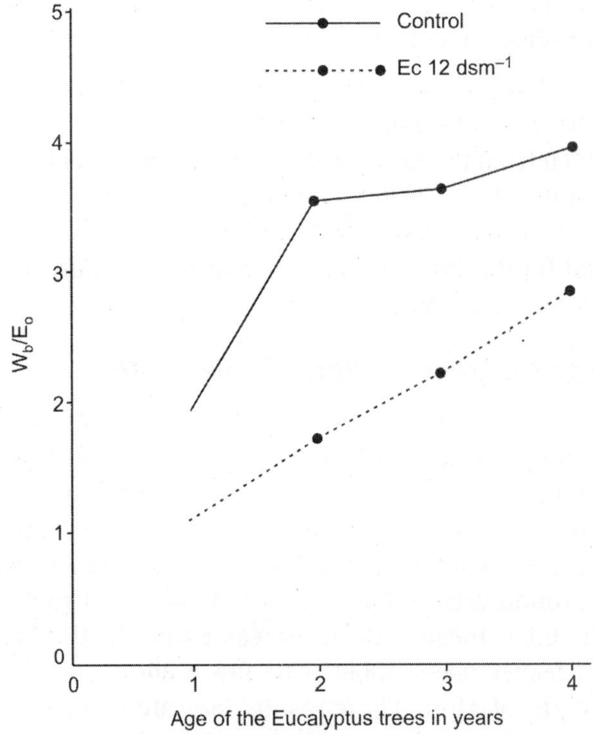

Figure 3.3 *Ratio of water biodrained (W_b) to pan evaporation (E_o)*

Effect of Seasons on the Biodrainage Capacity of Trees

In all the four years of the experiment, the amount of water lost through W_b, E_o and E_s was affected by the season. In the fourth year (Figure 3.4), the amount of water biodrained was maximum in summer (April to September) and minimum in winter (October to March), indicating that environmental factors like radiation and temperature are the most important factors controlling the biodrainage capacity of the plantations. Rawat et al. (1984) also observed that transpiration by Eucalyptus was a function of atmospheric aridity. Although the Eucalyptus tree is an evergreen plant, yet it has maximum leaf shedding in the month of September causing the lowest W_b as compared to that in the months of December and January when the temperature was the lowest. Contrary to that, E_o and E_s were more affected by climatic factors as compared to transpiration thus showing that the simple energy calculations are not applicable to the active transpiration process leading to effective control of the water table. Their values were lowest for the months of December and January and corresponded with the lowest temperatures of the year. Due to these reasons W_b/E_o (Figure 3.4) was more in the winter (October–March) than in summer (April–September). Sharma (1984) argues that the high transpiration rate by Eucalyptus even in the rainy season may be because the evaporation of intercepted water is higher than the transpiration from a dry canopy. In comparison to trees, the role of annual crops for bio-control of the water table is rather limited. The water uptake by annual crops is more variable and is mostly extracted from the surface shallow layers. Crops also do not need much water either during the early stages of plant growth or at maturity. There is also a clear lean period when no water is absorbed because there are no crops in the field.

Result of CSSRI Study

The CSSRI study shows that the Eucalyptus (*Eucalyptus tereticornis*) plant could biodrain 2880, 5499, 5578 and 5248 mm of water in the first, second, third and fourth years of the study period respectively from non-saline groundwater (EC 0.4 dS m^{-1}) and a water table depth of 1.5 m. In the fourth year, the biodrained quantity was 3.94 times of the standard pan evaporation. At groundwater salinity of 12 dS m^{-1}, the Eucalyptus plant biodrained 53 per cent of that under non-saline conditions.

The CSSRI experiment was conducted in lysimeters of 1.13 m^2 surface area. The growth of plant roots was restrained by the bounding walls. The volume of the transpired water per tree would have been much more under non-constrained conditions. The results of the experiment may correlate with those in the field in case the trees are planted in a grid pattern of 1 m × 1 m. The results of the experiment cannot be extrapolated and applied where the plantations are raised in the field in a grid pattern of 2 m × 2 m or 3 m × 3 m. More experimentation is needed to determine the optimum grid pattern for achieving the maximum possible rates of evapotranspiration by the trees.

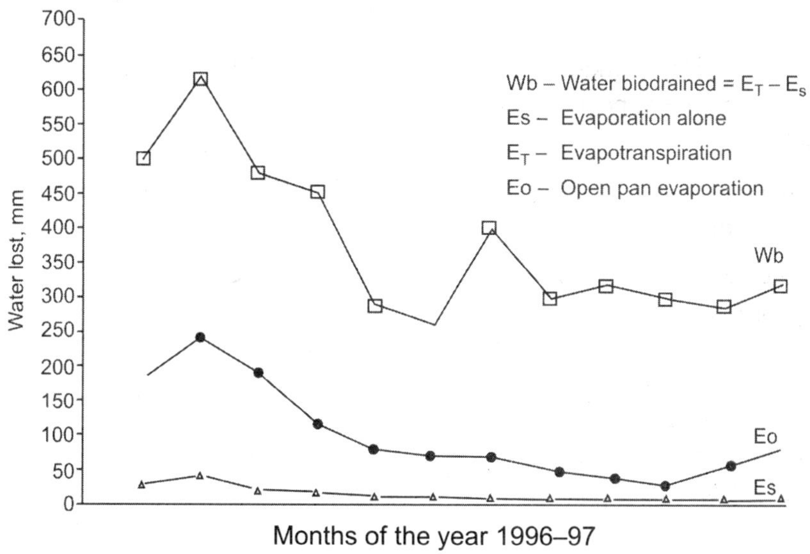

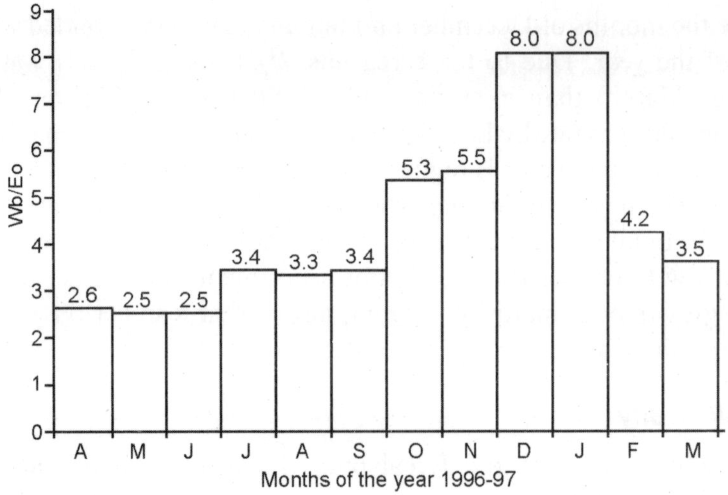

Figure 3.4 *Seasonal variations in W_b, E_o, E_s and ratio of W_b/E_o for eucalyptus plantation in the fourth year*

The study clearly demonstrates that Eucalyptus plants under conditions of abundant water supply in the root zone, can transpire huge quantities of water, which are much more than is generally recognised.

Sharma (1984), on the basis of measurements carried over a period of 5 years on Eucalyptus marginata and E. Calophylla species of trees in a region near Collie, West Australia with annual rainfall of 1100 mm/yr, concluded that during rainy winter periods, monthly evapotranspiration rates were up to three times that of pan evaporation.

FIELD STUDY ON PLANTATIONS, KARNATAKA, INDIA

Studies on Eucalptus plantations, their water use and sustainability were carried out for a number of years in Karnataka, South India. Calder (1994) have described the studies and their results in several papers and reports. The relevant descriptions and findings are discussed below.

Approximately, half of all plantation forestry in the tropics and subtropics is composed of the Eucalyptus species.

The higher water use of forests compared with the shorter vegetation is principally due to two processes. In the wet areas of the world, such as the uplands of the UK, the high aerodynamic roughness of forests leads to greatly enhanced evaporation rates in wet conditions (interception) and evaporation can, on an annual basis, be as much as twice of that for grass. In drier climates, the deep root systems of the forest and their greater water availability during the dry seasons leads to higher transpiration losses. The water-use studies carried out under the ODA-funded Eucalyptus project in the dry zone of southern India have established that the total evaporation (transpiration plus interception) from the forest is nearly 1.5 times greater than from agricultural crops.

Studies of the hydrology of Eucalyptus plantations, indigenous forests and an annual agricultural crop were initiated at four main sites in Karnataka, South India. Three of these sites were in the low rainfall zone (800 mm per annum) at the Devabal and Puradal experimental plantations near Shimoga and at the Hosakote experimental plantation near Bangalore respectively. The depths of the soils for the three experiments were different, at approximately 3 m at Devabal and Puradal and greater than 8 m at the Hosakote site. The fourth site is at Behalli in the high rainfall zone (2000 mm per annum) on deep soils (greater than 8 m).

In very low rainfall areas, the principal limiting factor for annual evaporation is the soil water availability. Studies (Harding et al., 1992) show that the available soil water capacity of both indigenous, dry deciduous forests and Eucalyptus plantation is of the order of 480 mm whereas, in the same region, the available water capacity for finger millet, an annual agricultural crop, is 150 mm. The annual evaporation from the indigenous and plantation forests is, within the errors of measurement (10%), equal to the rainfall of 800 mm/year and the evaporation from the finger millet, with a reduced soil water reservoir to exploit, is 500 mm/year. A conversion from forest to agricultural crops in this area will therefore, increase the annual flow (or catchment recharge) by this difference in annual evaporation. The studies also demonstrated the importance of tree size and age as limiting factors on evaporation. Measurements made on these (young) Eucalyptus plantations have established a new and surprisingly close correlation, as shown in Figure 3.5 (Calder et al., 1992), between the transpiration rate of an individual tree and its stem cross-sectional area (a better correlation than was found with the leaf area). The only meteorological data that is required is daily rainfall. Meteorological demand, though providing the driving force for evaporation, is not thought to be a limiting factor during most of the year, as the principal limitations on transpiration are thought to be soil moisture availability and tree size. These results obtained from semi-arid Karnataka, indicate that evaporation is limited principally by soil water availability and plant physiological controls.

This relationship implies that, irrespective of the size of the tree, sap velocities remain the same. As water in the soil becomes scarce and water stress develops, the same linear relationship between transpiration rate and basal cross-sectional area appears to hold, even though the sap velocity decreases. This finding enables the growth and water use of eucalyptus plantations in water-limited conditions to be modelled by assuming the volume growth is related to the volume of water transpired. Conventional methods for estimating the water use of vegetation are based on the assumption that evaporation is primarily determined by meteorological 'demand' – a function of solar radiation and atmospheric temperature, humidity and turbulence – with some regulation to the water 'supply' being imposed through stomatal controls exerted by the plants. In temperate

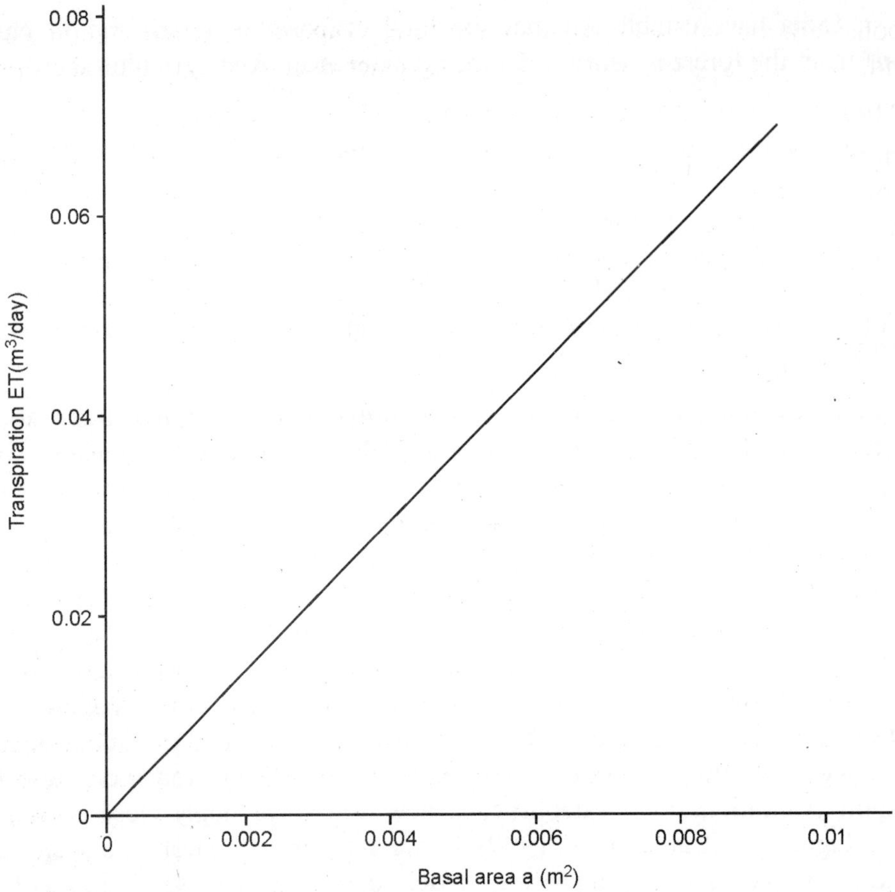

Figure 3.5 *Transpiration rate of eucalyptus tereticornis trees in conditions with little soil moisture stress at sites in southern India – measured using the deuterium tracing method plotted against basal (stem) cross sectional area of the tree measured at 1.2 m above ground level*

conditions where generally neither is the atmospheric demand very great nor the soil water totally limiting these methods are very appropriate and successful.

In the dry zones of India, on the other hand, it appears that the roles are reversed: evaporation is primarily determined by water supply rather than meteorological demand. The restrictions imposed by soil water availability – and for the young eucalyptus plantations, tree size – are so severe that, on a day-to-day basis, they limit the amount of evaporation that can take place. During the Indian dry season, the meteorological demand imposed by the hot dry climate is far in excess of the water supply. Although the meteorological demand provides the energy and driving force for evaporation, it does not – when viewed over time scales of a few days – control the evaporation rate. This implies that for these supply-control conditions, the evaporation rates can be estimated solely from consideration of the supply side of the equation without a detailed knowledge of the demand.

The main findings of the field study were:

- In the dry zone, the water use of young eucalyptus plantations on a medium depth soil (3 m depth) was no greater than that of the indigenous dry deciduous forest.
- At these sites, the annual water use of eucalyptus and indigenous forest was equal to the annual rainfall (within the experiential measurement uncertainty of about 10%).
- At all sites, the annual water use of forest was higher than that of annual agricultural crops (about 2 times higher than that of finger millet).
- At the dry zone deep soil (> 8 m depth) site, there are indications that the water use, over the three (dry) years of measurement, was greater than the rainfall. The model estimate of evaporation were 3400 mm as compared with a 2100 mm rainfall for the three-year period.
- At none of the sites was there any evidence of root abstraction from the water table. (Where this has happened in Australia, at a site where the water table is relatively shallow, there are reports of annual eucalyptus water use of 3600 mm in an area where the rainfall is only 800 mm).

FOREST RESEARCH INSTITUTE, DEHRADUN, INDIA

Gupta and Rawat (1984) state that when moisture is abundantly available as in case of plants which are maintained at field capacity the water is allowed to be transpired heavily as the tungor pressure of the leaf cells is not limiting. This is normally turned as 'luxury consumption' of water by plants. Dabral (1970) has shown that water utilised by E. Citriodora seedlings of 6–16 months age was 5526 mm per seedling per year when they were watered regularly to field capacity. In comparison, water use of E. Camaldulensis seedlings of same age was only 1200 mm per seeding per year with restricted water supply.

It was concluded that transpiration rate of Eucalyptus tereticornis was a function of atmospheric aridity under conditions of abundant soil moisture. On the other hand, under condition of restricted soil moisture, water conserving mechanism within the plant become operative resulting in much lower transpiration rates under similar aridity conditions.

The 'luxury consumption' behaviour of Eucalyptus has been successfully utilised to drain swamps in Israel.

AUSTRALIAN EXPERIENCE

Thorbunn (1999) describes results of measurements of evaporation of groundwater from plants from 20 diverse field experiments where groundwater was at least moderately saline (i.e., Ec > 5 dS/m). The studies covered trees, crops, shrubs and pastures in rainfed and irrigated areas. Water table evaporation rates ranged from 10 to 590 mm/year. The results indicate that plants have the potential to substantially increase evaporation compared with base soil, provided that (1) water tables are deep (e.g., > 3 m), (2) the plant's roots are initially close to the water table, (3) the relative water table salinity (the ratio of actual salinity to the threshold salinity at which plants can no longer extract water from the soil) is low, i.e., less than 0.5, and (4) soils have low permeability at field water contents (i.e., either fine or very coarse textured soils).

Greenwood et al. (1985) compared rates of evaporation from grazed pastures and Eucalyptus farm plantations in West Australia on saline farms. Annual evaporation from pastures was 390 mm and from trees upslope was 2300 mm for E. maculata and 2700 mm each for E. Globulus and E. Cladocalyx. It was concluded that planting of discrete areas of trees has the evaporation potential to assist in reclamation of saline farm land. Evaporation from plantations can be up to seven times that from the pasture.

▼ TREE WATER USE MEASURED BY HEAT PULSE METHOD

Rates of tree water use measured by the heat pulse method have been reported as follows :

$0.66\ A_{pan}$ and $0.83\ A_{pan}$	:	Western Australia, Eucalyptus camaldulenis trees, 9 year old at 1100 stems per hectare, water table depth 1.16 m, water table salinity 9 dS/m (Marshall et al., 1997; Akilan et al., 1997)
$0.189\ A_{pan}$	:	(Qld) Australia, *Eucalyptus camaldulensis* trees 5.7 m tall at 500 stems per hectare, water table depth 3.0 m, water table salinity 6.2 dS/m (Fraser et al., 1996).
$0.152\ A_{pan}$	:	WA, Australia, *Eucalyptus camaldulensis* trees 10 years old at 580 stems per hectare, water table depth 4.5 m, water table salinity 6.2 dS/m (Salama et al., 1994).
$0.471\ A_{pan}$	:	Tando Jam, Pakistan, *Acacia Nilotica* trees 3–4 years old at 800 stems per hectare, water table depth 1.3 m, water table salinity 20 dS.m (Khanzada et al., 1997).
$0.84\ A_{pan}$	:	Tando Jam, Pakistan, *Acacia Nilotica* trees 4–5 years old at 2500 stems per hectare. Water table depth 2 m, water table salinity 1.5 dS/m (Khanzada et al., 1997).
$0.72\ A_{pan}$	:	Pacca Anna, Pakistan, *Eucalyptus camaldulensis* trees 2–3 years old at 2500 stems per hectare, water table depth 1.5 m, water table salinity 5 dS/m (ACIAR Project 9316, 1996).
$0.66\ A_{pan}$	:	Pacca Anna, Pakistan, *Eucalyptus camaldulensis,* 2–3 years old at 2500 stems per hectare, water table depth 3.0 m, water table salinity 5 dS/m (ACIAR Project 9316, 1996).
$0.65\ A_{pan}$	:	NIAB Bio-saline Research Station Lahore, Pakistan, *Eucalyptus microtheca* trees 4–5 years old at 2000 stems per hectare, water table depth 1.5 m, water table salinity 3 dS/m (ACIAR Project 9316, 1996).
$0.252\ A_{pan}$	:	Vic., Australia, *Eucalyptus camaldulensis* trees 5–7 years old at 625 stems per ha., water table depth 2 m, water table salinity 6 dS/m (ACIAR Project 9316, 1996).
$0.259\ A_{pan}$	:	Vic., Australia, *Casuarina cunninghamiana* trees 5–7 years old at 625 trees per ha., water table depth 2 m, salinity 6 dS.m (ACIAR Project 9316, 1996).

EXPERIMENTAL METHODS

It is seen that different experimental methods of determining tree water use or transpiration rates give different results, sometimes the variations are quite large. Even with the same experimental method, scientists have reported differing results., There is a need to look into the matter and standardise methods and procedures for estimating transpiration rates.

Lysimeter Experiments

RCC or PVC chambers are filled with soil in which saplings of trees are transplanted. Water tanks supply water of desired salinity. Water table in the lysimeters is maintained at constant level and daily water use is measured. Another set of lysimeters with similar combinations of the water table and salinity levels but without trees, is used to monitor the loss of water through evaporation alone and to determine the net tree water use.

The merit of this method is that the volume of tree water use is measured directly and if there are no leakages in the experimental system, the experiment provides fairly accurate results.

The handicap in the method is limiting size of the lysimeter. Most cylindrical lysimeters are less than 1.5 m diameter and no more than 2.5 m deep. Rectangular lysimeters are of 2 m × 2 m size or less. Tree roots have a tendency to spread in all directions. The limiting size of the lysimeters puts a constraint on the spread of roots and therefore, on the growth of trees. That is why trees of the same specie and age exhibit better growth in the field in comparison to the trees planted in lysimeter. This in turn impacts the rate of transpiration and the lysimeter experiments, therefore, underestimate the potential rate of transpiration from the trees. In the CSSRI, Karnal experiment, the girth of Eucalyptus trees in lysimeters after 4 years was 27.8 cm while those growing outside in the field acquired a girth of 55 cm.

Another handicap with lysimeter experiments is that they are generally discontinued after a period of 4–5 years. The CSSRI Karnal experiment was carried out for a period of 4 years. The girth of Eucalyptus tree was 27.8 cm in the 4th year and the height was 1.35 m. In the Haryana Agricultural University experiment at Hissar, same specie of tree (E. Tereticernis) attained a height of 10–15 m and a girth, at breast height, of 50–60 cm after 6 years. To provide more dependable results, lysimeters should not only be large sized but the experiments should be continued till the trees attain full mature growth.

In the lysimeter experiments, the groundwater table is maintained at a constant level between 0–2 m. The rate of tree water use is found to decrease when the groundwater table is lowered. These experiments do not provide data on the rates of tree water use corresponding to the groundwater table depths of 3 m, 4 m and 5 m. Since groundwater table under tree plantations (in blocks or in parallel strips) for biodrainage may be needed to be maintained at deeper depths so that it does not encroach into the root zone of irrigated crops anywhere, it is desirable that lysimeter experiments generate relevant data.

In spite of these limitations, the lysimeter experiments provide very useful and dependable data for use in planning biodrainage plantations.

Sap Flow Rate Experiments

The sap flow system is based on the principle of thermal dissipation. The probe consists of two needles, one corresponding a (–) Reference T-type thermocouple, another one incorporating a (+) T-type thermocouple and a heating unit. Heated upper needle temperature is compared to lower ambient temperature needle (dT). Maximum dT occurs when the needle is hottest representing situation of no flow and correspondingly minimum dT occurs when the needle is coolest suggesting high flow. Measurement of apparent sap velocity through tree trunk is recorded using 30 and 50 mm TDP thermocule sensors each attached to a heater. The sap flow is recorded at regular intervals with the help of a multichannel data logger. The sap flow velocity is multiplied with sap wood area to arrive at the total quantum of flow.

This is an indirect method of measurement and the reported results do not match with the results provided by lysimeter experiments. They appear to underestimate the transpiration rates. Simultaneous testing by this technique and the lysimeter experiment is needed to correlate and calibrate the results. Steven J. Kalma et al. (1998) compared heat pulse and deuterium tracing technique for measuring sap flow in Eucalyptus grandis trees. The measurements were made on nine trees at two sites. The trees ranged in size from 4 to 14 m with leaf areas of 5 to 35 m^2. In all trees, sap flow estimated by the deuterium tracing technique was higher than sap flow estimated by the heat pulse method, with differences of 11 to 43 per cent in eight trees and 113 per cent in one tree.

In the experiments at water technology centre, Orissa (India) the reported results of transpiration rates of both eucalyptus and casuarina plants by porometry were about two times of those determined by the heat pulse method.

Other Methods

Other methods like the ventilated chambers method, energy balance method, the deuterium (a naturally occurring isotope of hydrogen) tracing method, the left behind salt method and the CO_2 gas analyser are sometimes used to estimate the tree water use. These are all indirect methods. Their validity and correctness of reported results needs to be reviewed and standardised.

In the porometry experiments in the laboratory evaporation rate from the surface of leaves is calculated on the basis of quantity of water consumed. The water used is generally non-saline. To provide useful results the experiments should the conducted with water of varying salinity. Also, observed data of leaf area index on trees growing in the field, needs to be made available.

EXPERIMENTAL RESULTS ON TREE WATER USE

These are tabulated in Table 3.3.

Table 3.3 *Estimated water use by trees on the basis of reported experimental results*

S. No.	Specie of tree and age	Experiment location and agency	Type of experiment	Tree spacing	Water table depth and salinity	Transpiration rate	TWU adjusted value corres-ponding to 2500 tree ha^{-1}	Remarks
1	2	3	4	5	6	7	8	9
1.	E. Globules 5 yrs.	Australia Heuperman (1995)	NR	Full canopy cover. 1100 trees ha^{-1}	NR	0.6 to 1.9 A_{pan}	1.3 to 4.3 A_{pan}	
2.	E. Globules 2 yrs. E. Camaldulensis 2 yrs.	Australia Dewan quoting Greenwood (1979)	NR	NR	NR	37 ld^{-1} 21 ld^{-1}	3376 mm = 2.4 A_{pan} 1916 mm = 1.36 A_{pan}	Assuming A_{pan} = 1403 mm (North victoria)
3.	Eucalyptus E. Camaldulensis	Gratten et al. (Stribbe) 1975) –Do–	NR NR	Full cover NR	NR 10 dS m^{-1}	NR NR	1.2 to 1.5 ET crop 0.83 ET crop	
4.	E. Camaldulensis (62%), Acalia Niltica (32%) others (6%) 6 yrs.	IGNP, Rajasthan, India Kapoor (2001)	Field measurement of volume of water transpired	1900 trees ha^{-1}	0–9 m 0.4–1.8 ds m^{-1}	3446 mm 1.2 A_{pan}	1.6 A_{pan}	
5.	Willow Poplar (3 years) Alder Tamarix Eucalyptus (full grown)	Western USA Bhutta et al.	NR	NR	NR	770–1350 mm^{-1} 45 ld^{-1} 1680 mm (annual) 60 ld^{-1} 143 ld^{-1}	0.5–0.9 A_{pan} 2.73 A_{pan} 1.12 A_{pan} 3.83 A_{pan} 8.69 A_{pan}	Assuming A_{pan} = 1500 mm. Some figures are very high. Perhaps they are maximum observed rates

S. No.	Specie of tree and age	Experiment location and agency	Type of experiment	Tree spacing	Water table depth and salinity	Transpiration rate	TWU adjusted value corres-ponding to 2500 tree ha^{-1}	Remarks
1	2	3	4	5	6	7	8	9
6.	Acacia Arabica E. Territicornis E. Camaldulenosis Moris Alba Delbergia Sisoo Tamaris Articulata	Pakistan Riotzch and Marsur. Reported by Bhutta et al. (2000)	NR	NR	Under luxury water consumption conditions	129 ld^{-1} 137 ld^{-1} 143 ld^{-1} 153 ld^{-1} 156 ld^{-1} 174 ld^{-1}	11771–15877 mm yr^{-1} 8–10 A_{pan}	Reported figures are very high. Perhaps they are maximum observed rates
7.	E. Grandis age – NR	South Africa Dye (1996): Reported by Lendsberg (1999)	Estimated water loss	NR	NR	38.5–43.7 ld^{-1} (average of 200 trees)	35154 – 39870 mm yr^{-1} = 2.5 A_{pan}	
8.	Red gum E. Camaldulensis E. Grandis, 4 yrs	Victoria (Australia Morris & Webner reported by Jason (1991)	NR	NR	NR	Over 4000 mm yr^{-1} Over 4000 mm yr^{-1} Over 3500 mm yr^{-1}	2.8 A_{pan} 2.8 A_{pan} 2.5 A_{pan}	Assuming A_{pan} = 1500 mm yr^{-1}
9.	E. marginata and E. Calophylla, age 5 yrs	West Australia reported by Sharma (1984)	NR	NR	NR	3 times A_{pan} (during rainy winter period)	3.0 A_{pan}	
10.	E. Citriodora, age 6–16 months	N.R. Dabral (1970) reported by Gupta (1984)	NR	NR	Irrigated regularly	5526 mm yr^{-1}	3.68 A_{pan}	Assuming A_{pan} = 1500 mm

S. No.	Specie of tree and age	Experiment location and agency	Type of experiment	Tree spacing	Water table depth and salinity	Transpiration rate	TWU adjusted value corresponding to 2500 tree ha^{-1}	Remarks
1	2	3	4	5	6	7	8	9
11.	E. Maulata E. Cladocalyx and E. Globulus age – NR	West Australia Greenwood (1985)	NR	NR	NR	2300 mm yr^{-1} 2700 mm yr^{-1} 2700 mm yr^{-1}	1.53 A_{pan} 1.80 A_{pan} 1.80 A_{pan}	Annual evaporation on saline farms and pastures 390 mm. From trees 2330–2700 mm
12.	C. Equisetiforbia E. Tereticornis M. Azedarach P. Juliflora T. Arjuna	Haryana (India) Haryana Agricultural University Hissar	Lysimeter dia 1.45 m	One tree in each lysimeter	80–90	47 ld^{-1} 46 ld^{-1} 18 ld^{-1} 38 ld^{-1} 16 ld^{-1}	4287 mm yr^{-1} = 2.73 A_{pan} 4147 mm yr^{-1} = 2.64 A_{pan} 1642 mm yr^{-1} = 1.04 A_{pan} 3467 mm yr^{-1} = 2.21 A_{pan} 1460 mm yr^{-1} = 0.93 A_{pan}	A_{pan} = 1570 mm yr^{-1}
13.	E. Camaldulunsis age 9 yrs	Western Australia Marshall (1997) Akilan (1997)	Heat Pulse	1100 ha^{-1}	1.16 m 9 ds m^{-1}	0.66 A_{pan} 0.83 A_{pan}	1.5 A_{pan} 1.89 A_{pan}	
14.	E. Camaldulensis (5.7 m tall)	Qld (Australia Fraser (1996)	Heat Pulse	500 ha^{-1}	3.0 m 6.2 ds m^{-1}	0.189 A_{pan}	0.945 A_{pan}	
15.	E. Camaldulensis (age 10 years)	Western Australia	Heat Pulse	580 ha^{-1}	4.5 m 6.2 ds m^{-1}	0.152 A_{pan}	0.655 A_{pan}	

S. No.	Specie of tree and age	Experiment location and agency	Type of experiment	Tree spacing	Water table depth and salinity	Transpiration rate	TWU adjusted value corres-ponding to 2500 tree ha^{-1}	Remarks
1	2	3	4	5	6	7	8	9
16.	Acacia Nilotica (age 3–4 years)	Pakistan Tando Jam Khanzada (1997)	Heat Pulse	800 ha^{-1}	1.3 m 20 ds m^{-1}	0.471 A_{pan}	1.47 A_{pan}	
17.	Acacia Nilotica (age 4–5 years)	Pakistan Tando Jam Khanzada (1997)	Heat Pulse	2500 ha^{-1}	2.0 m 1.5 ds m^{-1}	0.84 A_{pan}	0.84 A_{pan}	
18.	E. Camaldulensis (age 2–3 years)	Pakistan Pacca Anna ACIAR (1996)	Heat Pulse	2500 ha^{-1}	1.5 m 5 ds m^{-1}	0.72 A_{pan}	0.72 A_{pan}	
19.	E. Carmaldulensis (age 2–3 years)	Pakistan Pacca Anna ACIAR (1996)	Heat Pulse	2500 ha^{-1}	3.0 m 5 ds m^{-1}	0.66 A_{pan}	0.66 A_{pan}	
20.	E. Micothica (age 4–5 years)	Pakistan Lahore	Heat Pulse	2000 ha^{-1}	1.5 m 3 ds m^{-1}	0.65 A_{pan}	0.81 A_{pan}	
21.	E. Camaldulensis (age 5–7 years)	Victoria (Australia ACIAR (1996)	Heat Pulse	625 ha^{-1}	2.0 m 6 ds m^{-1}	0.252 A_{pan}	1.0 A_{pan}	
22.	Caspuarina Guminghamiana (*age 5–7 years)	Victoria Australia ACIAR (1996)	Heat Pulse	625 ha^{-1}	2.0 m 6 ds m^{-1}	0.259 A_{pan}	1.04 A_{pan}	

S. No.	Specie of tree and age	Experiment location and agency	Type of experiment	Tree spacing	Water table depth and salinity	Transpiration rate	TWU adjusted value corres-ponding to 2500 tree ha^{-1}	Remarks
1	2	3	4	5	6	7	8	9
23.	Eucalyptus 26–27 months	Orissa (India) Deltaic region, WTC Bubaneswar (2009)	Heat Pulse also porometer	625 ha^{-1}	at ground surface salinity NR	22 ld^{-1} (sap flow) 4.90 to 6.24 m mol m^{-2} s^{-1}	2007 mm yr^{-1} = 1.33 A_{pan} 2783 – 3545 mm yr^{-1} = 1.85 A_{pan} – 2.36 A_{pan} (porometer)	Annual rainfall 1400 mm A_{pan} (assumed) – 1500 mm yr^{-1} Leaf area index (assumed) = 3.0
24.	Casuarina (27 months)	Orissa (India) deltaic region WTC Bhubaneshwar (2009)	Heat Pulse Porometer	NR	Depth – NR Salinity up to 2.5 ds m^{-1}	16.5 ld^{-1} (sap flow) 5.20 m mol m^{-2} s^{-1} (porometer)	1506 mm yr^{-1} = 1.0 A_{pan} 2954 mm yr^{-1} = 1.97 A_{pan}	Assuming A_{pan} = 1500 mm yr^{-1} (Conversion factor: 1542 m mol m^{-2} s^{-1} = 1 lit m^{-2} hr^{-1})
25.	Casuarina (39 months)	–Do–	–Do–	NR	Depth NR Salinity up to 2.6 ds m^{-1}	16.5 ld^{-1} (sap flow) 5.23 m mol m^{-2} s^{-1} (av. porometer)	1506 mm yr^{-1} = 1.0 A_{pan} 2971 mm yr^{-1} = 1.98 A_{pan}	Leaf area index assumed 2.77 (av) on the basis of data given by Greenwood (1984) for 7 yrs old eucalyptus trees-range 0.96 to 4.34
26.	E. Tercticornis E. Camaldulensis	Tamil Nadu (India) Agricultural University	Portable photo synthesis system	2m × 2m	Very high water table, coastal area	14.32 m mol m^{-2} s^{-1}	8135 mm yr^{-1} = 5.4 A_{pan}	

S. No.	Specie of tree and age	Experiment location and agency	Type of experiment	Tree spacing	Water table depth and salinity	Transpiration rate	TWU adjusted value corres-ponding to 2500 tree ha^{-1}	Remarks
1	2	3	4	5	6	7	8	9
27.	E. Camaldulensis T. Aphylla	Rajasthan (India) AFRI, Jodhpur (2012)	Lysimeter		Water table depth 1.0 m salinity	In litres day^{-1}	mm day^{-1} A_{pan}	
					normal	31	2828 1.9	
					12 ds m^{-1}	14	1277 0.85	
					normal	39	3558 2.37	
					12 ds m^{-1}	10	912 0.61	
					normal	24	2190 1.50	
					12 ds m^{-1}	20	1825 1.20	
28.	Eucalyptus (4 years)	Karnal (Haryana) India (CSSRI)	Lysimeter 1.2 m dia	One tree in each Lysimeter	Water Table depth 2.0 m salinity ds m^{-1}			Assuming A_{pan} = 1500 mm yr^{-1}
					0.4		2.90 A_{pan}	
					3.0		2.44 A_{pan}	
					6.0		2.01 A_{pan}	
					9.0		1.80 A_{pan}	
					12.0		1.64 A_{pan}	
29.	Eucalyptus (hybrid)	Jabalpur (India) Tropical Forest Research Institute (2912)	Lysismeter		Water table depth 0–0.75 m salinity 0.368 ds m^{-1}	3558 mm yr^{-1}	2.14 A_{pan}	Assuming A_{pan} = 1500 mm yr^{-1}

TRANSPIRATION RATE NOT LIKELY TO BE LESS THAN A_{PAN}

An attempt is made to tabulate the results of various studies to estimate annual transpiration rates from tree plantations (trees spaced 2 m × 2 m or 2500 trees ha^{-1}) as multiple of A_{pan} (annual evaporation from open pan). The results are presented in Table 3.3. The results show variance. This is so because of different types of experiments, different water level salinities, different depths of groundwater table etc. For planning of biodrainage tree plantations, it would, therefore, be advisable to carry out experiments under conditions prevailing in the irrigated area facing the threat of waterlogging to determine the expected rate of transpiration from tree plantations. However, on the basis of available data and results it can be stated, that the transpiration rate of tree plantations from shallow groundwater table (0–5 m) with salinity levels of up to 5.0 dS m^{-1}, is unlikely to be less than 1.0 A_{pan}. Therefore, for making preliminary assessment, before more dependable data becomes available, it may be reasonable to assume transpiration rate from tree plantations equal to 1.0 A_{pan}.

CONCLUSIONS

On the basis of the reported data, study results and experiments, the following conclusions can be drawn:

Tree plantations, particularly Eucalyptus species, on maturity, with full canopy cover (density about 2500 trees/ha) with no constraint on water supply (such us under waterlogged conditions) have the potential of transpiring water at the following rates:

1. 2 to 3 times A_{pan} where EC of groundwater < 0.4 dS m^{-1}.
2. 1.2 to 1.5 times A_{pan} where EC of groundwater is 0.4 – 5 dS m^{-1}.
3. 1.0 times A_{pan} where EC of groundwater is 5 to 8 dS m^{-1}.

REFERENCES

Arid Forest Research Institute, Jodhpur, India (2009). *Study of Characteristic Features Pertaining to Biodrainage Potential of Some Selected Species*, Report Submitted to the Ministry of Water Resources, New Delhi, p. 25 and 56 (Tables 7 and 21).

Babla, A.M. and M.F. Soliman (1978). 'Evaluation of the Relative Specific Effect of Cations on Plant Growth and Nutrient Absorptions', *11th Intern. Soil Sci. Congr. Edmonton*, 1: 319–20.

Bhutta, M.N. and M.R. Chaudhry (2000). 'Biological Control of Waterlogging', Paper Presented at the 8th ICID International Drainage Workshop, February 4, New Delhi, India.

Calder, I.R. (1994). *Eucalyptus Water and Sustainability: A Summary Report*, ODA Series No. 6, A Joint Research Work of Karnataka Forest Department, Mysore Paper Mills and Institute of Hydrology, Wallingford, Oxon, UK.

— (1986). 'Water Use of Eucalyptus: A Review with Special Reference to South India', *Agricultural Water Management Journal*, 11: 333–42.

Chhabra, R. and H.S. Baddesha (1985). 'Sewage Water for Afforestation', Proceedings of the National Seminar on 'Man, Forest and Environment', Haryana Forest Department, Karnal, India.

Chhabra, R. and N.P. Thakur (1998). 'Biodrainage Using Trees to Control Waterlogging and Secondary Salinisation in Canal Irrigated Areas', Proceedings of the National Seminar on 'Man, Forest and Environment', Haryana Forest Department, Karnal, India.

Dwivedi, A.P. (1992). *Principles and Practices of Indian Silviculture*, Surya Publications, Dehradun.

Food and Agriculture Organisation (FAO) (1992). 'Crop Water Requirements', Irrigation and Drainage Paper No. 24.

Grattan, S.R., M.C. Sharma et al. *Production Functions of Eucalyptus for the Design of Saline Drainage Water Re-use Systems.* quoted by Kapoor A.S. in 'Biodrainage – A Biological Option for Controlling Waterlogging and Salinity' (2001). ISBN 0-07-0110231-0, p. 88.

Greenwood and Beresford (1979). Quoted by P.L. Diwan in paper presented at the 7th ICID Workshop in Malaysia.

Greenwood, E.A.N. and J.D. Beresford (1979). 'Evaporation from Vegetation in Landscapes Developing Secondary Salinity Using the Ventilated-chamber Technique: Comparative Transpiration from Juvenile Eucalyptus above Saline Ground Water Seeps', *J. Hydrol.*, 42: 369–82.

Greenwood, E.A.N., L. Kelvin, J.D. Beresford and G.D. Watson (1985). 'Differences in Annual Evaporation between Grazed Pasture and Eucalyptus Species in Plantation on Saline Farm Catchment', *J. Hydrol.*, 78: 261.

Gupta, B.B. and J.S. Rawat (1984). *Transpiration as Affected by Soil Moisture in Eucalyptus Tereticornis Seedlings*, Forest Research Institute, Dehradun, India, January, p. 38.

Handa, A.M. and A.E. El-Enany (1994). 'Effect of NaCl Salinity on Growth and Mineral Element Contents and Gas Exchange of Plants', *Biologia Plantarum*, 36 (1): 75–81.

Haryana Agriculture University, Hissar, India (2009). *Study of Characteristic Features Pertaining to Biodrainage Potential for Waterlogged Soils of Haryana*, Report Submitted to the Ministry of Water Resources, New Delhi, pp. 15–17.

Hutton, T., P. Reece, P. Taylor and K. McEwan (1999). *Does Leaf Water Efficiency Vary Among Eucalyptus in Water Limited Environments*, Published in RIRDC (Australia) Publication No. 99/37, titled 'The Ways Trees Use Water', p. 32.

Heinrich, P.A. and R. Sands (1990). 'Water Use by Trees', Woodlots Workshop, Proceeding No. 32, Victoria, Australia, pp. 47–52.

Heuperman, A.F. (1985). *Salt and Water Dynamics Beneath a Tree Plantation Growing on a Shallow Water Table*, Department of Agriculture, Energy and Minerals, Victoria, Australia.

Jason, Alexandria (1991). 'The Role of Trees in the Sustainable Management of Irrigated Land', Proceedings of the Symposium held at the Institute of Sustainable Agriculture at Tatura, Australia, July 23, p. 89 and 93.

Kapoor, A.S. (1997). 'Biodrainage: A Case Study', Proceedings of a Workshop Held at New Delhi, India.

Karajeh, F.F. and K.K. Tanji (1991). *Agroforestry Drainage Management Model*, University of California, Davis, USA.

Landsberg, J.J. (1999). *Tree Water Use and Its Implications in Relation to Agroforestry System*, Published in RIRDC (Australia) Publication No. 99/37, titled 'The Ways Trees Use Water', p. 3, 12, 16 and 19.

Marshal and Chester (1991). Morris and Wefner (1987). Quoted by A.F. Heuperman (1995). *Salt and Water Dynamics Beneath a Tree Plantation Growing on a Shallow Water Table*, Department of Agriculture, Energy and Minerals, Victoria, Australia.

Morris, J.D. (1991). *The Water Use Process and Salinity Management in Irrigation Areas*, Proceedings of the Symposium Held at the Institute of Sustainable Agriculture at Tatura, Australia, July 23, pp. 18–20.

Pandey, S.N. and B.K. Sinha (1995). *Plant Physiology*, 3rd Edition, Har Anand, New Delhi.

Rawat, P.S., D.S. Negi, J.S. Rawat and K. Gurumurti (1985). 'Transpiration, Stomatal Behaviour and Growth of Eucalyptus Hybrid Seedlings Under Different Soil Moisture Levels', *Indian Forester*, 111 (12): 1097–112.

Robert, J. (1983). 'Forest Transpiration: A Conservative Hydrological Process', *J. Hydrology*, 66: 131–41.

Sharma, M.L. (1984). 'Evaporation from a Eucalyptus Community', *J. Agric. Water Man.*, 8: 41–56.

Stebbing (1952). Quoted from A.P. Dwivedi (1992). *Principles and Practices of Indian Silviculture*, Surya Publications, Dehradun.

Sumayao, C.R., G.T. Kenemasu and T. Hodgest (1977). 'Soil Moisture Effects on Transpiration and Net CO_2 Exchange of Sorghum', *Agric. Meteorology*, 18: 401–408.

Tamil Nadu Agricultural University (2012). *Screening Superior Genotypes of Eucalyptus for Biodrainage Through Ecophysiological Approaches*, Report Submitted to the Ministry of Water Resources, New Delhi, p. 25 and 26 (Tables 2 and 3).

Tanji, K.K. and F.F. Karajeh (1993). 'Saline Drain Water Re-use in Agroforestry System', *Journal of Irrigation and Drainage Engineering*, 119 (1): 170–80.

Thorburn, Peter J. (1999). *The Limits to Evaporation from Shallow Saline Water Tables: The Big Picture?* Published in RIRDC (Australia) Publication No. 99/36, titled 'Agroforestry Over Shallow Water Tables', pp. 18–20.

Tropical Forest Research Institute, Jabalpur, India (2012). *Identification of Suitable Tree Species and Other Vegetation for Biodrainage in Bargi Command Area*, Report Submitted to the Ministry of Water Resources, New Delhi, p. 70 (Table 7).

Water Technology Centre, Bhubaneswar, India (2009). *Fitting in and Evaluation of Biodrainage in Different Agroecological Situations of Deltaic Region of Orissa*, Report Submitted to the Ministry of Water Resources, New Delhi.

Wood (1925). Quoted from S.N. Pandey and B.K. Sinha (1995). *Plant Physiology*, 3rd Edition, Har Anand, New Delhi.

4

Mineral Absorption by Plants

▼ MINERAL CONTENTS IN PLANTS

Under normal conditions of growth, all green plants are autotrophic and they require the supply of only inorganic substances from outside. With the exception of hydrogen, carbon and oxygen, all other inorganic plant requirements are obtained directly or indirectly from the soil. Salts of nitrogen, phosphorus and potassium (NPK) are used in agriculture because these are generally deficient in arable soils. The micronutrients added to the soils are absorbed by crop plants from soil water through the root system. Alkaline soil, with pH higher than 7.5, commonly found in arid regions, can be corrected by the addition of elemental powdered sulphur to the soil.

Earlier it was assumed that mineral salts are carried into the plants passively with the absorption of water. Pfeffer (1900) stated that the nature of plasma membrane is such as to render it possible for a substance to combine chemically with the plasmatic elements, thus being transmitted internally and set free again. The work done in the 1930s finally recognised that the uptake of salts is predominantly active and not passive as was thought earlier. The soil serves as a main source of mineral salts in which crystals with a central nucleus called 'micelle' are present in the colloidal form. The micelle are negatively charged and in order to maintain a balance, these attract and hold positively charged ions on the surface of colloidal clay crystals. The balance is always maintained. When disturbed due to salt absorption by the roots, the equilibrium is restored by the transfer of some of the absorbed ions in the solution. Minerals of soil, occurring in ionic forms become available to plant roots only when they are broken down physically or chemically. The process described by Pandey and Sinha (1995) is as below:

Soil Solution

The available minerals of soil occur in ionic forms. The common catonic forms are K, Mg, Ca, Fe, Mn, Cu, Zn, Co while the anionic forms are P, B, S and Cl. These ions are found either in the form of loosely absorbed ions or as firmly absorbed ions. The loosely absorbed ions can be easily displaced on decreasing their own concentration in the soil solution while the firmly absorbed ions can be replaced by other ions which have more affinity for the colloid or ion exchange. The ion exchange may be either cation exchange or anion exchange. Although not fixed, for some of the cations the order for cation retentive capacity of colloids is as follows:

$$H^+ > Ca^{++} > Mg^{++} > K^+ > NH_4^+ > Na^+$$

In plants these electrolytes penetrate in the form of ions and in general, the entry of one type of ion is accompanied with the entry of another type of ion of equal electrostatic charge. The monovalent cations such as K and Na are absorbed more rapidly as compared to the bivalent or polyvalent cations such as Ca and Mg.

In case of an unbalanced ratio of cation and anion uptake, the equilibrium is maintained by certain changes in the ionic composition of the cell sap and salt solution. For example, when the cation uptake is more, the cell produces organic acids resulting in more anions to balance the excess cations. The hydrogen ions of the organic acids, however, pass out into the external medium to compensate for the cations which have

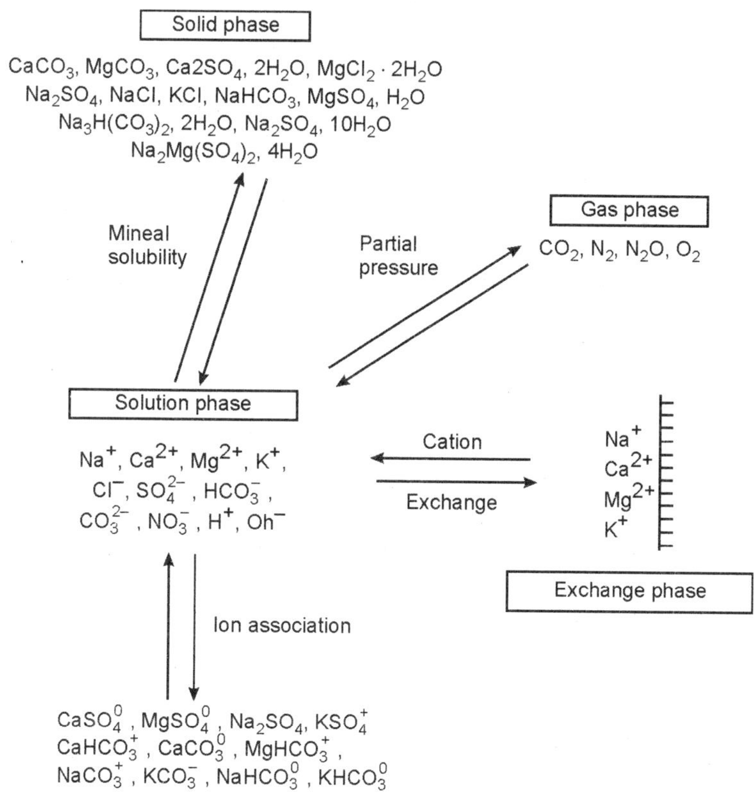

Figure 4.1 *Interactive chemical reactions in soil water systems (Tanji, 1972)*

been removed. The readily available ionic forms of water, the H ions, and OH ions, also compensate for the shortage of cations and anions wherever required.

The complex chemical interactions that take place between the solution, solid, exchanger and gas phases are illustrated in Figure 4.1 (Tanji, 1972). A change in soil water content by irrigation and rainfall or evapotranspiration by plants will cause the equilibrium to shift due to mineral precipitation or dissolution, association or dissociation of ion pairs, adsorption or desorption of cations, and emission or absorption of gases (Paul et al., 1966). Moreover, the free ions and ion pairs are subject to transport by diffusion and convection, e.g., water movement.

Salt Precipitation in Soils

Some precipitation of less soluble salts such as gypsum and carbonates occurs from soil water solutions due to complete reactions involving plant roots and the minerals present in the irrigation water. The relative amounts of salt that may chemically precipitate would

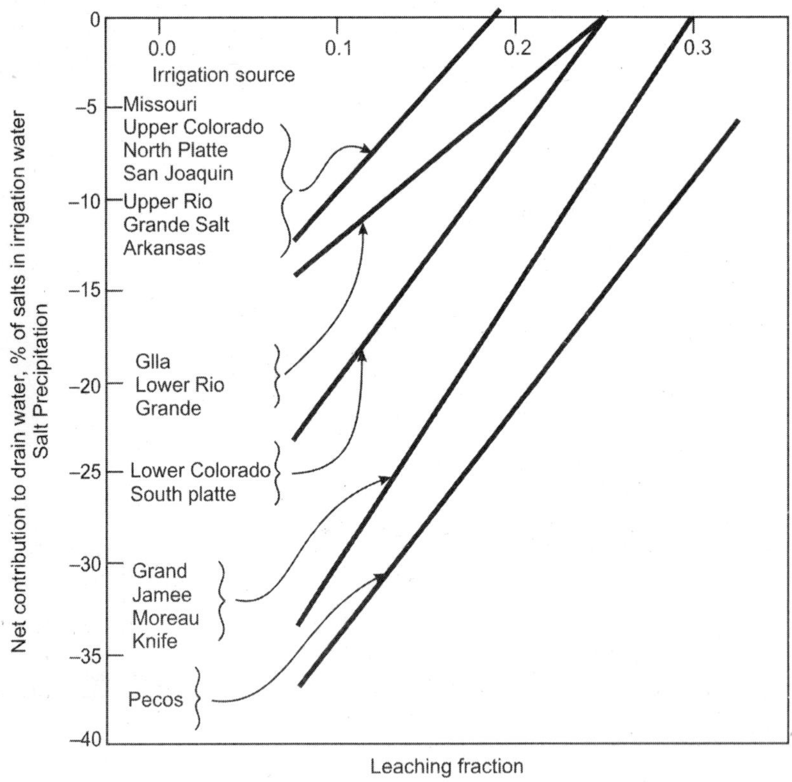

Figure 4.2 *Net contribution of salt precipitation to salt content of drainage water, expressed as a per cent of the salt applied from various river waters (each line represents an average of the percentages for calcareous and noncalcareous pachappa sandy loam soil). (Rhoades et al., 1974)*

depend upon the salt content in irrigation water and in the soil, and the leaching fraction. Rhoades (1974) demonstrated in a three-year lysimeter study that the amount of salt leached below the root zone may be less than that applied when leaching fraction is less than 0.25 and salt content in irrigation water is more than 100 mg/*l* as some salts precipitate in the root zone and become stored in the soil profile.

Figure 4.2 shows net contribution of salt precipitation to salt content on drainage water, expressed as a per cent of the salt applied from various river waters. All irrigation waters illustrated had salt content exceeding 500 mg/l. At low leaching fraction (L = 0.1) 20 per cent or more of the salt in irrigated water was found to precipitate. At moderate leaching fractions (L = 0.2), about 15 per cent (average) of the salt in irrigation water was found to precipitate. Salt precipitation is a significant part in calculating the salt balance, particularly when the leaching fraction is low.

In the soil water, the salts occur in the form of ions, and plant roots also absorb them in the form of ions. Electrostatic balance is always maintained in the soil water solution as well as inside the plant body. Plant roots have greater affinity to absorb HCO_3 anions in comparison to the cations of calcium, magnesium, sodium etc. To restore the electrostatic balance equivalent hydroxide (OH) anions are secreted within the plant body and excreted out into the soil solution, where they combine with the unabsorbed cations of Ca, Na Mg etc., to form hydroxides (Figure 4.3). Carbon dioxide (CO_2) produced by roots during respiration on reaction with hydrocides produces carbonates, that being less soluble (particularly those of calcium and magnesium) are precipitated, and thus removed from the soil solution.

Rajkumar (1998) describes how irrigation of saline sodic soils with waters of low salt content can promote Na_2CO_3 formation as a result of hydrolysis of exchangeable Na as follows:

$$\text{Ca x} + 2\text{Na Cl} = \text{Na}_2\text{ x} + \text{CaCl}_2$$

$$\text{Na}_2\text{ x} + 2\text{H2O} = \text{H}_2\text{ x} + 2\text{NaOH}$$

$$2\text{NaOH} + \text{CO}_2 = \text{Na2CO3} + \text{H}_2\text{O}$$

where x denotes soil cation exchange sites (Donner and Lyn 1989).

In a study in the IGNP, India, the canal water contained 260 mg/l total soluble solids (cations 24% and anions 76%). The HCO_3 anions constituted 79 per cent of the total anions. The groundwater samples drawn from piezometers in a tree planted area adjacent to the canal were found to contain 920 mg/l total TSS (cations 30% and anions 70%). The HCO_3^- anions constituted 60 per cent of the total anions. The Ca^+ + Mg^+ cations, which constituted 73 per cent of total cation weight in the canal water were found to be only 13 per cent in the groundwater. The weight of Na^+ cations which was 24 per cent in the canal water was found to be 85 per cent in the groundwater. The study results show significantly less content of Ca^+, Mg^+ cations and HCO_3 anions in the groundwater as compared to those in the canal water. This indicates that there occurs significant precipitation of Ca^+ and Mg^+ cations from the soil water solution.

According to USSLS (1954), in waters containing high concentrations of bicarbonate ions, there is tendency for calcium and magnesium to precipitate as carbonates as the soil solution becomes more concentrated. This reaction does not go to completion under ordinary circumstances, but in so far as it does proceed, the concentrations of calcium and magnesium are reduced and the relative proportion of sodium is increased.

The quantity of Ca^+ and Mg^+ cations that are precipitated from the soil water solution in this manner can be quite significant. Research is needed to find reliable methods of quantitative assessment. It is possible that when natural river waters of low salinity are used for irrigation, the total cations of Ca^+, Mg^+, K^+ absorbed by the plants together with the cations precipitated from the soil water solution in the form of insoluble carbonates may equal or exceed the total cations contained in the irrigation water, the excess cations coming from the natural soils. Calcareous soils can provide sufficient dissolved calcium and at the same time decrease the sodium hazard appreciably.

Availability of Mineral Salts

Mineral salts are found either as soluble fractions of soil solution or as adsorbed ions on the surface of colloidal particles. Many physiologists believe that the uninterrupted supply of mineral nutrients from the adsorbed fractions is possibly caused by ionic exchange. So far, two theories have been put forward to explain the ionic exchange; these are: (i) carbon dioxide hypothesis, and (ii) cation exchange hypothesis.

(i) Carbon Dioxide Hypothesis

According to this theory, CO_2 produced by roots during respiration, reacts with water to produce carbonic acid which in turn dissociates into hydrogen ions and bicarbonate ions. The hydrogen ions change places with cations adsorbed on the colloidal particles and the bicarbonate ions release the adsorbed anions to make both cations and anions available to the roots of plants close by. However, the most remarkable point in this hypothesis is the movement of ions between adsorption sites in the soil and roots.

(ii) Cation Exchange Hypothesis

This theory, by eliminating the shortcoming of ionic movement between adsorption sites in the soil and roots, postulates that ions are not completely static. They are always oscillating around their adsorption surface and when the oscillation volumes of the ions on the roots and those on the colloidal particles overlap with each other, ion exchange occurs. An equilibrium is, however, always maintained between the dissolved fractions as any depletion in the soil solution is immediately covered by the movement of ions.

Although the attempts of these hypotheses explaining ionic availability in soil are not universally accepted, yet the forces involved in both the theories are considered to supply additional quantities of mineral salts to roots.

The various theories proposed to explain the mechanism of mineral salts absorption can be placed in the following two broad categories:

(i) Passive Absorption

It includes the theories of diffusion, ion exchange, Donnan equilibrium and Mass flow.

(ii) Active Absorption

It includes theories related with the carrier concept such as cytochrome pump hypothesis, ATP theories, protein lecithin as carrier theories, etc.

There are many theories to explain the mechanism of mineral salt absorption but no conclusion has as yet been arrived at. The passive absorption concept has had to face a lot of criticism while active absorption concepts have been widely supported. However, there are certain experimental evidences in support of the passive absorption concept such as the effect of transpiration on salt absorption and experiments of Lopushinksy (1960) which remain to be explained by the supporters of the active absorption concept. On the other hand, evidences in the support of active absorption concept are quite high in number which could not be explained by supporters of the passive absorption concept. Therefore, it is believed that one or both of these mechanisms may be involved in salt absorption. According to Wilkins (1969), the knowledge is in a quite primitive stage and there is need to fill this gap by future investigators.

CHEMICAL ELEMENTS IN PLANTS

The essential chemical elements generally found in plants are: carbon, hydrogen, oxygen, nitrogen, sulphur, phosphorus, potassium, calcium, magnesium and iron (C, H, O, N, S, P, K, Ca, Na, Fe). Carbon and oxygen enter ordinary green plants as carbon dioxide (CO_2) while hydrogen and oxygen enter as water (H_2O). Small amounts of oxygen are absorbed from the soil by active roots. Green plants form carbohydrates from carbon dioxide and water using the energy of sunlight. They derive energy as well as material from chemical compounds.

Four-fifths of the atmospheric air is free nitrogen, but ordinary green plants do not assimilate nitrogen from air and instead receive their supply of nitrogen through organic compounds in the soil.

When plants are completely burnt, all of the carbon, hydrogen, oxygen and nitrogen are given off as gases, but there remain small amounts of many other essential and non-essential elements in the form of incombustible ash. The total ash of ordinary plants constitutes only about 5 per cent of the total dry weight. The other essential elements (S, P, K, Ca, Na, Fe) of the ash are absorbed by ordinary plants from the soil in the form of inorganic salts. Little is known about how the small amounts of essential ash constituents are used in the plant. Sulphur occurs in proteins, phosphorus in nucleins, magnesium in chlorophyll and iron is essential for the formation of chlorophyll. Grasses accumulate silica in the epidermis.

Different parts of the same individual plants differ in ash content. Leaves are generally richer in ash than the stems and roots. The ash content alters with the age of the organ or tissue. A soil may be unproductive if it contains too much or too little of soluble mineral salts or because it contains highly injurious substances in toxic amounts.

All substances supplying essential and non-essential elements are absorbed by diffusion in solution. When the transpiration rate is high, these may enter the roots from the soil by a mass streaming and filtration of the soil solution through the peripheral cells, to the xylem vessels. To be able to enter plant cells, these substances must be dissolved in water. They diffuse through the peripheral, water-impregnated cell walls, into the protoplasm. Carbon dioxide and oxygen diffuse through the suberin or lignin of cell walls that are impregnated with one of these substances, as well as through the imbibed water.

MINERAL CONSTITUENTS IN A PLANT

Plant physiology text books describe mineral contents in plants. According to Palladin, the bulk of dry weight of a plant is made up of:

Carbon	45.0 per cent
Oxygen	42.0 per cent
Hydrogen	6.5 per cent
Nitrogen	1.5 per cent
Mineral constituents	5.0 per cent
	100 per cent

Pandey and Sinha give the average chemical composition of plant body as follows:

	By weight
Carbon	20 per cent
Oxygen	62 per cent
Hydrogen	10 per cent
Nitrogen	3 per cent
Mineral constituents	5 per cent
	100 per cent

Some other reported results are shown in Table 4.1.

Table 4.1 *Mineral contents in plants*

Reference	*Plant component*	*Mineral content (% of dry weight)*
Principles of Plant Physiology by Walter Stiles (1995)	*Aethalium Septicum Plasmodium*	Calcium salts except calcium carbonate 0.5 Other salts 6.5 Total 7.0
	Myxomycete Plasmodium	Mineral matter 4.4
	Cabbage leaves Cytoplasm	Ash 6.45
	Zea mais	
	Leaves	7.71
	Stems	4.85
	Grain	0.87
	Cobs	0.62
	Roots	2.12
Plant Physiology by H.S. Srivastava (1996)	Maize	3.13 (excluding silicon)
		4.30 (including silicon)

EXPERIMENT AND RESEARCH RESULTS

Several researchers have conducted experiments to determine the mineral contents in plants. The reported results of some studies are detailed below:

Central Arid Zone Research Institute (CAZRI), Jodhpur, India

Dhir et al. (1984) determined the mineral content in the foliage of 35 commonly found species of trees, shrubs and grasses after collecting and analysing 294 plant species from 33 locations spread all over arid Rajasthan (India). The reported results are given in Table 4.2.

Table 4.2 *Concentrations of macro-elements in tree, shrub and grass species (mean values) in % (dry weight basis)*

Species	Ca	Mg	K	Na	P	Total
			Tree species			
Prosopis cineraria	1.85	0.31	1.40	0.058	0.169	
Acacia nilotica	1.96	0.22	1.11	0.066	0.178	
Acacia senegal	4.80	0.69	1.33	0.065	0.153	
Acacia tortilis	2.16	0.35	1.44	0.058	0.137	
Salvadora oleoides	12.67	0.54	1.10	0.377	0.091	
Tecomella undulata	2.24	0.49	1.28	0.078	0.169	
Ailanthes excelsa	3.16	0.75	2.05	0.068	0.414	
Prosopis juliflora	3.18	0.53	1.51	0.085	0.208	
Albizzia lebbek	2.67	0.40	1.23	0.035	0.147	
Azadirachta indica	3.05	0.50	1.49	0.056	0.194	
Mean	3.77	0.48	1.39	0.095	0.203	5.938
			Shrub species			
Zizyphus nummularia	1.84	0.29	1.15	0.072	0.163	
Capparis aphylla	0.85	0.23	1.81	0.037	0.186	
Calligonum polygonoides	3.48	0.71	1.33	0.073	0.124	
Balanites aegyptiaca	3.33	0.60	1.62	0.272	0.162	
Leptadenia pyrotechnica	2.72	0.56	1.36	0.077	0.140	
Maytenus emarginata	4.46	0.92	1.22	0.171	0.242	
Lycium barbarum	1.43	0.58	1.24	0.072	0.202	
Tamarix auriculata	4.56	0.54	1.04	0.618	0.101	
Acacia jacquemontii	1.19	0.59	0.51	0.128	0.128	
Mean	2.65	0.56	1.25	0.168	0.161	5.59
			Grass species			
Cenchrus biflorus	0.72	0.42	1.23	0.021	0.148	
Cenchrus ciliaris	0.69	0.28	0.79	0.115	0.155	
Cenchrus setigerus	0.59	0.36	1.34	0.071	0.159	
Lasiurus sindicus	1.16	0.33	0.66	0.026	0.112	
Panicum turgidum	0.87	0.33	0.50	0.071	0.127	
Panicum antidotale	0.63	0.29	1.12	0.193	0.146	
Brachiaria ramosa	0.88	0.64	1.37	0.049	0.224	
Digitaria Marginata	0.72	0.84	1.11	0.035	0.163	
Dicanthium annulatum	0.51	0.26	0.66	0.027	0.112	
Aristida funculata	0.50	0.24	0.35	0.020	0.143	
Eragrostis tremula	0.61	0.21	0.32	0.018	0.203	

Species	Ca	Mg	K	Na	P	Total
Eceusine compressa	0.71	0.26	0.35	0.228	0.127	
Cymbopogon jawarancusa	0.63	0.20	0.49	0.031	0.146	
Cynodon dactylon	1.22	0.30	0.65	0.037	0.140	
Dactyloctenium sindicum	0.86	0.32	0.41	0.149	0.146	
Dactyloctenium aegyptium	1.12	0.49	0.61	0.167	0.178	
Mean	0.80	0.36	0.75	0.082	0.152	2.144

Dhir et al. (1985) also studied the mineral content in the foliage of 20 arid zone tree and shrub species under similar soil and climatic conditions. The reported results are shown in Table 4.3.

Table 4.3 *Average concentration (dry matter basis) of chemical elements in different species*

Species	Family	Ca (%)	Mg (%)	K (%)	P (%)	Na (ppm)	Mn (ppm)	Zn (ppm)	Cu (ppm)	Total Ca + Mg + K + P
Acacia aneura	Mimosaceae	2.25	0.18	0.53	0.07	677	84.6	19.0	7.7	3.03
Acacia calycina	Mimosaceae	4.19	0.25	0.91	0.15	386	28.8	17.9	7.7	5.50
Acacia ciliata	Mimosaceae	4.31	0.37	0.90	0.12	243	63.4	17.8	7.7	5.70
Acacia ligulata	Mimosaceae	2.00	0.30	0.50	0.09	228	50.0	17.9	7.7	2.89
Acacia nilotica	Mimosaceae	2.69	0.30	0.92	0.15	254	50.0	32.4	16.9	4.06
Acacia senegal	Mimosaceae	3.62	0.76	1.19	0.15	465	53.8	33.3	14.6	5.72
Acacia tortilis	Mimosaceae	2.62	0.37	1.44	0.15	375	71.0	28.3	8.4	4.58
Ailanthes excelsa	Simaroubaceae	2.97	0.64	1.60	0.15	534	63.5	26.2	8.0	5.36
Albizzia lebbek	Mimosaceae	2.61	0.37	0.81	0.11	353	63.4	30.4	9.6	3.90
Azadirachta indica	Meliaceae	2.92	0.55	0.92	0.14	376	42.4	24.8	8.1	4.53
Cassia auriculata	Caesalpiniaceae	1.90	0.50	1.25	0.11	347*	48.1	26.2	14.1†	3.76
Delbergia sissoo	Papilianaceae	2.86	0.49	1.37	0.15	337	164.0	58.5	13.8	4.87

Species	Family	Ca (%)	Mg (%)	K (%)	P (%)	Na (ppm)	Mn (ppm)	Zn (ppm)	Cu (ppm)	Total Ca + Mg + K + P
Dichrosta-chys nutans	Mimosaceae	1.90	0.39	1.65	0.17	263	44.7	19.7	7.7	4.11
Eucalyptus camaldulen-sis	Myrtaceae	1.47	0.21	1.11	0.14	482	54.0	40.5	9.0	2.93
Moringa oleifera	Moringaceae	1.47	0.16	1.72	0.23	573	44.2	25.0	10.3	3.58
Prosopis cineraria	Mimosaceae	2.34	0.28	1.36	0.10	406	31.7	23.9	11.6	4.08
Salvadora oleoides	Salvadoraceae	6.27	0.41	0.90	0.09	2362	59.6	21.4	10.3	7.67
Tecomella undulata	Bignoniaceae	2.06	0.41	1.45	0.13	1045	46.1	27.4	14.1	4.05
Zizyphus mauritiana	Rhamnaceae	1.75	0.40	0.88	0.11	521	59.6	38.2	9.0	3.14
Zizyphus nummularia	Rhamnaceae	1.44	0.31	1.68	0.12	543	65.7	44.5	11.5	3.55
Mean		2.68	0.38	1.16	0.13	534	61.9	28.9	10.4	4.35
									Average: 4.35	

Indian Council of Agriculture Research (ICAR)

ICAR in its report entitled, "Desertification and its Control" (1977), highlights the chemical analysis of some desert feed species as shown in Table 4.4.

Table 4.4 *Palatability ratings and chemical analysis of individual top-feed species*

(Figures Represent the Percentages of the Constituent on Dry Wt. Basis)

Botanical name	Common name (Local)	Palatability ratings	Crude protein	Crude fibre	Nitrogen free extract	Ash	P	Ca	Mg	Total P+Ca + Mg
Acacia arabia Wild	Babul	Good	13.9	9.2	69.8	7.1	0.1	2.6	0.4	3.1
Acacia senegal Wild	Kumat	Moderate	10.3	9.5	65.7	16.4	0.05	6.9	0.6	7.55
Albizzia lebbek Beneath	Siris	Moderate	29.2	25.3	43.8	7.5	0.2	1.8	0.5	2.5
Anogeissus pendula edgew	Dholkra	Moderate	7.6	19.0	65.3	8.1	0.1	3.5	0.3	3.9
Anogeissus rtoundifolia	—	Moderate	8.9	20.0	62.2	8.6	0.1	3.7	0.5	4.3
Azadirachta indica A.juss.	Neem	Fair	12.4	11.4	66.6	9.4	0.1	3.4	0.6	4.1
Calligonum	Phog	Moderate	7.4	20.7	64.2	7.6	0.1	2.4	0.8	3.3

Botanical name	Common name (Local)	Palatability ratings	Crude protein	Crude fibre	Nitrogen free extract	Ash	P	Ca	Mg	Total P+Ca + Mg
Polygonoides linn.										
Cassia auriculata linn.	Tarwad	Poor	10.9	12.5	67.3	9.1	0.1	3.7	0.7	4.5
Grewia tenax Forsk	Gangan	Fair	10.3	18.9	61.1	9.5	0.1	3.8	0.5	4.4
Gymnosporia spinosa Forel	Kankora	Fair	9.3	19.0	63.5	8.1	0.2	1.8	0.8	2.8
Prosopis juliflora	Bilayathi babool	Fair	21.4	20.8	50.0	7.7	0.2	1.5	0.5	2.2
Prosopis spicigera Linn.	Khejri	Good	13.9	20.3	59.2	6.5	0.2	1.9	0.5	2.6
Salvadora oleoides dene	Pilujal	Good	9.6	9.3	40.2	40.8	0.1	11.9*	0.7	12.7
Salvadora persica linn.	Kharajal	Good	14.2	9.4	44.9	31.4	0.2	8.8	0.7	9.7
Tecomella undulata seem	Rohida	Fair	12.2	15.8	63.0	8.9	0.2	3.4	0.9	4.5
Zizyphus nummularia rox	Bordi	Good	11.7	16.0	65.0	7.2	0.2	1.6	0.3	2.1
								Average: 4.64		

*Contains free Ca_2SO_4

Note: Palatability ratings of 'good', 'moderate', 'fair' and 'poor' were arrived at by conducting trials with sheep, following the 'cafeteria technique'.

Source: Ganguli, Kaul and Nambiar, 1964.

The phosphorus + calcium + magnesium content in the reported species ranges from 2.1 to 12.7 per cent, with a mean value of 4.46 per cent. The potassium and sodium contents are not reported. Allowing for these, the range of the total macro-element content would be from about 3 to 14 per cent, with a mean value of about 6 per cent.

In "Chemical Composition and Nutritive Value of Indian Feeds and Feedings of Farm Animals", ICAR gives the mineral contents of some feeds on a dry matter basis as shown in Table 4.5.

Table 4.5 *Mineral contents of Indian feeds on dry matter basis*

International feed name	International feed no.	Ash (%)	Calcium%	Phosphorus %	Mg %	Na %	K %	Total minerals
(1)	(2)	(3)	(4)	(5)	(6)	(7)	(8)	(9)
1. DRY FORAGES AND ROUGHAGES								
Cicer arietinum, Chickpea								
–hay, sun–cured, midbloom	—	11.2	1.13	0.07	0.29	0.25	2.48	4.22
–hay, sun–cured, straw	1–01–217	6.2	0.34	0.12	0.22	0.06	2.42	3.16

International feed name	International feed no.	Ash (%)	Calcium%	Phosphorus %	Mg %	Na %	K %	Total minerals
Horgeum vulgare, Barley	1–00–498	6.1	0.33	0.08	0.21	0.20	1.31	2.13
–straw		8.1	0.31	0.07	0.09	0.22	2.23	2.92
Phaseolus aconitifolius, Moth bean								
–hay, sun–cured	1–00614	12.0	2.04	0.26	0.72	0.22	1.68	4.92
Triticum aestivum, Wheat								
–hay, sun–cured	1–05–172	7.2	0.21	0.22	0.24	0.04	1.49	2.20
–straw	1–05–175	8.2	0.26	0.09	0.07	—	—	
–straw, ground	1–10–873	13.3	0.30	0.07	0.07	0.21	4.1	4.75
–straw, ground	1–10–873	14.1	0.48	0.16	—	—	7.5	8.14
zea mays, Maize								
–aerial part w/o ears, w/o husks sun–cured, mature	1–12–179	10.5	0.56	0.11	—	—	—	0.67

Indian Council of Medical Research (ICMR)

The Indian Council of Medical Research in 'Nutritive Value of Indian Foods', gives approximate values of minerals in common foods (grams per 100 grams of edible portion) as shown in Table 4.6.

Table 4.6 *Minerals in common foods all the values are per 100 gm of edible portion*

S.No.	Name of the food stuff	Moisture (g)	Minerals (g)	% on dry weight basis	Average%
(1)	(2)	(3)	(4)	(5)	(6)
CEREAL GRAINS AND PRODUCTS					
1.	BAJRA	12.4	2.3	2.62	
2.	BARLEY	12.5	1.2	1.37	
3.	ITALIAN MILLET	11.2	3.3	1.81	1.99
4.	JOWAR	11.9	1.6	2.43	
5.	MAIZE, tender	67.1	0.8	1.72	
6.	WHEAT, whole	12.8	1.5		
PULSES AND LEGUMES					
7.	BENGAL GRAM, whole	9.8	3.0	3.32	
8.	BLACK GRAM, dal	10.9	3.2	3.59	
9.	COW PEA	13.4	3.2	3.65	
10.	FIELD BEAN, dry	9.6	3.2	3.54	3.56
11.	GREEN GRAM, whole	10.4	3.5	3.92	

S.No.	Name of the food stuff	Moisture (g)	Minerals (g)	% on dry weight basis	Average%
(1)	(2)	(3)	(4)	(5)	(6)
12.	MOTH BEANS	10.8	3.5	3.88	
13.	PEAS green	72.9	0.8	2.95	
14.	RAJMAH	12.0	3.2	3.64	
LEAFY VEGETABLES					
15.	BENGAL GRAM LEAVES	73.4	2.1	7.89	
16.	BROAD BEAN LEAVES	77.6	1.3	5.55	
17.	CABBAGE	91.9	0.6	7.41	
18.	CARROT LEAVES	76.6	2.8	11.96	
19.	CAULIFLOWER, Green	80.0	3.2	16.00	
20.	CORIANDER LEAVES	86.3	2.3	16.79	
21.	COW PEA LEAVES	89.0	1.6	14.54	
22.	LETTUCE	93.4	1.2	18.18	14.31
23.	MUSTARD LEAVES	89.8	1.6	15.68	
24.	RADISH LEAVES	90.8	1.6	17.39	
25.	RAPE stem	91.4	1.4	16.28	
26.	RAPE leaves	84.9	2.5	16.55	
27.	SAFFLOWER Leaves	91.1	1.3	14.60	
28.	SPINACH	92.1	1.7	21.52	
ROOTS AND TUBERS					
29.	ONION, small	84.3	0.6	3.82	
30.	POTATO	74.7	0.6	2.37	6.67
31.	RADISH, pink	90.8	0.9	9.78	
32.	RADISH, white	94.4	0.6	10.71	
OTHER VEGETABLES					
33.	BROAD BEANS	85.4	0.8	5.48	
34.	CAULIFLOWER	90.8	1.0	10.87	
35.	CLUSTER BEANS	81.0	1.4	7.37	8.10
36.	CUCUMBER	96.3	0.3	8.11	
37.	TOMATO, green	93.1	0.6	8.69	

Note: The reported values under columns (3) and (4) are per 100 gms of green matter. Converting these on dry weight basis, the mineral content values would be as shown under columns (5) and (6).

Other Results

Some other reported results on mineral content in crop are shown in Table 4.7.

Table 4.7 *Mineral contents (% of dry weight) in some crops*

Reference	*Wheat*		*Cotton*		*Groundnut*		*Gram*	
	Grain	By-product	Grain	By-product	Grain	By-product	Grain	By-product
K.C. Sen (ICAR) (Total ash)	1.9 to 3.337			15.85 to 30.60 (bran)		7.14 (husk) 8.02 (stem)		5.55 to 6.65
S.L. Chopra & J.S. Kanwar		6.8 (bran) 12.9 (bhusa) 6.4 (hay)	4.0 to 5.3 (seed)			13.7	2.4	9.5 to 10.1 (bhusa)
R.S. Reddy ICAR (1988)						4.5 (shell)		
H.L.S. Tandon FDCU (1993)							2.5 to 3.2	
National Institute of Nutrition ICMR (Hyderabad) (1989)	1.5 to 3.5				2.5			

United States, Experience

The US Department of Agriculture mentions, in Handbook No. 60 "Consider alfa alfa growing in the Imperial Valley, California where 6 tonnes per acre of sun-cured hay is a common yield. The salt added to the soil in the irrigation water consumed by this crop would be about 4 tonnes. Of this salt, not more than 0.4 tonne would be removed by the harvested crop." Two inferences can be drawn from this – first that the salt content in irrigation water approximated 1000 mg/1 (assuming total depth of irrigation water as 100 cm), and second the salt content in harvested crop of Alfa Alfa (sun-cured) was estimated to be up to 6.67 per cent.

ASCE reports that the average amount of salt contained in mature crops of Alfa Alfa, barley, corn silage, Sudangrass, and sweet clover grown in Texas Rio Grande area was 3.6 per cent (Lyerly and Longenecker, 1962). At intermediate levels of salinity, Chapman (1966) reported that the salt content of Alfa Alfa, corn and sorghum was about 3 per cent of the dry tissue weight. In one study, water from the Pecos river with an electrical conductivity, *Ec* of 3.3 ds/m was applied to Alfa Alfa grown on sandy loam soil. The amount of salt removed in the harvest was 3% of the dry forage for leaching fractions varying from 0.1 to 0.3 (Rhoades et al., 1974). Francois (1981) reported a salt content of 3 to 4 per cent for Alfa Alfa grown under saline conditions.

International Laboratory Tests

Martin et al. (1987) report that the methods and apparatus used for plant analysis are not always fully comparable. Monitoring procedures were evolved to compare analytical results obtained by a number of laboratories on the same samples. An international group, the Inter-institute committee on Analytical Techniques for Foliar Diagnosis set up in 1959, provides such laboratories with samples of fifteen species which have been analysed by its members and for which the mean analytical values and their statistical properties have been published. The determination of P, K, Na, Ca, Mg was done by the method of ashing. The reagents used were concentrated HCl (Sp.gr. 1.9), HF and demineralised water. Ca and Mg were determined by atomic absorption spectrometry, K and Na by flame photometry and P by calorimetry. The number of laborations taking part was between 18 to 23 for different crops. The analytical results are reported along with 'normality' 'arithmetic mean' 'median', 'variance' and 'standard variation' values. The reported results (median values) of K, Ca and Mg content in the crops are shown in Table 4.8.

Table 4.8 *Mineral content (%) in crops by ash analysis*

Name of crop	*Mineral content %*				
	K	*Ca*	*Mg*	*K + Ca + Mg*	*Overall average*
Artichoke (Italy)	0.950	2.960	0.400	4.310	
Codia discolor (New Caledonia)	0.363	0.830	0.281	1.471	
Cotton (Africa)	1.845	3.100	0.413	5.358	
Eucalyptus (Spain)	0.680	1.583	0.207	2.470	
Rubber (Ivory coast)	0.920	1.060	0.330	2.310	
Maize (Spain)	1.930	1.120	0.361	3.311	3.58
Olive (Spain)	0.500	2.610	0.130	3.240	
Orange (Spain)	1.500	4.890	0.320	6.710	
Palm (Africa)	0.890	0.802	0.296	1.988	
Peach (France)	2.180	2.160	0.475	4.815	
Apple (golden delicious, Belgium)	1.440	1.270	0.283	2.993	
Apple (coes orange Pippin, France)	1.040	2.150	0.296	3.986	
Vine (France)	1.230	2.550	0.280	4.060	

INDIRA GANDHI NAHAR PROJECT, RAJASTHAN, INDIA TEST RESULTS

The mineral content and composition in plants may differ not only from species to species but also within the same species growing under different soil water and climatic conditions. Different organs of the same plant differ markedly in their selectivity in accumulating various mineral ions. Samples of tree plantations along the main canal RD (952–957), IGNP, Rajasthan, India were collected (1998) and tested for chemical contents. Samples of different parts of trees, viz., the bark, stems, twigs and leaves were collected and analysed separately. Samples were taken from different species of trees growing in the area. A similar analysis was done on important crops grown in the area.

The collected samples were washed with liquid detergent to remove the waxy coating on the surface and soil particles with N/10 hydrochloric acid to remove metallic contamination and finally with de-ionised water to wash the previous two solutions. After wiping the extra moisture, the samples were dried in an oven at 70°C. The dried samples were cut into small pieces by a scatter before grinding in a pestle and mortar and finally sieved.

The diacid wet digestion method was followed to determine the mineral contents. One gram powdered material was placed in a 50 ml conical flask to which 20 ml of nitric acid was added and the solution was allowed to stand still until the initial reaction subsided. The solution was heated till all solid particles were dissolved. After cooling, 10 ml of percholoric acid (72%) was added and the solution was heated, gently at first and then more intensely till the solution became clear and colourless. Heating was continued till the volume of the solution reduced to 3–4 ml. After cooling, the solution was transferred to a 100 ml volumetric flask and deionised water was added to it to make the desired volume.

From the diacid digested extract, sodium and potassium were determined by the 'flame photometric' method, calcium by titration with ethylene diaminete tetra acetate [(EDTA) (Versenate)] in the presence of sodium hydroxide using ammonium purpurate indicator. Calcium + magnesium were determined in a similar manner in the presence of ammonium chloride, with ammonium hydroxide as a buffer solution and erichrome black T indicator. Magnesium was estimated by substracting calcium from calcium + magnesium. Sulphur was determined calorimetrically after precipitation as barium sulphate using the Klett Summerson calorimeter.

The results of the experiment and analysis are shown in Tables 4.12 and 4.13.

Table 4.12 *Plant analysis of trees along Indira Gandhi main canal (RD 952-957)*

Tree species	*Tree component*	*Mineral Content %*					
		Na^+	*K^+*	*Ca^{++}*	*Mg^{++}*	*S^-*	*Total*
Eucalyptus	Stem-bark	0.092	0.506	4.700	0.420	0.036	5.754
	Stem-wood	0.083	0.274	1.300	0.120	0.073	1.850
	Branch-bark	0.092	0.498	5.800	0.180	0.056	6.626
	Branch-wood	0.245	0.432	1.600	0.060	0.069	6.626
	Twig	0.046	0.365	2.200	0.060	0.066	2.737
	Leaves	0.027	1.328	1.500	0.300	0.106	2.261

Tree species	Tree component	Mineral Content %					
		Na^+	K^+	Ca^{++}	Mg^{++}	S^-	Total
Accacia Nilotica (Babool)	Fruits	0.073	1.519	1.400	0.240	0.109	3.341
	Stem-bark	0.009	0.265	2.400	0.120	0.043	2.837
	Stem-wood	0.073	0.407	1.500	0.060	0.089	2.069
	Branch-bark	0.101	0.448	2.200	0.120	0.099	2.968
	Branch-wood	0.135	0.232	0.600	0.120	0.076	1.163
	Twig	0.085	0.500	0.500	0.180	0.102	1.367
	Leaves	0.105	0.888	2.600	0.420	0.290	4.303
Ziziphius (Beri)	Fruits	0.100	1.345	1.000	0.060	0.610	3.115
	Stem-bark	0.150	0.373	2.900	0.420	0.095	3.938
	Stem-wood	0.235	0.398	1.200	0.240	0.089	2.162
	Branch-bark	0.190	0.432	2.400	0.480	0.046	3.548
	Twig	0.195	0.714	1.600	0.060	0.056	2.625
	Leaves	0.060	0.996	2.500	0.780	0.165	4.501
Prosopis Cineraria (Khejri)	Stem-bark	0.070	0.830	4.600	0.480	0.191	6.171
	Stem-wood	0.070	0.664	1.100	0.120	0.274	2.228
	Branch-bark	0.070	0.830	3.000	0.660	0.148	4.708
	Twig	0.060	0.655	1.700	0.120	0.925	3.460
	Leaves	0.070	1.245	2.600	0.600	0.584	5.099
Tecomella Undulata (Rohida)	Stem-bark	0.070	0.921	2.100	0.180	0.043	3.314
	Stem-wood	0.090	0.116	0.460	0.420	0.300	1.386
	Branch-bark	0.070	0.664	1.500	0.360	0.056	2.656
	Branch wood	0.030	0.232	1.100	0.120	0.019	1.501
	Twig	0.030	0.722	0.900	0.240	0.033	1.925
	Leaves	0.030	0.838	2.100	0.540	0.294	3.802
Delbergia Sissoo (Sheesham)	Fruits	0.010	1.303	0.100	0.300	0.106	1.819
	Stem-bark	0.210	0.489	2.600	1.260	0.139	4.698
	Stem-wood	0.180	1.477	1.900	1.260	0.290	5.107
	Branch-bark	0.080	0.448	2.800	1.320	0.092	4.740
	Branch wood	0.060	0.290	0.800	0.120	0.066	1.336
	Twig	0.050	0.805	2.100	0.780	0.099	3.834
	Leaves	0.110	1.826	1.900	1.260	0.056	5.152
Azatirachta Indica (Neem)	Fruits	0.070	1.220	0.700	1.020	0.089	3.099
	Stem-bark	0.410	0.382	0.500	0.240	0.191	1.723
	Stem-wood	0.300	0.274	1.000	0.120	0.076	1.770
	Branch-bark	0.240	0.614	1.100	0.420	0.224	2.598
	Branch-wood	0.310	0.249	0.300	0.060	0.053	0.972
	Twig	0.270	1.594	1.200	0.240	0.076	3.380
	Leaves	0.290	0.954	1.800	0.540	0.488	4.072

Table 4.13 *Plant analysis of crops in Indira Gandhi Canal command (Mitharia Minor RD 945)*

Crop	Crop component	Mineral content %					
		Na^+	K^+	Ca^{++}	Mg^{++}	S^-	Total
Wheat	Stem	0.090	0.488	2.100	0.420	0.155	3.253
	Leaves	0.320	1.594	2.000	0.840	0.336	5.090
	Grains	0.250	0.315	0.200	0.120	0.188	1.073
Groundnut	Stem	0.320	1.345	2.300	1.080	0.205	5.250
	Twig	0.170	1.286	1.700	1.860	0.063	5.079
	Leaves	0.110	1.104	1.700	1.920	0.046	4.880
	Seeds	0.050	0.664	0.400	0.060	0.049	1.223
Mustard	Seeds	0.020	0.523	0.700	0.300	0.066	1.609
Guar	Seeds	0.070	0.855	6.500	0.240	0.069	7.734
Gram	Grain	0.100	0.822	0.700	0.180	0.099	1.901
Cotton	Stem	0.040	0.706	0.750	0.750	0.096	2.342
	Twig	0.020	0.191	0.300	0.420	0.057	0.988
	Leaves	0.070	1.336	1.500	1.980	0.822	5.708

The variation in the reported results of mineral contents in plants can be due to the following reasons.

In ash analysis, the residue left after burning is described as plant ash. This analysis does not reveal the exact percentage of mineral content of a plant. The mineral matter is present in a highly oxidised form in the ash. A part of the mineral content undergoes sublimation or vaporisation, e.g., in case of chlorine, sulphur and to some extent in phosphorus, calcium and potassium.

- The mineral content in a plant depends upon the prevailing site and soil conditions. If a particular mineral element is absent or is in short supply in the soil water solution in the root zone, the actual mineral content in the plant would be less than its capacity to absorb and retain the particular mineral element.
- There is significant variation in the mineral content in different parts (stem, leaf, bark etc.) of the same plant. A change also takes place in the content with the age of the plant.

It would therefore be advisable to conduct tests and an analysis on plants growing in a particular area and repeat them periodically for the correct estimation of mineral absorption by plants.

Salt Tolerant Grasses and Halophytes

Cervinka et al. (2001) have identified some suitable salt tolerant grasses and halophytes. These include Alkali sacaton (Salado) (Sporobolus airiodes), Beardless wild rye (Rio) (Leymus triticoides), (Elymus triticoides), Cordgrass (Spartina species), Five hook bassia (Bassia hyssopifolia), Iodine bush (Alenrolfea Occidentalis), Jose Tall wheat grass (Elytrigia elongata) (Agropyron elongatum), Koleagrass (Perla) (Phalaris tuberosa var.Hirtiglumis) , Pickleweed (samphire) (Salicornia bigelovii), Purslane (Portulaca oleracea) (Sesuvium portulacastrum), Saltbush (Atriplex species), Saltgrass (Distichlis spicata) and Tall Fescene (Alta or Goar) (Festuca arundinacea). The most vigorous halophytes were found to be five-hook bassia, salt grass and salicornia.

Salicornia is a low growing very succulent annual plant that is 15 to 38 cm tall with green scale like leaves. The plant is established from seed by broadcast or drilling on a well prepared firm seed bed, similar to establishing alfa alfa stands. Seeding should be done after the frost period in the spring. The stand can be flood or sprinkler irrigated. The plant requires salty water of EC 28 to 30 dS/m. Surface soil in this stand may have ECe about 50 dS/m. Due to its ability to grow under such salty conditions, salicornia produces an excellent plant canopy for evaporating salt water. Salicornia was reseeding itself. Salicornia may have multiple uses. One of its main uses is for seed production. When processed it produces oil which contains polyunsaturated fat close to the level of safflower oil and better than soyabean oil. The meal from the oil processing can be used as a feed source for poultry and livestock. The young top portions of the plant are used as salad green and a tasty vegetable. Salicornia was recommended by the HEI company from Arizona which has been growing this crop on several hundred hectares in Saudi Arabia and India. Salicornia dry biomass production was 1.05 kg/m^2 (10.5 t/ha). Plant samples ware collected for nutrients, chloride, sulphate, sodium and trace elements. Per cent of ash is an indicator of mineral content, and was quite high at 43.5 per cent. Selenium and boron uptake is also high, at 892 mg/kg and 3.8 mg/kg respectively.

From this description, the Salicornia harvest has the potential of evacuating as much as 4.5 t/ha/year of salts from the soil and water where it grows.

Some plants like *Salvadora oleoides* (Ca, Na), *Tecomella undulata* (Na), *Acacia calycina* and *Acacia ciliata* (Ca), *Acacia senegal* (Mg), *Zizyphus nummulania* (K) *Moringa oleifera* (P) have a special affinity for particular elements and offer a choice for the selection of plants to be grown to suit the prevailing conditions.

Bhutta and Chaudhry (2000) report that Naik and Manjunatha studied the salt scavenging capacity of three tree species, namely *Sesbania grandiflora* pers, *Prosopis juliflora* DC and *Acacia tortilis* forsk, in field and laboratory experiments. The potential uptake of toxic level of ions from the root zone of the field crops was found to be maximum in *Sesbania grandiflora*, followed by *Prosopis juliflora*. The accumulation of salts in the shoot parts of *Sesbania grandiflora*, and *Prosopis juliflora* help in lowering the salinity during the active growth of field crops. In another experiment, the co-cultivation was carried out using local crop cultivars of sugarcane, sunflower and groundnut alongwith the salt scavenging tree species namely *Sesbania grandiflora*. The association of Sesbania boosted the productivity of salt-sensitive cultivars under saline conditions by

10 to 30 per cent. This increase was seen in terms of biomass, number of tillers, leaf area and chlorophyll contents.

MINERAL CONTENT IN BIOMASS

The mineral content in biomass varies from plant species to species and depends upon the prevailing conditions of soil, water and climate under which the plants grow. The mineral content also varies in different parts of the plant body. Plant physiology literature indicated an average mineral content of about 5 per cent in biomass on a dry weight basis.

The mineral content in plants is generally determined by ash analysis or by the wet digestion method. In the 'diacid wet digestion' method, the plant sample is washed with hydrochloric acid, treated and heated with nitric acid and then with perchloric acid. The mineral element (Na^+, K^+, Ca^{++}, Mg^{++}, ...) content in the sample is then determined in the laboratory by standard methods. The method determines the content of mineral elements in the biomass by weight. Some of the elements like chlorine and sulphur volatilise and escape on heating or on treatment by acids and are not fully estimated in the analysis. The chemical analysis of plant biomass, therefore, gives lower values of the mineral content than the actuals.

CATION-ANION EXCHANGE BETWEEN PLANTS AND SOIL SOLUTION

The plants absorb most of the mineral elements from the soil solution through their root system. The plant roots absorb the electrolytes in the form of ions, and in general the absorption of one type of ion is accompanied with the absorption of another type of ion of equal but opposite electrostatic charge. This ensures the maintenance of electrostatic balance both inside the plant cell as well as in the soil solution outside [Figure 4.3 (a)].

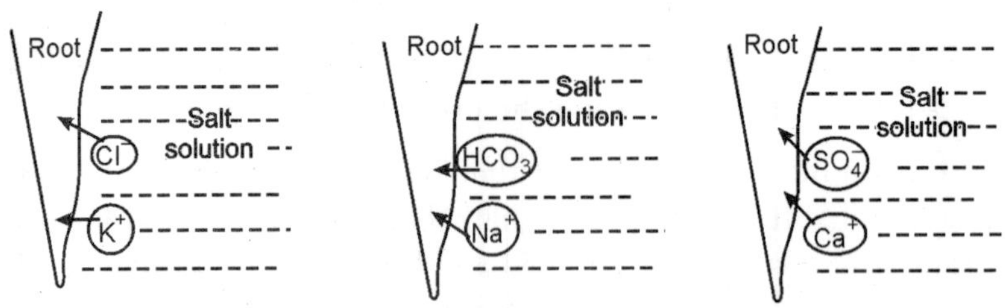

Figure 4.3 *(a) Both cations and anions absorbed by the plant root on equivalent basis*

It is believed that sometimes the plant cells absorb cations and anions in different proportions. In such cases the electric balance between cations and anions both inside the plant cells and in the outside soil solution is maintained by changes in the ionic composition of the cell and the soil solution. If the uptake of cations in the plant cell is more than that of anions, hydroxide anions (OH^-) and hydrogen cations (H^+) are produced in the plant cell, the OH^- cations help maintain balance within the plant cell while the H^+ cations are excreted from the plant cell and passed out into the soil solution to restore the balance in the salt solution outside. The excreted hydrogen cations can combine with the excess anions in the soil solution to form acids [Figure 4.3 (b)]. In case the anion uptake in the plant cell is more than that of cations, equivalent hydroxide (OH^-) anions are secreted and passed out into the soil solution to maintain the balance [Figure 4.3 (c)].

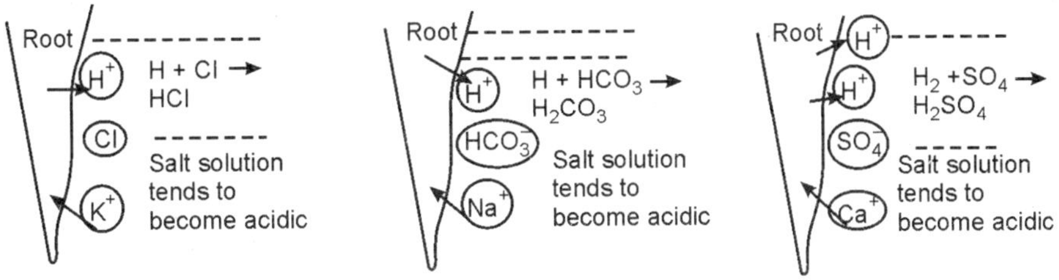

Figure 4.3 *(b) More cations (than Anions) absorbed by the plant root. Balance maintained by excretion of H^+ Ions*

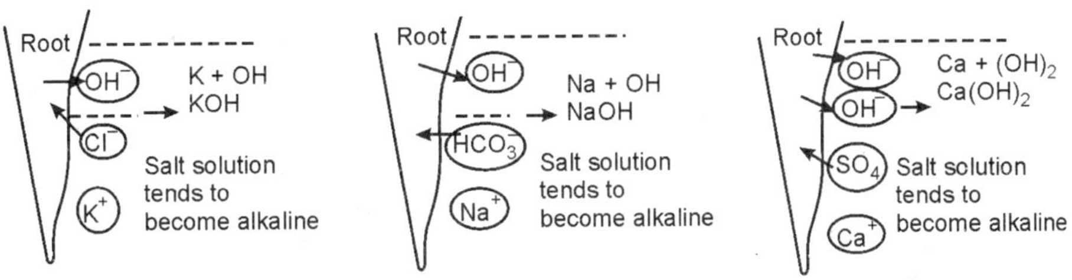

Figure 4.3 *(c) More anions (than cations) absorbed by the plant root. Balance maintained by excretion of OH^- ions*

There is another way in which the electrostatic balance can be restored when the plant cells absorb cations and anions in non-equivalent proportions. The excess unabsorbed anions or cations in the soil solution would be leached out of the soil solution rapidly. At the same time the plant cell would produce hydrogen (H^+) or Hydroxide (OH^-) ions within the cell to balance the excess absorbed anions or cations.

ESCAPE OF CHLORINE AND SULPHUR FROM GROUNDWATER INTO ATMOSPHERE

Chlorine

Verma explains that the dispersed particles of all hydropholic solutions are electrically charged. A colloidal solution is electrically neutral because every charge carried on a micelle is neturalised by a charge of opposite value carried by ion in the dispersion medium. The movement of positively charged ions towards the negatively charged poles is termed cataphoresis or electrophoresis.

Sodium chloride in water solution results in ionisation of Na^+, Cl^-, H^+ and OH^- ions

$$NaCl \rightleftharpoons Na^+ + Cl^-$$

$$H_2O \rightleftharpoons H^+ + OH^-$$

According to Tulsi et al. (1996), on electrolysis the Na^+ and H^+ ions move towards the cathode but H^+ ions are able to accept electrons much more easily then Na^+ ions as a result of which hydrogen is liberated with the decomposition of more water molecules while Na^+ ions remain in solution. At the anode, the Cl^- ions give up their electrons in preference to OH^- ions and Cl_2 gas is therefore liberated there.

At the cathode, $$H^+ + e = H^0$$

$$H^0 + H^0 = H_2$$

or $$2H_2O - 2e \rightarrow H_2 + 2OH^-$$

At the anode, $$Cl^- - e = Cl^0$$

$$Cl^0 + Cl^0 = Cl_2$$

The solution on electrolysis can become richer in sodium hydroxide, the overall ionic equation being

$$2Na^+ + 2Cl^- + 2H_2O \rightarrow 2Na^+ + 2OH^- + H_2\uparrow + Cl_2\uparrow$$

Some chlorine contained in soil water solution may escape into the atmosphere.

Sulphur

Plant roots have a tendency to uptake more of cations than anions from sulphate solutions. In such cases electrostatic balance is maintained by secretion of hydrogen cations in the plant body and their excretion into the soil solution (Figure 4.3) resulting in formation of H_2SO_4.

Tulsi et al. describe that carbon is a good reducing agent. It reduces sulphuric acid to sulphur dioxide.

$$C + 2H_2SO_4 \rightarrow CO_2 + 2H_2O + 2SO_2\uparrow$$

Plant roots contain abundance of carbon and they on reacting with H_2SO_4 in soil solution can cause formation of SO_2 and its eventual escape into the atmosphere.

Soils along with water interact continuously both with the atmosphere and with the biological populations that they support. Cresser et al. (1993) describe that in biogeochemical cycling of nutrient elements, sulphur occurs in the form of SO_4^{2-}, SO_2

and H_2S. The input of sulphur may be as sulphates to the atmosphere from oceanic sprays or from emission in the form of SO_2 from industries. Sulphur is transferred to the atmosphere from marshes as sulphur dioxide, especially in esturies and hydrogen sulphide although the transfer is difficult to quantify.

It follows from this discussion that some portions of chlorine and sulphur contained in the groundwater may escape into the atmosphere in gaseous form, some are absorbed by the plants and some remain in the groundwater solution.

In all above explained scenarios and regardless of whether the cation and anion absorption by the plant cells is equal or unequal, the absorption of a certain quantity of cations by the plant cell is always accompanied by the evacuation of an equivalent quantity of anions from the salt solution surrounding the plant roots.

MINERAL SALTS ABSORBED BY PLANTS

The commonly found soluble salts in soil solutions are the salts of calcium, sodium and magnesium. The percentage gravimetric weight of the respective elements in the salts compounds is as shown in Table 4.9.

Table 4.9 *Percentage in gravimetric weight of element in the salt compound*

Calcium		*Sodium*		*Magnesium*	
In Calcium Sulphate ($CaSO_4$) (20.04 ÷ 68.07) × 100	29.4	In Sodium Chloride (NaCl)	39.3	In Magnesium chloride ($MgCl_2$)	25.5
In Calcium Chloride ($CaCl_2$)	36.1	In Sodium bicarbonate ($NAHCO_3$)	27.4	In Magnesium sulphate ($MgSO_4$)	20.2
		In Sodium Sulphate (Na_2SO_4)	32.4		

Note: The equivalent weights of respective ions are

Ca^{++} = 20.04	Na^+ = 23.00	Mg^{++} = 12.16
$CaSO_4$ = 68.07	NaCl = 58.45	$MgCl_2$ = 47.62
$CaCl_2$ = 55.50	Na_2SO_4 = 71.03	$MgSO_4$ = 60.19
	$NaHCO_3$ = 84.01	

On an overall average basis, the gravimetric weight of mineral elements (Ca, Na, Mg...) in the mineral salt compounds may be taken as about 33 per cent. In other words, the total weight of mineral salts absorbed by a plant, or evacuated from the soil solution, is about three times the total weight of mineral elements (cations) found in the plant.

SALT BALANCE USING NATURAL RIVER WATER FOR IRRIGATION

The salts in irrigation water are mostly bicarbonates, chlorides and sulphate of calcium, sodium, potassium and magnesium. Natural unpolluted river waters generally contain less than 200 mg/l of total soluble solids of which about one-third proportion is made up of cations and two-thirds by anions. A major component of the anions is made up of the bicarbonate (HCO_3) anions. In the Sutlej-Beas rivers, India, in the canal water used for irrigation, the HCO_3 anions constitute about 80 per cent of the total anions. The USSLS (1954) gives data on chemical composition of some river waters used for irrigation in Western United States from which it is found that on an overall average basis the HCO_3 anions formed about 55 per cent of the total anions.

The normal average requirement of water for irrigation for agriculture crops may be about 500–600 mm. The data on salt inflow with irrigation water, and salt uptake by plants is presented in the Tables 4.10 and 4.11.

Table 4.10 *Salt inflow with irrigation water (t/ha)*

Irrigation water (mm)	*Salts in irrigation water mg/l*				
	100	*200*	*300*	*500*	*1000*
500	0.5	1.0	1.5	2.5	5.0
600	0.6	1.2	1.8	3.0	6.0
700	0.7	1.4	2.1	3.5	7.0
800	0.8	1.6	2.4	4.0	8.0
1000	1.0	2.0	3.0	5.0	10.0
1500	1.5	3.0	4.5	7.5	15.0

Table 4.11 *Salt evacuation* by plants (t/ha)*

Plant yield t/ha (dry biomass)	*'Cation' (%) content in plant body on ash analysis*			
	2	*3*	*5*	*10*
5.0	0.30	0.45	0.75	1.50
7.5	0.45	0.67	1.12	2.25
10	0.60	0.90	1.50	3.00
15	0.90	1.35	2.25	4.50
20	1.20	1.80	3.00	6.00
30	1.80	2.70	4.50	9.00

**Note:* Assuming that quantity of salt compound evacuated is 3 times the quantity of 'Cations' found in the plant body.

Normal crops grown by using natural river water (salinity less than 200 mg/l), can evacuate almost as much salt from soil as is brought in by the irrigation water. This is without taking into account the salts that are precipitated (about 15 to 20%) and thus removed from the soil water solution due to complex chemical reactions involving the soil, plant roots and the salt ions.

CONCLUSIONS

1. A general perception that plants do not take any minerals alongwith the soil water, is not totally correct. About 5 per cent of the dry weight of a plant is constituted by minerals, the source of all of which is the soil in which the plants grow.
2. Total mineral uptake by plants, where irrigation water is of good quality, can be quite significant and equal to the quantity of minerals brought in by the irrigation water.
3. About 15–20 per cent of salts contained in irrigation water are precipitated.
4. The generally adopted ash analysis or wet digestion methods for determining the mineral content in biomass give only approximate results, as the elements that sublime or vaporise during the test procedure are not accounted for.
5. The total weight of the mineral salt compounds absorbed by a plant and or evacuated from the soil solution is approximately three times the total weight of cation elements found in the biomass of the plant.

REFERENCES

American Society of Civil Engineers (ASCE) (1990). 'Agricultural Salinity Assessment and Management', ASCE Manuals and Reports on Engineering Practice No. 71.

Bhutta, N.N. and M.R. Chaudhry (2000). 'Biological Control of Waterlogging', Paper Presented at the 8th ICID International Drainage Workshop, February 4, New Delhi, India.

Cervinka, V., C. Finch, M. Martin, F. Menezes, D. Peters and K. Buchnoff (2001). *Drainwater, Salt and Selenium Management Utilizing IFDM/Agroforestry Systems*, Report to the US Bureau of Reclamation, Sacramento, CA.

Chopra, S.L. and J.S. Kanwar. 'Groundnut', Indian Council of Agricultural Research (ICAR). Quoted by Kapoor A.S. in 'Biodrainage – A Biological Option for Controlling Water Logging and Salinity (2001), ISBN 0-07-040231-0, p. 120.

Cresser, M., K. Killham and T. Edwards (1993). *Soil Chemistry and Its Applications*, Chapter 4, Cambridge University Press, Cambridge, p. 95 (Figure 4.14).

Dhir, R.P., B.K. Sharma and B.K. Dutta of CAZRI, Jodhpur (1984). *Mineral Nutrient Elements in Natural Vegetation of Arid Rajasthan: I Macro-elements.*

Dhir, R.P., B.K. Sharma, B.K. Dutta and S.K. Saxena of CAZRI, Jodhpur (1985). *Concentration of Mineral Elements in Some Arid Zone Species from a Common Habitat: 1 Tree and Shrub Species.*

Indian Council of Agricultural Research (ICAR) (1997). *Desertification and Its Control.*

— (1997). *Chemical Composition and Nutritive Value of Indian Feeds and Feedlings of Farm Animals.*

Indian Council of Medical Research, Hyderabad, India (1989). *Nutritive Value of Indian Foods.*

Jackson, M.L. (1967). *Soil Chemical Analysis*, Prentice Hall of India Pvt. Ltd., New Delhi, pp. 321–38.

Lopunshinksy (1960). Wilkins (1969). Quoted from S.N. Pandey and B.K. Sinha (1995). *Plant Physiology*, 3rd Edition, Har Anand, New Delhi.

Martin, P., J. Gagnard and P. Gautier (1987). *Plant Analysis as a Guide to the Nutrient Requirements of Temperate and Tropical Crops*, Lavoisier Publishing, New York, pp. 143–54 (Tables 7.1–7.13).

Palladin, V.I. (1992). *Plant Physiology*, Edited by B.L. Livingston, 2nd American Edition, Arihant Publishers, Jaipur.

Pandey, S.N. and B.K. Sinha (1995). *Plant Physiology*, 3rd Edition, Har-Anand Publications, New Delhi.

Paul et al. (1966). As reported in the American Society of Civil Engineers Report on Engineering Practice No. 71, titled 'Agriculture Salinity Assessment and Management' (1990).

Rajkumar (1998). *Mineralogy and Mineral Transformation in Agricultural Salinity Management in India*, CSSRI, Karnal, India, p. 116.

Rhoades et al. (1974). Quoted by Glenn J. Hoffman (1990). 'Leaching Fraction and Root Zone Salinity Control', ASCE Manual No. 71, titled 'Agricultural Salinity Assessment and Management', pp. 239–40.

Richard, L.A. (ed.) (1954). *Diagnosis and Improvement of Saline and Alkali Soils*, USDA Handbook No. 60, Oxford Publishing Co., New Delhi, pp. 127–34.

Sen, K.C. Indian Council of Agricultural Research (ICAR). Quoted by Kapoor A.S. in 'Biodrainage – A Biological Option for Controlling Water Logging and Salinity (2001), ISBN 0-07-040231-0, p. 120.

Shekhawat, Dr et al. Departments of Agriculture and CAD, State of Rajasthan (1998). Laboratory Tests for the Determination of Mineral Contents in Plant Samples from IGNP and Their Results.

Srivastava, H.S. (1996). *Plant Physiology and Biochemistry*, Foundation for Science and Society, Lucknow.

Stiles, Walter (1995). *Principles of Plant Physiology*, Atlantic Publishers & Distributors, New Delhi.

Tandon, H.L.S. (ed.) (1993). *Methods of Analysis of Soils, Plants, Waters and Fertilisers*, Fertiliser Development and Consultation Organisation, New Delhi.

Tanji, K.K. (1972). As reported in the American Society of Civil Engineers Report on Engineering Practice No. 71, titled 'Agriculture Salinity Assessment and Management' (1990).

Tulsi, G.D., B.S. Bahl and Amba Prasad (1996). *Intermediate Inorganic Chemistry*, 18th Edition, Jodhpur, p. 411 and 462.

US Department of Agriculture (1954). *US Salinity Laboratory Diagnosis and Improvement of Saline and Alkali Soils*, Handbook No. 60, p. 38 and 77 (Table 12).

Verma, V. (1991). *A Textbook of Plant Physiology*, 7th Edition, Chapter 3, Atma Ram & Sons, Delhi, p. 107.

5

A Case Study on Biodrainage

DESERT LAND OF RAJASTHAN, INDIA

Rajasthan is the second largest State in India with 25.6 million hectares (m ha) cultivable area. Nearly 60 per cent of its area is desert land. During summer, the desert area experiences intense heat, strong winds, and temperature of up to 45–50°C with the average maximum temperature being 42°C. The coldest month is January when the average minimum temperature is 6°C. The mean annual rainfall, occurring mostly during mid-July to September, varies from less than 100 mm in the west to 360 mm in the east. The annual mean potential evapotranspiration varies between 2000 mm in the east to 2500 mm in the west. The subsoil water level in the desert area is 60–100 m below the ground level (bgl). The groundwater is highly saline and unfit for irrigation.

The area used to be very thinly populated. As against an average population density of 112 numbers per square kilometre of India, and 48 numbers per square kilometre of Rajasthan State in the year 1951, the population density in the desert area varied between 1.5 numbers per square kilometre in Jaisalmer district to 16 numbers per square kilometre in Bikaner district. The main source of livelihood of the people was rearing of cattle for butter, wool and leather. The grasses growing in the area provided grazing lands and fodder for the animals.

INDIRA GANDHI NAHAR PROJECT (IGNP)

After the partition of the erstwhile Indian State in 1947, two independent sovereign States, namely India and Pakistan, came into being. The Indus river basin also got divided between the two countries and after protracted negotiations and with the assistance of the World Bank, an agreement on the river water rights was reached in 1960 in the form of the Indus Water Treaty. India got the right of exclusive use of the waters of three eastern rivers, namely the Ravi, Beas and Sutlej. The water of these rivers is used in the States of Punjab, Haryana, Rajasthan and Jammu & Kashmir of the Indian Union. An

Inter-State water agreement determines the share of water of each State. Out of a total available quantity of 21.12 billion cubic metres (bcm) of Ravi and Beas waters in an average year, Rajasthan State is entitled to receive 9.36 bcm annually.

The head work of the project comprises of two storage reservoirs, one each on the rivers Ravi and Beas. The Beas river has been dammed (Pong dam) near Talwara (Punjab) to store 6.9 bcm of water. The dam on the river Ravi can store 2.3 bcm water. The rivers have been interlinked.

The waters of both Ravi and Beas rivers are picked up at Harike Barrage and carried through a 204 km long Rajasthan feeder canal (with a capacity of 524 m^3/sec) to the border of Rajasthan, from where the water is carried through a 445 km long main canal to Mohangarh. The index plan of this project is seen in Figure 5.1.

Water from the main canal is distributed through a large number of secondary and tertiary canals, of a total length of 9180 km. The project provides irrigation facility to 1.87 million hectares of land. The water travels to a distance of about 1500 km to irrigate the farthest land.

For convenience in planning, construction and operation, the project was divided into two stages, viz., stage I and stage II. The two stages are distinguished by their soil, climatic and settlement characteristics, with stage II implying greater challenges in almost all respects.

Stage I

Stage I of the project starts from Harike Barrage and terminates after a length of 189 km of the main canal. The total culturable command area (CCA) is 553 thousand hectares, (th ha) out of which 492 th ha are irrigated by flow and the remaining 61 th ha by lift (up to 60 m). The canal system has been constructed with a water allowance of 0.36 lit/sec/ha which is proposed to be reduced to 0.21 lit/sec/ha ultimately. The actual area irrigated in (1999) in Stage I was 699 th ha, with an overall average intensity of 126 per cent.

Stage II

Stage II of the project starts immediately downstream after 189 km of the main canal and extends up to Gadra Road in Barmer district. The total CCA is 873 th ha under flow and 443 th ha by lift, with the total being 1316 th ha. The intensity of irrigation and rate of water allowance for the flow area are 80 per cent and 0.21 lit/sec/ha respectively. The annual irrigation in (1999) was 221 th ha.

Irrigation Development in the Area

The IGNP had been receiving water for irrigation since the year 1961. The figures of the actual annual irrigated area are shown in Table 5.2.

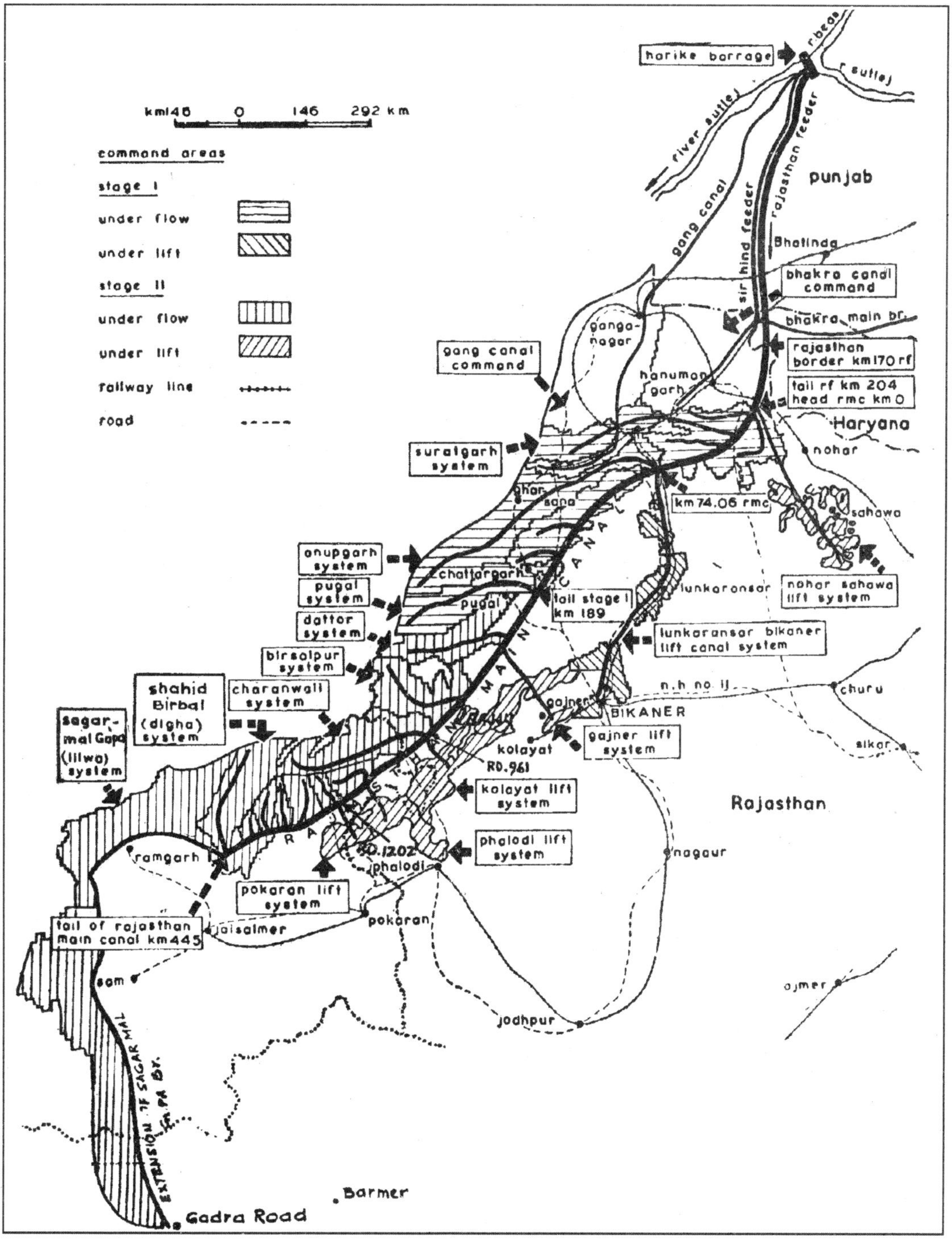

➲ **Figure 5.1** *Index plan of Indira Gandhi Nahar Project (IGNP)*

Table 5.2 *Development of irrigation in IGNP*

Year	Area irrigated (in thousand hectares)		
	Stage I	Stage II	Total
1975–76	289	—	289
1985–86	463	2	465
1995–96	664	137	801
1996–97	682	159	841

So long as the downstream areas were not fully opened for irrigation, the quantity of available water for the limited opened area was quite liberal. During the period 1988–1995, the average rate of water use, released at the head of the feeder canal, was 1260 mm as against the designed value of 560 mm. The values for the neighbouring project areas of Gang and Bhakra commands, during the same period, were 575 and 515 mm respectively.

Waterlogging

The depth of the water table in the command of Stage I in the year 1952 generally varied between 40 to 50 m below ground level (bgl). With the introduction of irrigation the groundwater table started rising. During the decade 1981–82 to 1991–92, the average annual rate of rise of water table was 0.92 m.

In Stage II of the project, the groundwater table generally ranged between 20 to 100 m bgl before the advent of irrigation. With irrigation it has been rising, though not at the same rate as in Stage I.

The position of the waterlogged area in the IGNP command area during the period 1992–1997 is shown in Table 5.3.

Table 5.3 *Position of waterlogged area in IGNP command area*

(area in hectares)

Category	Year				
	1992–93	93–94	94–95	95–96	96–97
STAGE I					
(i) Area inundated	13750	9680	10192	14750	17220
(ii) Area with water table up to 1.5 m bgl	22000	17760	18970	20670	24140
(iii) Area with water table between 1.5 m to 6 m bgl.	292960	160760	198643	207300	297820
(iv) Area with water table more than 6m bgl.	224290	364800	325195	310280	213820
Total	553000	553000	553000	553000	553000

Category	Year				
	1992–93	93–94	94–95	95–96	96–97
STAGE II					
(i) Area inundated	1000	526	1000	800	800
(ii) Area with water table up to 1.5 m bgl	4062	NA	4500	5470	5740
(iii) Area with water table between 1.5 m to 6m bgl.	17900	NA	19000	19000	19800
(iv) Area with water table more than 6m bgl.	1293038	NA	1291500	1290730	1289660
Total	1316000	1316000	1316000	1316000	1316000

Note: Under Stage II the total area with completed canal network was 5,60,000 ha up to March 97 and, the actual irrigation was 1,59,000 ha.

IGNP Main Canal, RD 750-1365 (km 228-416)

The construction work on the main canal, Stage II was commenced in the year 1975 and completed up to tail (RD 1458) in the year 1986. The release of water in the canal was done as follows:

Main canal reach	Year of first filling of canal
Up to RD 620	1975
Up to RD 961	1983
Up to RD 1458	1986

Soon after the release of water in the main canal, it was found that seepage water collected at the natural ground surface on both sides of the canal at many places. This occurred more in reaches where the canal embankment was in filling and the earth for making canal embankments had been taken from borrow areas along the canal alignment. The canal is lined with conventional clay tile lining in cement mortar, and the appearance of pools of water, so soon, along the canal was unexpected.

The longitudinal section of the main canal from RD 950 to RD 980 is shown in Figure 5.2. A typical cross-section of the canal in filling reach is shown in Figure 5.3.

A survey conducted in the year 1991 indicated that the total number of locations where pools of water were formed on both sides of the canal was 127 and the total area where the water appeared on the surface was 900 hectares. Because of the plantations that were subsequently raised along the canals, and around the pool areas, there was progressive reduction in the affected area and in June, 1997, there were only 8 locations of water pools with a total area of 20 ha. The position of progressive reduction in affected area is shown in Table 5.4.

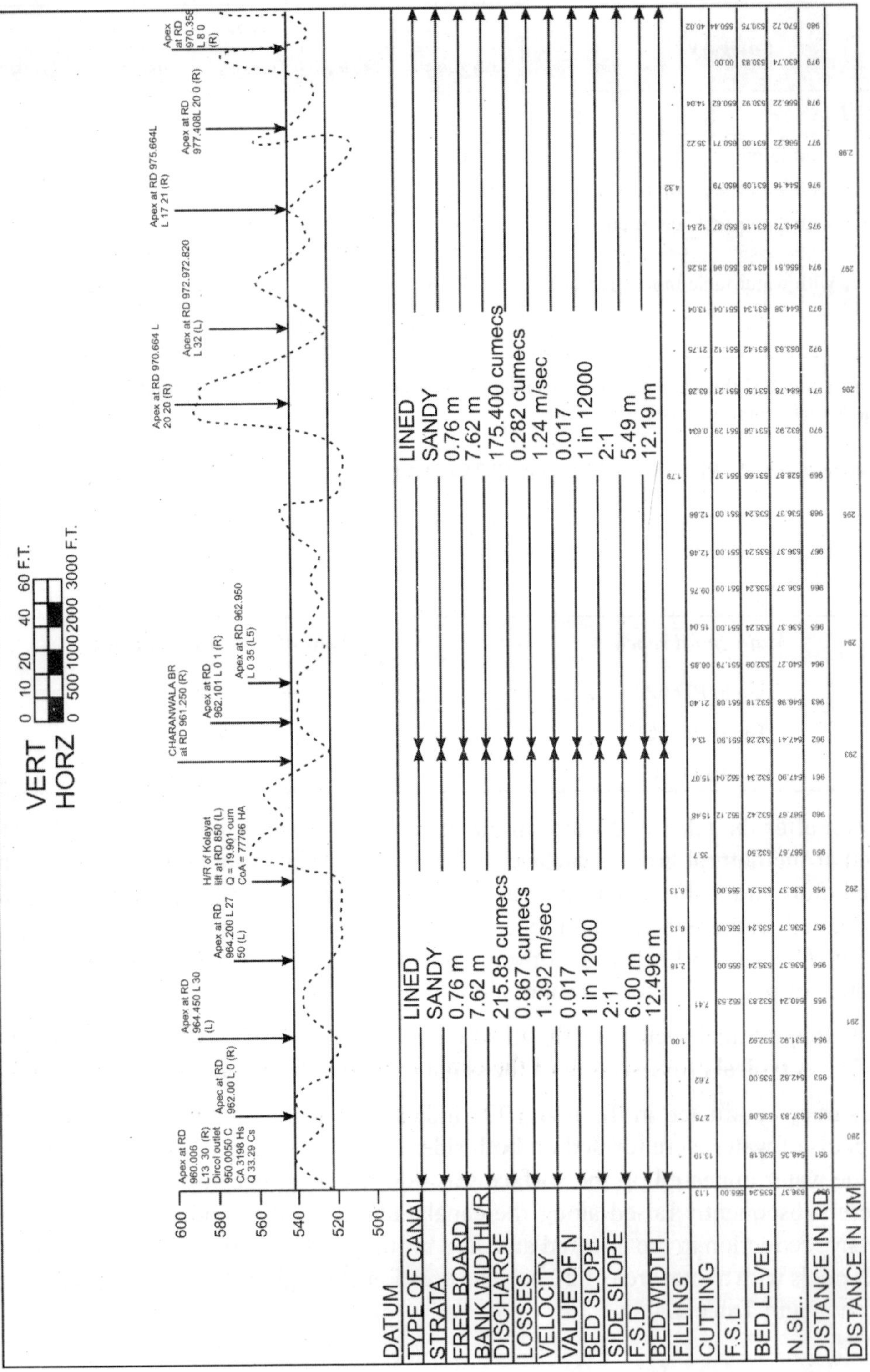

Figure 5.2 *Longitudinal section of Indira Gandhi main canal from RD 950 to 980.*

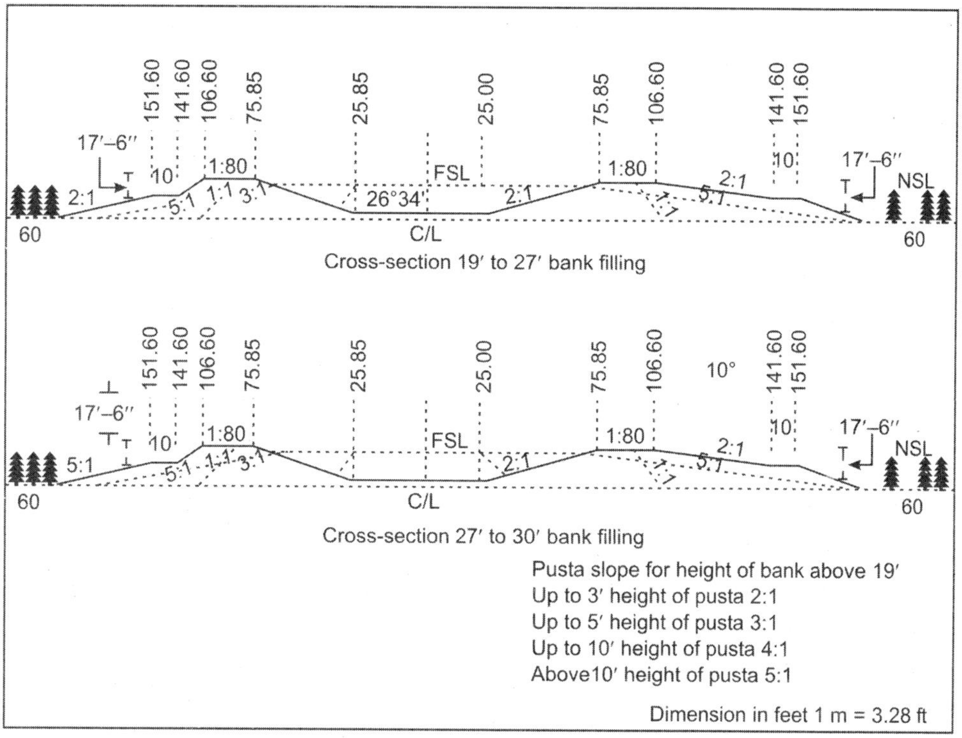

Figure 5.3 *Typical cross-section of Indira Gandhi main canal*

Table 5.4 *Indira Gandhi main canal RD 750-1365, area with groundwater at surface*

(Area in ha)

Location	*June 91*	*June 93*	*June 95*	*June 97*
RD 750–861	254	35	24	10
RD 860–961	83	—	—	—
RD 961–1121	533	471	83	4
RD 1154–1365	30	24	20	6
Total	900	530	127	20

RD – unit of 1000 ft = 305 m

Afforestation

Afforestation work on the IGNP was undertaken to protect the canals from wind-blown sand deposits by creating effective shelter belts, to produce timber, fuel and fodder to meet local needs and to improve the general environment. The afforestation activity in Stage I had been started in the year 1962 but was taken up on a large scale in 1974.

The position of categories of plantation and area covered up to March 1997 is shown in Table 5.5.

Table 5.5 *Different categories of plantations and area covered up to March 1997*

S.No.	Type of plantation	Area covered (ha)		
		Stage I	Stage II	Total
1.	Canal side plantation	11703	12552 (Main canal only)	24255
2.	Road side plantation	2582	340	2922
3.	Sand dune stablisation & pasture development	94908	11660	106568
4.	Fuel wood (Abadi) plantation	5270	5277	10547

Canal Side Plantations

Strips of 100 m width on either side of the main canal and branch canal, of 50 m width on either side of the distributary and of 25 m width on either side of the minor canal were initially planted at 3m × 2m spacing, which was later changed to 3m × 3m spacing. The irrigation frequency was generally 1, 9, 6, 6, 3, 3, 2 and 2 times from the first to the eighth year. In command areas, *Eucalyptus camaldulensis*, *Delbergia sisso* and *Acacia nilotica* have been planted whereas in the non-command areas, *Prosopis cineraria*, *Tecomnella undulata* and *Ziziphus spp.* and on naked dunes, *Acacia tortillas* have been planted. No irrigation was done in the plantation areas where the groundwater had come up to the surface.

Block (Abadi) Plantations

These are command area plantation on 12.5 ha *abadi* lands which are earmarked for afforestation for each *chak* (village). The species grown are mainly *Delbergia sisso*, *Eucalyptus camaldulensis* and *Acacia nilotica* with some mixing of *Prosopis cineraria*, *Tecomnella undulata* and *Ziziphus spp.* The spacing of plantations and frequency of irrigation is the same as in the case of canal side plantations.

Sand Dune Stabilisation

Planting is done at a spacing of 3m × 3m. The active dunes are mulched, using local material. The mulch lines are perpendicular to the prevailing south-west direction and the spacing between two lines is 5m. The seeds of *Lasirus sindicus* (Sewan) and some local bushes are also sown in the area before the rainy season. The irrigation frequency is 4, 6, 6, 3, 2, 2, 1 and 1 from the first to the eighth year. The species involved in this activity are *Acacia tortillas* on high and shifting dunes and *Tecomnella undulata*, *Prosopis cineraria* and *Ziziphus spp.* on low semi-stabilised dunes and in the inter-dunal areas.

Roadside Plantations

Seedlings are planted at 3m × 3m spacing except for shady and ornamental trees which are planted at a spacing of 5m × 5m. The irrigation schedule is similar to that of canal side plantations.

Plantation in the Reach RD 952-957 (km 290-291.7) Left Side

A 1524 m long and 261 m wide strip along the left side of the main canal from RD 952–957 (km 290 to 291.77), was selected for detailed study. The plantation work was carried out during the years 1987 to 1994. A field census was carried out and the distribution of different species of actually growing trees in July 1997 is shown in Table 5.6.

Particulars of Main Canal RD 952-957

The hydraulic data of the main canal at RD 953 (km 290) is as follows:

Full supply discharge	—	216 m^3 /s
Full supply depth	—	6 m
Bed width of canal	—	12.5 m
Side slopes	—	2:1
Longitudinal bed slope	—	1 in 12000

Table 5.6 *Distribution of different species of growing trees along left bank of main canal RD 952-957*

S.No.	Location	*Number of growing trees of species*							
		Eucalyptus camaldulensis	*Acacia nilotica*	*Azadirachta indica*	*Ziziphus spp.*	*Delbergia sisso*	*Prosopis cineraria*	*Tecomnella undulata*	*Total number of trees*
1.	RD 952.00 to 952.200	396	1077	—	134	—	—	—	1607
2.	RD 952.200 to 952.450	566	902	—	93	—	—	44	1605
3.	RD 952.450 to 953.00	3296	3411	36	27	—	133	—	6903
4.	RD 953.00 to 953.250	1465	325	—	—	—	—	—	1790
5.	RD 953.250 to 953.500	1527	494	—	115	—	—	—	2136
6.	RD 953.500 to 953.800	1090	643	—	380	—	—	—	2113

S.No.	Location	Number of growing trees of species							
		Eucalyptus camaldulensis	*Acacia nilotica*	*Azadirachta indica*	*Ziziphus spp.*	*Delbergia sisso*	*Prosopis cineraria*	*Tecomnella undulata*	*Total number of trees*
7.	RD 953.800 to 954.00	659	580	—	—	—	—	—	1248
8.	RD 954.00 to 955.00	6950	3352	—	—	506	648	—	11456
9.	RD 955.00 to 955.300	1220	870	—	153	22	—	83	2348
10.	RD 955.300 to 955.500	995	285	—	414	3	—	—	1697
11.	RD 955.500 to 956.00	2932	1238	—	219	—	—	159	4545
12.	RD 956.00 to 957.00	7137	2022	—	—	—	214	386	9759
	Total	28233	15208	36	1535	531	995	672	47210

The canal is fully lined and the canal embankment is in filling. The canal bed level is about 1.2 to 1.4 m above the natural ground level. At its full supply depth, the water surface level in the canal is about 7 m above the ground level.

The canal has not been run with full supply discharge but the water level is maintained at about the full supply level to run the off-taking channels, by regulation at the control structure at RD 961 (km 293).

The canal cross-section is shown in Figure 5.3.

Formation of Water Pools Along the Canal RD 952-957

The canal was first filled with water in 1983. Soon afterwards, pools of water were formed on both sides of the canal. The maximum water pool area on the left side of the canal in this reach was 25 ha (in 1988). The plantation work was taken in hand in 1987 along the canal and around the water pool area. With the growth of trees, the water pool area reduced progressively as follows:

Year	*Water pool areas (ha)*
Up to year 1988	25
1989	23
1990	20
1991	15
1992	9
1993	2
1994	—

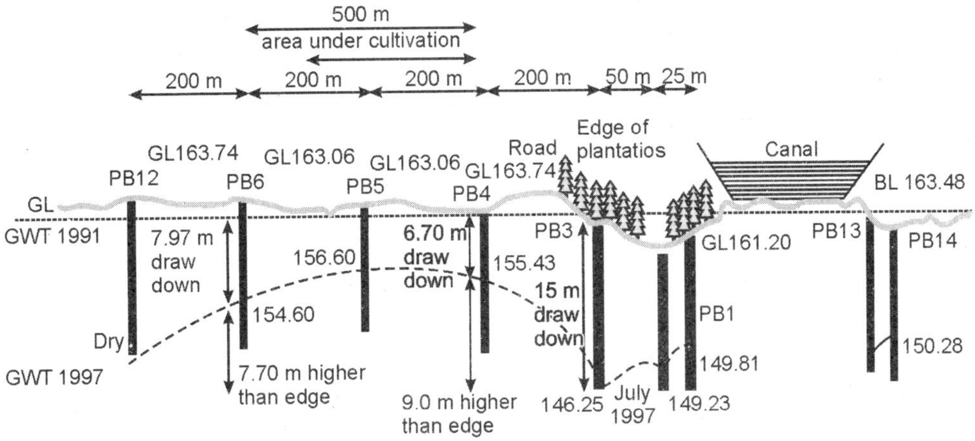

➲ **Figure 5.4** *Groundwater level fluctuations in piezometers (at RD 953-212 of IGMN)*

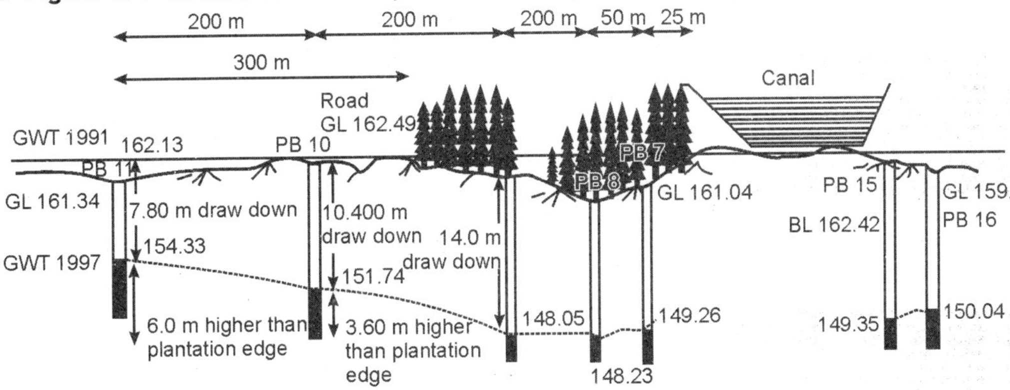

➲ **Figure 5.5** *Groundwater level fluctuations in Piezometers at RD 953-900 of IGMN*

The deepest pool bed was about 3.5 m lower than the natural ground surface level (NSL). In April 1994, the groundwater went down to 4.9 m below the NSL, in April 1996 to 8.8 m below the NSL, in September 96 to 10.3 m below NSL and in July 1997 to 12.9 m below the NSL. At the cross-section at RD 953.2, the groundwater level has gone down by as much as 15 m.

Peizometers were installed in 1997 to determine the groundwater profile at two cross-sections across the canal. The position of the observed water levels and other details are shown in Figures 5.4 and 5.5.

GROUNDWATER BALANCE

Recharge

(a) From Canal Seepage

The canal length in the reach under consideration is 1524 m. The average side slope length and half bed width together total 19 m. The design value of the rate of loss from lined canals in the IGNP is 0.6 m^3/sec per million square metre of the wetted surface area. Special tests were done in the reach to assess the quantum of seepage loss by the tracer-technique and the seepage loss was estimated to range between 0.57 to 1.78 m^3/sec/million sq m. By analytical approach the seepage loss was estimated as 1.05 m^3/sec/million sq m.

(b) From Rainfall

The observed annual rainfall at the nearest rain-gauge site (which is about 50 km away) during the period 1991–1997, varied between 85 mm (1991) to 525 mm (1996), with the average annual figure being 335 mm. Assuming a net recharge to groundwater at the rate of 5 per cent, over the area (1524 m × 800 m), the volume of recharge due to rainfall during the six-year period was $(1524 \times 800) \times 05 \times 335 \times 10^{-3} \times 6\ m^3 = 12.24$ ha m, say 12 ha m.

(c) From Irrigation

The area receiving irrigation in the study area was 33 ha. Assuming the annual water application as 500 mm and deep percolation loss as 20 per cent, the volume of recharge to groundwater during six years was $33 \times (0.5 \times 0.2) \times 6 = 19.8$ ha m, say 20 ha m.

Evapotranspiration

The removal of surface water from water pools in the non-planted area is mostly by surface evaporation. This area in the year 1991 was 14.69 ha. The volume of pool water was 54 ha m. The maximum depth was about 7m. It is assumed that all this water was removed by surface evaporation during the period 1991 to 1994.

The bulk of the groundwater in the land strip 1524 m long and about 800 m wide was removed by transpiration from the plantations. The average depth by which the groundwater table was lowered during the six year 1991–1997, is about 10 m. It is assumed that surface evaporation accounted for removal of water from the top 1 m layer and groundwater from a 9 m thick layer was removed by transpiration from plantations.

The average dry bulk density of soils in the top 1.5 m layer is about 1.6 and the moisture content 5 per cent (1997). On the basis, the fillable porosity of the top layer is 0.3. But lower down, the soil strata is much more dense and hard. The soils are mostly sandy loams. An average fillable porosity of 0.15 is considered for the whole depth of 10 m to estimate the volume of subsoil water removed by transpiration.

Transpiration by Plantations

The volume and rate of transpiration from the plantation area (25 ha) during the six-year period, 1991–1997, is estimated as follows:

Plantation area

Length = 1524 m

Width = 261 m

Total area = 1524 × 261

= 39.77 ha

Less water pool area (1991) = (–) 14.69

Net plantation area = 25.08 ha, say 25 ha.

Volume of groundwater removed (extracted)

Length = 1524 m

Width = 800 m

Depth (av.) = 9 m

Fillable porosity = 0.15

Volume = (15.24 × 8.0 × 9) ha m = 1097 ha m

Less water pool volume (–) 54 ha m

Net 1043 ha m

Net volume of groundwater

removed in 6 years = 1043 × 0.15 = 157 ha m

Recharge by Seepage from Main Canal

Seepage from half section of the main canal at the rate of 0.6 m^3/ sec per million square metre of wetted area:

$= 1524 \times 19 \times 0.6 \times 1/10^6\ m^3/sec$

$= 17.37 \times 10^{-3}\ m^3/sec$

$= (17.37 \times 10^{-3}) \times 24 \times 60 \times 60 = 1500\ m^3/day$

$= 1500 \times 365 \times 10^{-4}$ ha m/year

= 54.7 ha m/year

Total seepage in six years = 6 × 54.7 ha m = 328.2 ha m

Total Recharge and Transpiration

Recharge from rainfall = 12 ha m

Recharge from irrigation = 20 ha m

Total volume of water

removed by plantation on 25 ha area

in 6 years (1991–1997) = (157 + 328 + 12 + 20) = 517 ha m

Therefore, annual rate of transpiration = 517/25 × 6 m

= 3446 mm per year

It can be seen that with the possibility of fillable porosity in soil being more than 0.15, the rate of seepage from canal exceeding 0.6 m^3/sec per million sq m, and lowering of the groundwater table beyond the 800 m boundary of the strip, would all tend to increase the calculated rate of transpiration. The groundwater flow to the outside study area, which is small, has been neglected. The estimated value of 3446 mm is more than the annual rate of evaporation from the free water surface in the area, which at one station in the eastern part of the project was measured as 2971 mm. Heuperman, on the basis of the study in Northern Victoria, Australia suggests that the water use of a five-year old irrigated plantation may range between 0.6 A_{pan} to 1.9 A_{pan} and that with a closed canopy (1100 trees per ha) approximates A_{pan} evaporation. The conservative estimated value in the IGNP reach under study is 1.2 A_{pan}.

In the IGNP, with an estimated annual groundwater recharge of 2.6 billion cubic meters (bcm), a forest cover over an area of 86700 ha (with a 3000 mm annual rate of transpiration) would be needed to transpire the whole quantity of recharge. This would be about 5 per cent of the project area. By considering an overall average annual transpiration rate of 1500 mm from the forest area, a requirement of 1,77,000 ha or 10 per cent of irrigable area would be needed to be put under tree plantations to maintain the water balance.

Transpiration by Comparing with Evaporation

There is another simple way to estimate the rate of transpiration by the tree plantations. The total study area was 40 ha (approximate) which was all waterlogged (with groundwater almost at surface). Before the tree plantations came up, the water levels had stabilised at about ground level. This shows that the annual rate of groundwater inflow (from all sources) into the area equalled the rate of evaporation from the surface (without trees) and was therefore nearly equal to 40 × A_{pan}, assuming rate of evaporation from water surface and waterlogged ground equalled A_{pan}.

As the trees grew up (in 25 ha area), and started transpiring water, the groundwater levels started going down. After the groundwater disappeared from the surface and went sufficiently deep, direct evaporation from the ground and water pool surface almost ceased, and the removal of water from the ground in the area thereafter was almost wholly by virtue of transpiration from the plantation trees. There was little change in the conditions of canal water flow and irrigation during the period and in the rate of groundwater inflow into the area. Therefore, the volume of annual transpiration from trees in the plantation area (25 ha) exceeded or at least equalled the volume of annual inflow which was $40 \times A_{pan}$. The rate of annual transpiration, was therefore, approximately equal to $40 \times A_{pan} \div 25 = 1.6\ A_{pan}$.

DRAW-DOWN EFFECT OF PLANTATIONS

At cross-section 953.2, the groundwater under the plantation has gone down by 15.7 m in a period of six years and it is continuing to go down further. At a distance of 100 m from the edge of the plantation, the groundwater level was about 9 m higher than at the edge, with a draw-down of 6.7m. At a distance of 500 m from the edge the groundwater level was 7.6 m higher that at the edge, with a draw-down of 8 m. The higher ground level at 100 m from the edge as compared to that at 500 m from the edge is apparently due to recharge from irrigation from the area under cultivation.

At cross-section 953.9, the groundwater under the plantation has gone down by 14 m in a period of six years. At a distance of about 100 m from the edge of the plantations, the groundwater level was 3.6 m higher than at the edge, with a draw–down of 10.4 m. At a distance of about 300m from the edge, the groundwater level was 6.2 higher than at the edge with a draw-down of 7.8 m.

It can be said that the plantations are acting like shallow wells, withdrawing water from the ground at the rate of 3.446×10^4 m^3/year per ha of plantation. This works out to 3.93 m^3 / hr per ha of plantation or 98 m^3/hr for the 25 ha plantation in the reach under consideration. This would be equivalent to three tube wells with a withdrawal rate of 33 m^3/ hr working continuously 24 hours a day, 365 days a year. In other words, the plantations are acting like wells, spaced 500 m apart with a pumping capacity of 33 m^3/hr. The maximum draw-down during a period of six years has ranged from 13 m to 15 m and was yet to stabilise.

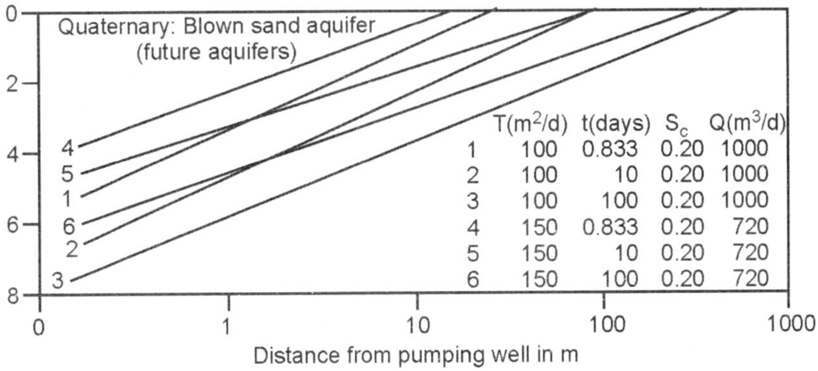

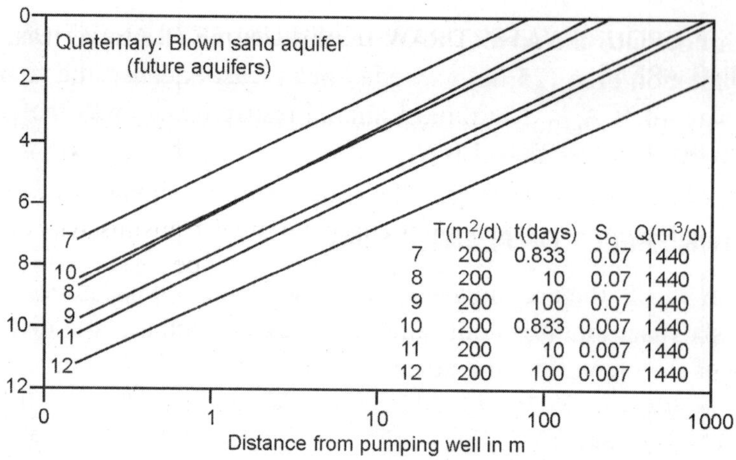

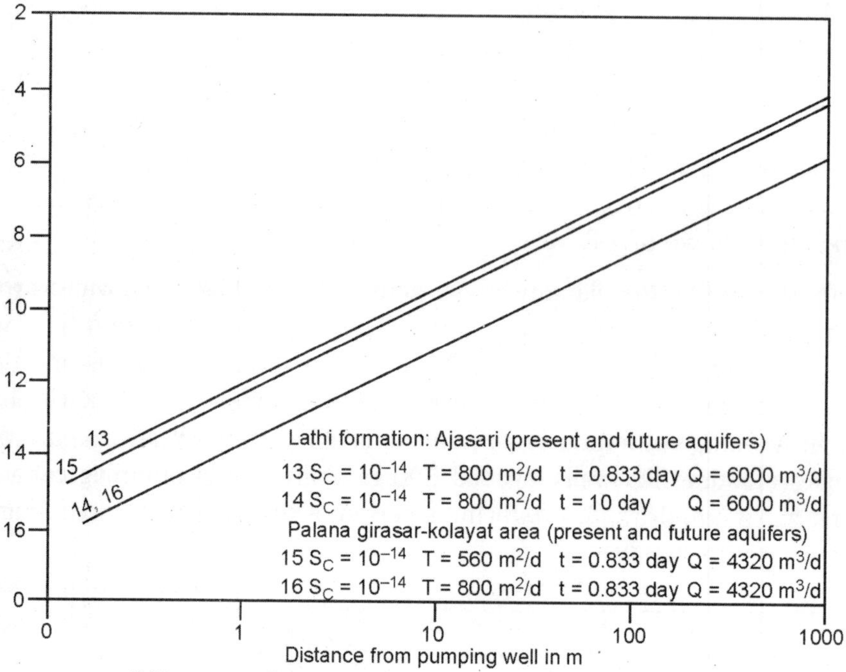

$$S_C = \frac{4UT_t}{r^2} \qquad Z = \frac{Q}{4\pi T} W(u)$$

Theoretical draw-down versus distance from pumping wells for selected combination of aquifer parameters, duration and rate of pumping for different aquifers

z = Draw-down, in m, t = Time since pumping started, in days

r = Radial distance from discharge well to the point of observation, in m

$u = r^2 S_c / 4Tt$ T = Transmissivity, in m^2/day

S_c = Storage coefficient, dimension-less Q = Constant rate of discharge of well, in m^3/day

$w(u)$ = Weil function u

Figure 5.6 *Theoretical draw-down (in metres)*

IMPACT OF GROUNDWATER DRAW-DOWN UNDER PLANTATION, ON ADJOINING AREAS

The draw-down effect on the groundwater would be maximum directly under the plantations and would decrease with increasing distances from the plantation sites. The shape of the groundwater surface profile can be expected to be similar, as would occur in the case of pumping from wells. The aquifer characteristics, transmissivity of strata and rates of recharge and pumping would determine the shape of the water table profile. WAPCOS, on the basis of the non-equilibrium formula, suggest relationships between rate of pumping, draw-down and distances as shown in Figure 5.6.

The observed data shows that plantations, during a period of six years have caused a draw-down of 8 m at a distance of 500 m from the edge of the plantations at the cross-section at RD 953.2, and of 7.8 m at a distance of 300 m from the edge of plantations at cross-section 953.9. The magnitudes of draw-down are quite significant even at considerable distances from the edge of the plantations.

In any biodrainage scheme, the maximum draw-down (under the plantations) cannot exceed the maximum depth up to which the plantation roots can draw water. In plantation at RD 953 of the Indira Gandhi Main Canal, the trees were surviving and healthy even after the groundwater table had gone down by 15.7 m. Furthermore, for the drainage scheme to be successful the groundwater surface, at the farthest distance from the plantations, must be at least 2 m below the ground surface.

If in a large area with a high groundwater table, the plantation is done in a small area, and the quantity of water transpired through trees is relatively small in comparison to the total quantity of water going into the water reservoir as recharge, then the draw-down impact on the groundwater table is likely to be limited and confined to the area directly underneath the plantations. The groundwater levels even at a short distance from the edge of the plantations may not be depressed. But if the plantation area is sufficiently large so that the quantity of the transpiring water through the trees exceeds the total quantity of recharge to the groundwater reservoir from the adjoining areas, then the groundwater table in the region will progressively go down till a balance is reached between the recharge and the withdrawal. In such case, though the maximum

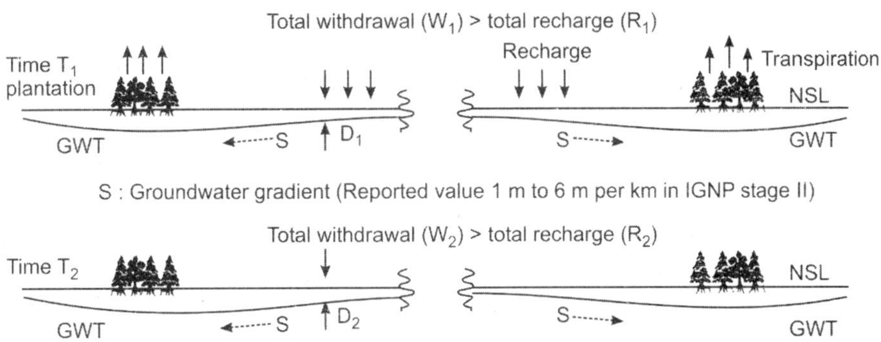

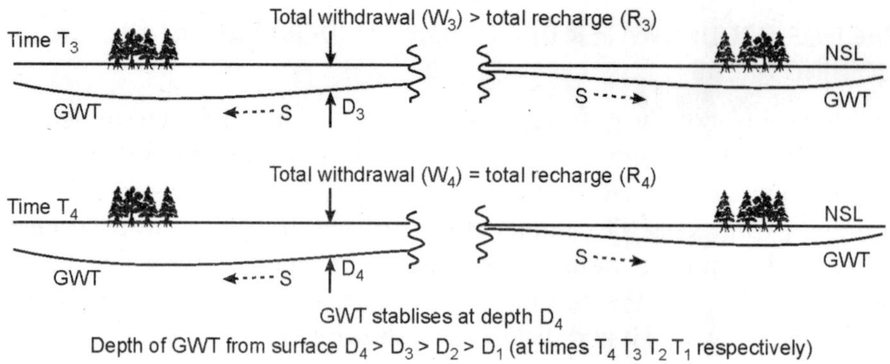

Figure 5.7 *Groundwater table movement with biodrainage*

draw-down would occur underneath the plantations, there would be significant draw-down of the groundwater table in adjoining areas. This position is illustrated in Figure 5.7.

Distance up to which Plantations can Provide Effective Biodrainage

There is another way of looking at the depression of the groundwater surface under the plantations and induced groundwater flow from the adjoining areas. If there are two plantation areas separated by distance *L*, the depression of the water table underneath them would result in groundwater flow behaviour similar to that as in the case of a flow towards parallel ditches penetrating an unconfined aquifer. On equilibrium, the position would be as shown in Figure 5.8, and the relationship between the depression of the groundwater table, rate of recharge, hydraulic conductivity, depth to barrier layer and distance between plantations can be expressed by the Donnan equation (Donnan, 1946).

$$L^2 = \frac{8kY_o h}{R} + \frac{4kh^2}{R}$$

Stirzaker et al. (1999), using Dupuit-Forchhlimer theory (Kirkham, 1967) give an equation to calculate the maximum spacing of parallel belts of trees to hold the water table just below the crop root zone. It can be shown that this equation is identical to the Donnan equation given in Figure 5.8.

In the IGNP command, a total recharge of 2.6 bcm from canals and irrigation spread over 1.8 m ha culturable command area, would give a recharge rate of 0.4 mm/day. The recharge due to rainfall is insignificant. The total recharge *R*, may be taken as 0.5 mm/day.

In the equilibrium stage when the groundwater surface at the highest point is 2 m below natural ground surface and the groundwater table underneath the plantations is, say 12 m, the value of *h* would be 10 m.

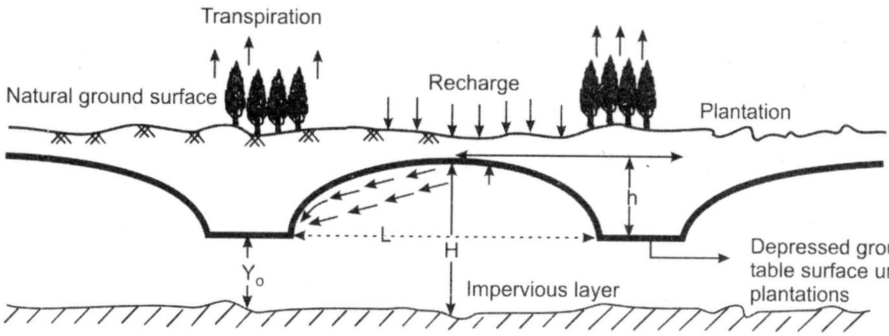

Donnan equation $L^2 = \frac{8KY_0h}{R} + \frac{4Kh^2}{R}$

L = Distance between plantations | K = Hydraulic conductivity

R = Rate of recharge | h = Head difference

Y_0 = Height of water level above barrier layer

With R = 0.5 mm/day, h = 10.0 m, Y_0 = 10.0 m

and K = 100 mm/day, L = Works out to 500 m

Figure 5.8 *Flow towards depressed groundwater table under plantations*

The values of hydraulic conductivity of ground strata in IGNP generally vary between 250 mm/day to more than 1000 mm/day.

The depth to the barrier layer varies from place to place. In the following example it is assumed as 10 m.

The equation may be re-written as:

$$L^2 = \frac{8\{k\}Y_oh}{\{R\}} + \frac{4\{k\}h^2}{\{R\}}$$

Corresponding to the hydraulic conductivity values of 1000, 100 and 10 mm per day, the value of L (distance between plantations) works out to 1500, 500 and 150 m respectively. In other words, it can be said that where the hydraulic conductivity of sub-strata is 1000 mm/day, and depth to the barrier layer is 22 m below the natural ground surface, plantations can provide effective biodrainage up to a distance of 750 m from the edge of the plantations.

For satisfactory control of the water table by biodrainage with the help of plantations, it would be necessary to plan so that:

(a) total withdrawals from groundwater by plantations equal or exceed the total net recharge to groundwater; and

(b) the minimum depth of the water table from the ground surface, at the farthest distance from the plantations, is not less than 2 m.

Objective (a) can be achieved by deciding on the size of the plantation area and (b) by appropriate distribution of the plantation area within the command area.

▼ SALT BALANCE

The total volume of soil water removed by plantations during a period of six years is estimated to be 517 ha m. The average salt content in the canal water is 125 mg/l. The total salt content in 517 ha m water would be 646 tonnes. If the trees were not to take any salt from the soil water during transpiration, there should have been an increase in the salt content in the soil mass (1524 m × 261 × 10 m with an average bulk dry density of 1.7) of 0.009 per cent (by weight) during the period of six years.

An approximate relationship between the salt content in the soil and the corresponding *ECe* value of saturation extract is given in Figure 5.9 and Table 5.7.

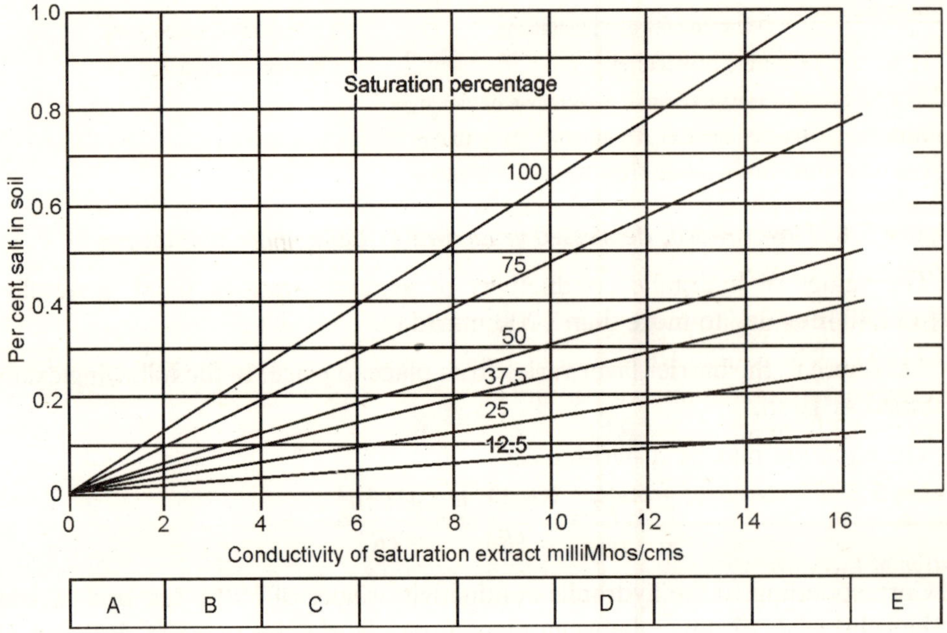

A = Salinity effects mostly negligible

B = yields of very sensitive crops may be restricted

C = Yields of many crops restricted

D = Only tolerant crops yield satisfactorily

E = Only few tolerant crops yield satisfactorily

Source: United States Department of Agriculture (1954) Handbook No. 60.

Figure 5.9 *Salinity and crop response*

Table 5.7 *Salt content in saline soils and ECe values*

Name of soil by degree of salinisation	*Toxic salt content*	*ECe × 10^{-3} in m.mhos/cm of saturation extract*
C_1 – non-saline	less than 0.2	Less than 4
C_2 – slightly saline	0.2 – 0.4	4–6
C_3 – Medium saline	0.4 –0.8	6–12
C_4 – Strongly saline	0.8 – 1.0	
	1.0 – 1.5	12–16
		16 – 30
C_5 – Very strongly saline	1.5 –3.0	30 –60
C_6 – Intensely saline	3.0 –6.0	60 –120

It would appear that the increase in salinity of soils underneath plantations due to canal water is quite small and it might take several hundred years for the soil underneath the plantations to become agriculturally unproductive due to high salinity. Therefore, if the groundwater table is maintained adequately deep, there would be little threat of salinisation of soils in the root zone underneath the plantations for a long time to come.

EC Values of Groundwater under Plantations and in Adjoining Area

The EC values of groundwater were determined for water taken out from piezometer holes (1997). The results are shown in Table 5.8. Out of the test results at 11 locations, the EC values at six locations were found to be less than 1 ds/m, at four locations between 1 ds/m and 2ds/m at one location it was 3 ds/m.

Table 5.8 *EC values of groundwater from Piezometers IG Main Canal (L) RD 952-957*

Location of Piezometer	*Depth of water table from NSL (m)*	*EC value micro siemens/cm at 25°C*
RD 953.212 (L)		
25 m from Canal	11.4	700
75 m from Canal	12.9	690
175 m from Canal	14.9	450
375 m from Canal	7.1	610
575 m from Canal	6.4	3000
775 m from Canal	9.6	1800
RD 953.900 (L)		
25 m from Canal	11.7	710
75 m from Canal	10.8	1290
175 m from Canal	13.2	430
375 m from Canal	10.8	1350
575 m from Canal	7.0	1650

The plantations extend up to a distance of about 260 m from the canal. As expected, the lowering down of the water table was maximum in the middle of the plantation area. But there appeared to be no concentration of salts in the groundwater under the plantation area.

The *EC* values of groundwater, though less than 2 ds/m, are still higher than those of canal water, suggesting that some salts brought by the canal water have remained in the groundwater. For a period of about 10 years (1984–94), there was a standing pool of water extending over an area of 15 ha, from whose surface evaporation was continuously taking place. The volume of evaporated water (at 2500 mm/yr) was about 375 ha m and all salts contained in this volume of water were left behind by the evaporating water. This has, perhaps, contributed to the higher salt content in groundwater as compared to that in the canal water.

EC Values in Root Zone and in Groundwater under Plantations

The groundwater table in the IGNP command is on the rise and would continue to do so till the withdrawals by drainage from groundwater do not equal the net recharge to groundwater. The net recharge to groundwater can be reduced by avoiding an excess application of irrigation water and adopting improved water saving and management practices.

The *EC* of deep groundwater, before the introduction of irrigation was quite high. With the rise of the groundwater table to shallow depths below the ground surface, the *EC* values of the groundwater table at surface have been found to be moderate in Stage I of the project. It is expected that the quality of water at the surface of the groundwater table when it comes up to shallow depths would be similar under stage II. At that stage, the plantations would be able to draw most of their water requirement from the groundwater surface and some concentration of salts will occur in soil in the root zone as the transpiring water leaves behind most of the salts contained in the groundwater in the root zone. The soils in the area are mostly sandy loams. The capillary zone may be expected to extend up to about 1 m above the groundwater table surface. This is the zone in which the maximum concentration of salts is likely to occur.

Trees would draw very little water from the zone above the upper limit of the capillary zone and, therefore, there should not be any significant concentration of salts above the capillary zone. In the saturated zone below the groundwater table surface, the groundwater flow would continuously occur from all directions towards the area underneath the plantations, to restore the imbalance caused by transpiration through trees. At the ultimate stage, when the quantity of drainage matches the quantity of net recharge to the groundwater and if the total salts removed equal or exceed the total salt contained in the irrigation water, both the groundwater table as well as the salinity levels, in the groundwater would stabilise. Thereafter, there should therefore be no progressive increase in groundwater salinity. Somewhat higher salinity levels may occur in the groundwater immediately underneath the plantation area, but there should be no progressive increase in salinity levels because of the continuous groundwater flow from the adjacent areas, which would be relatively less saline.

As and when the salinity level in soils in the capillary zone above the water table surface reaches high levels, inhibiting water uptake by trees, the roots of trees should be able to extend to the groundwater table surface to draw water. This would be feasible so long as the salinity level of the groundwater table does not reach the threshold level (say more than 12 ds/m), and the groundwater table does not go down beyond the reach of tree roots.

CONCLUSIONS

The following conclusions can be drawn from the study:

(a) The conservative transpiration rate from the plantation in the IGNP command is estimated as 3446 mm/yr which is about 1.2 A_{pan}.

(b) The draw-down of the groundwater table underneath the plantation was found to be 15 m and more.

(c) When the plantations are grown on sufficiently large areas, the draw-down effect on groundwater is not confined to areas immediately underneath the plantations but extends to distance of 500 m and beyond the edge of plantations. Knowing the values of hydraulic conductivity, the rate of recharge and depth to barrier layer, it is possible to estimate the draw-down effect of plantations on the groundwater surface at varying distances from the edge of plantations.

(d) No abnormal increase in salinity levels of soils and groundwater was observed underneath the plantations.

(e) Afforestation/plantations hold good promise of providing biodrainage to irrigated lands to overcome the threat of waterlogging and salinisation.

REFERENCES

Central Ground Water Board, India (1995). *Studies on Conjunctive Use of Surface Water and Ground Water Resources in IGNP Stage–I*, Rajasthan.

Donnan, W.W. (1946). *From Hydrogeology of Different Types of Plains*, Vol. I, International Institute for Land Reclamation and Improvement, The Hague.

Forest Department of the State Government of Rajasthan. Data on Plantations Along Indira Gandhi Main Canal RD 750–1300 (Unpublished).

Indira Gandhi Nahar Board (1997). Project Profile.

— (1997). Indira Gandhi Canal Project: Experiences and Lessons.

Kapoor, A.S. (1997). 'Biodrainage of Lands Under Irrigation in Dry Arid Regions with Particular Reference to IGNP, Rajasthan' (Unpublished).

— (1997). *Prevention of Soil Salinisation of Irrigated Lands in Dry Arid Regions with the Help of Plantations*.

Stirzaker, R.J., F.J. Cook and J.H. Knight (1999). 'Where to Plant Trees on Cropping Land for Control of Dryland Salinity: Some Approximate Solutions', *Agricultural Water Management Journal*, 39: 115–33.

US Department of Agriculture (1954). *US Salinity Laboratory Diagnosis and Improvement of Saline and Alkali Soils*, Handbook No. 60.

Upadhyaya, A.K. and R.G. Soni (1997). 'Growth, Biomass Production and Dry Matter Distribution Pattern of Eucalyptus Camaldulensis Grown Under Irrigated Conditions in IGNP Area of Rajasthan', *Indian Forester*, 123 (3).

Upadhyaya, A.K. (1994). *Irrigated Forestry in IGNP Stage–II: Planning for Sustainability in Irrigation.*

WAPCOS (1995). *Feasibility Report, IGNP Stage–II.*

WAPCOS/SANYU Consultants (1993). *Remedial Measures Project (Phase–I) for Indira Gandhi Main Canal*, Rajasthan, Final Report.

6

Some Other Experiences

THE ARAL SEA BASIN - CENTRAL ASIA

Irrigation and Drainage Development

A programme to utilise the waters of rivers Amu Darya and Syr Darya for irrigation on a large scale to transform barren desert into productive farm land covering an area of about 8 million hectares was undertaken in the Central Asia in the 1960s. The growth of irrigated area according to Dukhovny and Umarov (2000) was as follows:

Year	*Annual irrigated area (m ha)*
1990	2.5
1960	4.5
1970	5.2
1980	6.9
1990	7.5
2000	8.0

The five Sea Basin States are Turkmenistan, Uzbekistan, Tajikistan, Kirgizistan and Kazakhstan. The average annual water resource of the basin is estimated as 120 billion cubic metres (bcm). Dukhovny and Yaku Bov (2000) give annual water used figure as follows:

Water Use in Cubin Km (or bcm) in Aral basin

	Year (1990)	*Year (1995)*
Water resource	117.5	110.5
Return water	32.8	30.0

From the total volume of return water (30–36 cu km), 55–60% is used for irrigation. The total irrigated area has expanded up to 8 m ha.

The withdrawal of water from the rivers resulted in progressively reduced inflow into the Aral Sea into which the rivers were ultimately discharging. As a result, the Aral Sea started shrinking in the 1970s; salinity of the Aral Sea rose and several phenomena of environmental degradation such as the destruction of the ecosystems around the sea, extinction of many fish species, and pollution of groundwater were discovered. The occurrence of illness among the residents rose because of salty dust sprays in the surrounding areas, as high winds picked up salts from the drying bed of the sea. Takano (1994) stated that the cause of the Aral Sea crisis is the same one that destructed the Mesopotamia and Mohenjodaro civilisations. The same catastrophic phenomena that mankind experienced those days are recurring today due to inadequate counter measures against salinisation and other failures of arid region irrigation. If we fail today to solve the environmental crisis because of our ignorance and arrogance, the next generation will lose their place to live and suffer. The cause of the Aral Sea crisis is apparent and its counter measures technically attainable. We must solve the crisis since it is a crisis universal to us, the human beings.

The ill-effects of the extensive irrigation-cum-drainage programme are not confined to the Aral Sea alone. The groundwater table in the irrigated areas is rising progressively along with an increase in the salinity of soils. Dukhovny and Umarov (1990) give the following picture:

Table 6.1 *Rise in groundwater table and soil salinity*

Year	*Irrigated area (m ha)*	*Depth of groundwater level (% area)*			*Degree of soil salinity (% a year)*		
		< 2 m	*from 2 to 3 m*	*> 3 m*	*Nil*	*Low*	*High*
1990	7.47	24.8	32.7	42.5	48	29.4	22.6
1994	7.96	31.1	29.9	39	42.2	29.9	27.9

The waterlogging and salinity conditions have caused decline in agriculture produce. As an illustration, Dulhovny and Umarov give the example of Makpaaral zone of South Kazakhstan in the middle reach of Syr Darya river, where an irrigated area of 40.6 thousand hectare was a fertile land with cotton yields of 3.2 to 3.5 tonnes per hectare up to the year 1985 which progressively declined to the low level of 1.9 tonne per ha. by the year 1996.

It is estimated that out of an irrigated area of 8 m ha, 5.7 m ha area needs drainage. The drainage measures have so far been carried out over a total area of about 4.5 m ha; an area of 1.0 m ha is provided with subsurface horizontal drainage, 0.8 m ha with vertical drainage and the rest with 'mufriga' and 'zauras' which are open drains and canals. It has not been possible to maintain and operate the drainage systems properly as a result of which these have deteriorated and do not provide the intended relief. The main reasons of neglect of drainage systems are high costs, shortage of electric power (for pumping), lack of participation by water users, low land productivity and unavailability of investment funds. Most existing drainage facilities now require reconstruction.

Dukhovny and Umarov (2000) describe to approach the problem in the following three principal directions:

- completion of drainage facilities, where they have not been, as required by design;
- reconstruction of the existing drainage facilities;
- proper maintenance of the existing drainage facilities.

To start these it is necessary to:

- work up a more complete understanding of the present structure of land use;
- improve the methods and assessment criteria of the state of land melioration and fertility, considering the following: condition of irrigation and drainage networks, shortage of water resources and deteriorating quality of irrigation water. Assess lands in Central Asia using the new criteria and methods;
- develop a system to assess the state of drainage. Identify the operational state of existing drainage facilities and necessary rehabilitation measures;
- determine standards for drainage system design with consideration for on-farm irrigation demands, water quality and the transition to new regulations on water and lands use;
- design standards for the operation of drainage systems that will allow responsibility for operation and maintenance to be transferred to Water User Associations.

In principle, the proposed actions will minimise water loss in irrigation network, but requires a great capital investment. It will require the following steps:

- implementation of crop irrigation regimes in accordance with planned yields;
- a sharp improvement in the scale of leaching regimes by adoption of intensive methods of crop cultivation (deep soil loosening, chemical ameliorants and organic fertilisers, deep ploughing, crop rotation);
- to secure uniform moistening and desalinisation of soil and minimisation of groundwater infiltration through optimisation of the size of irrigated plots and land levelling;
- revision of crop selection with regard for the environmental, economic and social conditions in the region;
- exclusion of strongly saline lands from cropping.
- raising the efficiency of inter-farm and on-farm irrigation network by anti-filtration lining;
- organising of the manufacture and large scale introduction of better drainage machinery and improved irrigation technology to achieve a uniform supply of water along furrows and consistent moistening of the root zone.

- organisation of regular cleaning and maintenance of inter-farm and on-farm collectors and water intakes, to prevent further deterioration of technical conditions of primary drains and the appropriate drainage of irrigated lands;
- to initiate repair and rehabilitation works of on-farm drains, drainage collectors and tube wells, as well as expansion of the modern drainage technologies;
- to define proper government support to drainage network, especially for vertical drainage and inter-farm collectors;
- to organise local base for maintenance and repair of drainage system on contract base;
- to rehabilitate the local production of tubes and drainage techniques;
- to create the training and education of young specialists in drainage (construction, operation, maintenance).

It would be an uphill task to carry out the proposed measures successfully. A fresh look is needed on the two vital aspects, of water management and drainage methodology to search alternative approaches and methods that may be effective.

Water Resource Management

Assuming that 96 bcm (80% of the available annual water resource of 120 bcm) is used for irrigation and an additional quantity of return water, 18 bcm (55–60% of 30–36 bcm) is also used for irrigation, the total water volume of 114 bcm is presently used to irrigate an area of 8 m ha annually. This gives the overall irrigation water use figure of 1.45 m. This is far too excessive and is an important cause of the prevailing problems.

Dukhovny and Umarov (2000) mention that in the Makpaaral zone of South Kazakhstan, cotton yields of 3.2 to 3.5 tonnes/ha were being obtained up to 1970s, with irrigation water use (including leaching requirement) of 650 mm. In the Hunger Steppe area, the recommended water delivery for optimal reclamative regime is 895 mm against the old practice of delivering water at the rate of 1344 mm. Kitamura and Yasuda (2000) report average irrigation efficiency for the Aral Basin as 60% before water reached the field. In a study conducted over a 716 ha irrigated block (rice and alfa alfa) the losses within the block were 27.7% of the water delivered to the block.

Usman et al. (2011) analysed the situation prevailing at Shoma–Khulum Water Users Association (irrigated area 1885 ha) and estimated the crop water requirement, simulated with the SIMIS model, with major crops of cotton perennial crops and wheat as actually grown in the year 2006, as 590 mm.

In the Indus basin in India and Pakistan, irrigation over an area of 26 m ha is based on a water delivery of about 600 mm.

The reasons for the use of excessive water for irrigation appear to be (i) liberal availability of water to the water users particularly in upper reaches; and (ii) increasing salinity in the irrigation water as well as in the soils requiring greater fractions of leaching water. The remedial measures should aim to control and regulate irrigation water supply at rates no more than actually needed by the crops.

In case the rate of water application for irrigation is restricted and regulated to 650 mm then the water use for irrigating 8.0 m ha area would be about 52 bcm (against 114 bcm as at present) resulting in a possible saving of 62 bcm. The volume of return water may reduce to about 10 bcm (20% of 52 bcm), that is 26 bcm less than at present (which is 36 bcm). The net saving in water use would work out to 62 – 26 = 36 bcm. This quantity of water can then be used for irrigating additional lands or allowed to flow into the Aral sea or shared between the two options. This arrangement would help to:

- overcome the environmental and other problems being faced due to shrinking of Aral Sea.
- reduce the problem of waterlogging and salinity in irrigated areas.
- improve the quality of water in the rivers.
- save water for additional productive use.

Salinity Management

It is reported by Dukhovny and Yakubov (2000) that 55 to 70 per cent of return water in Syr Darya basin and 40–48 per cent of that in Amu Darya basin is discharged into the rivers. The total annual return water volume being 30–36 cu km, the net volume of return water being discharged into the rivers is therefore about 18 cu km.

The reported levels of salinity in the return waters for period 1990–97 are as follows:

	Salinity in mg/l
Upper reaches	1500–3000
Middle reaches	3500–6000
Lower reaches	5000–7000

Taking the overall average salinity of the return water discharge into the river as 3500 mg/l, the total salt load discharged into the river annually by the return waters works out to about 63 million tonnes.

The volume of drainage water from one hectare of irrigated land in Hunger steppe under the old zone of irrigation (unoptimal reclamative regime) was 7877 cum containing 27.57 tonnes of salt. Under the new zone of irrigation (optimal reclamative regime) the volume of drainage water from one hectare of irrigated land reduced to 2130 cum containing 10.22 tonnes of salt. It was concluded that if irrigation in Central Asia is practised in accordance with the recommended optimal reclamative regime, the salt discharge into the rivers due to return waters would reduce to 40 per cent of the present value. But even under the recommended optimal reclamative regime, discharge of 10 tonnes of salt from one hectare of irrigated land would result in a total annual discharge of 80 million tonnes from 8 mha and if 60 per cent of this finds its way into the rivers, the net annual discharge of salts into the river would still amount to 48 m tonnes.

Dukhovny and Yakubov mention that in the Syr Darya basins, the salts content in the return water in the upper reaches is 1500 to 3000 mg/l and in the lower reaches 3500–7000 mg/l. Similar picture is found in the Amu Darya basin. The collector drainage water salinity is 1500–4800 mg/l. While discussing the reuse of drainage water for irrigation they classify the water with salinity 600–1090 mg/l as good (I Class), that with salinity 1000–2500 mg/l as satisfactory (II class), that with salinity 2500–6000 mg/l as weakly satisfactory (III class) and that above 6000 mg/l as bad (IV class).

Comment

In the Aral Sea basin an annual river flow of 120 bcm of salinity 250 mg/l and return flow of 18 bcm with average salinity of 3500 mg/l would give a total annual salt contribution of 93 million tonnes. When drainage water is discharged into the rivers, the salinity of the mix-water in the river increases. The quantity of increase would depend upon the salinity of drainage water and its volume. Table 6.2 shows the average salinity of mix-water in the river corresponding to different salinity levels of the return water and different mix ratios. It is seen that there is manyfold increase in river water salinity due to the discharge of saline drainage water into the rivers. Kitamura et al. (2000) report that the salinity of water in lower Syr Darya in the Kzyl–Orda region increased from 400–600 mg/l to 1300–2000 mg/l during the last three decades.

Table 6.2 *Salinity of mix-water in the rivers*

Salinity of natural river water (mg/l)	*Salinity of return water (mg/l)*	*Average resultant salinity of mix-water of ratio (natural: return)*			
		4:1	*5:1*	*6:1*	*7:1*
	1500	500	458	429	406
	2000	600	542	500	469
250	3000	800	708	643	594
	4000	1000	875	786	719
	5000	1200	1042	929	844
	6000	1400	1208	1071	969
	7000	1600	1375	1214	1094

When mix-water from the river is used for irrigation, the need for leaching fraction increases with increase in salinity in the irrigation water. The direct use of saline drainage water for irrigation results in even greater requirement of water for leaching and irrigation. The volume of net drainage water discharged into the river may be reduced but not the quantity of salts. Therefore, the salinity in river water is not reduced. Discharge of drainage water from irrigated fields into the river, can provide relief from salinity to the drained areas on regional or local basis, but in the overall terms it creates more problems than it solves. Horizontal subsurface or open drainage systems can be no solution to the overall problem being faced in the Aral Sea Basin, unless the drainage water can be safely disposed off into evaporation tanks or discharged into the oceans through dedicated outfall drains without being put into the rivers. The experience on evaporation tanks, so far, is not very encouraging. The outfall drains are very expensive and difficult to construct and maintain.

The scenario of discharge of drainage water into the river thereby increasing river water salinity, then using greater volumes of saline water for irrigation and again discharging back salts into the river result in a vicious cycle in which the demand for irrigation water keeps on growing and the salinity of river water keeps on increasing. The Aral Sea Basin, unfortunately, is caught in this vicious cycle. The traditional drainage measures, the horizontal subsurface or open drainage systems or the vertical drainage systems cannot solve the problem of the Aral Sea Basin area, in the long term on a sustainable basis. The viable solution can only be that in which the salts from irrigated areas do not find their way into the rivers.

Biodrainage Feasibility

Dukhovny and Umarov classify reclamative regimes as follows:

automorphous regime	–	irrigation water not connected with groundwater. Infiltration is free.
semi-automorphous regime	–	groundwater backs up irrigation water but almost doesn't feed plants
semi-hydromorphous	–	groundwater feeds plants prevailing over irrigation water
hydromorphous	–	plants are mainly fed by groundwater

Under the biodrainage scheme, the irrigated crops (cotton etc.) can grow under automorphous regime while the tree plantations to provide drainage would be under the hydromorphous regime. The tree plantations are to be raised in blocks and strips, adjacent but outside the blocks of irrigated areas. Alternatively, they can be raised in parallel strips within the irrigated areas or in a scattered pattern.

Water Balance

Dukhovny and Umarov, for the optimal reclamative regime in respect of Hunger Steppe estimate the volume of drainage water as 2130 cubic metres corresponding to total water application of 11530 cubic metres. The volume of drainage or recharge water is approximately 20 per cent of the applied water. They give the figure of annual potential evaporation as 1200–1700 mm.

The tree plantations can transpire quite large volumes of groundwater at shallow depth. To estimate the area required for tree plantations for transpiring as much quantity of groundwater as is added in the form of recharge from irrigated area [Kapoor (2001) gives the following relationship, Figure 8.1].

$$P/C = \frac{R_F \times A_F \times I_R}{T_R}$$

where P/C = represents the fraction of culturable area that must be under afforestation

R_F = is the recharge factor, i.e., ratio of net recharge (to groundwater) to total irrigation water supply

A_F = is the area intensity factor of irrigated agriculture

T_R = is the transpiration rate from tree plantations

A_{Pan} = is surface evaporation from a standard pan

If $T_R = A_{Pan}$,

Then $$P/C = \frac{R_F \times A_F \times I_R}{A_{Pan}}$$

For the Aral Sea Basin, if $R_F = 0.2$, $A_F = 1.0$, $I_R = 650$ mm and $A_{Pan} = 1300$ mm, then P/C = 0.10, that is, tree plantations over 10 per cent of the area can provide the needed biodrainage to check the rise of groundwater table and stabilise it at safe deep level. Tree plantation in blocks or strips, suitably distanced and dispersed in the irrigated area can help achieve a water balance to overcome the threat of waterlogging.

The use of Donnan equation can be made to decide on the spacing of plantations (Figure 5.8):

$$L^2 = \frac{8KY_o h}{R} + \frac{4Kh^2}{R}$$

where L = distance between plantations

R = rate of recharge

Y_o = is height of water level above barrier layer underneath plantations

K = is hydraulic conductivity

h = is head difference

Salt Balance

The annual salt input with 650 mm of irrigation water containing 250 mg/l salts would be 1.65 tonnes/ha/yr.

The harvested raw cotton produce of 3 tonnes/ha along with 7 tonnes/ha of foliage containing 3.33 per cent average minerals (cations – Na + Ca + K + Mg), would enable extraction and removal of salt compounds to an extent of 1.0 tonne/ha/year (The weight of salt compounds of sodium, calcium etc., is on average about three times the weight of cation elements found in the plant by ash analysis). The data of cotton/foliage produce and mineral content in them is to be checked for the relevant site and if it is different from the figures given above, then the actual field data should be used to estimate mineral removal by harvested agriculture crops.

Tree plantations over 10 per cent area with utilisable biomass produce of 30 tonnes/ha/year, containing 2.5 per cent minerals (cations), would enable evacuation of another 0.2 tonnes of salt compound per ha per year.

Total extraction and removal of salts by crops and plants would be about 1.2 tonnes/ha/year. This would be 0.45 tonnes/ha/year less than the salt input of 1.65 tonnes/ha/year. This would result in an increase in the salinity of groundwater. The rate at which the groundwater salinity may be expected to increase can be estimated (Figures 2.5 and 2.6). This rate is likely to be too slow to be a matter of concern while using irrigation water of low salinity. Reducing irrigation water application to 600 mm (from 650 mm) and river water salinity to 200 mg/l (from 250 mg/l) can help in achieving salt balance.

There can be methods to achieve full salt balance. Cervinka et al. (2001) describe experiment on growing Salicornia plant giving a biomass produce of 10.5 tonnes/ha with as much as 43.5 per cent mineral content. Growing of Salicornia plant on one hectare area by irrigating from pumped groundwater can help remove about 4.5 tonnes of salts from the groundwater. Such plantations over relatively small area can help in achieving a total salt balance.

An acceptable salt balance is feasible with the help of crops and plantations so long as the irrigation water is of good quality. Therefore, a feasible approach, for overcoming the problem being faced in the Aral Sea Basin, is to prevent the pollution and increase in salinity of the river water. If the discharge of drainage water from irrigated lands is totally stopped along with elimination of discharge from other polluting sources, the salinity level in the river water can be restored to the original natural level of 250 mg/l or even lower. Biodrainage can help to achieve this.

Other Benefits

In addition to providing drainage, the tree plantations would give economic and environmental benefits of great value, so much so that the net cost of providing drainage may be near zero. Good deal of data is available on suitable plantations for different conditions of soils and climates. A start can be made on the basis of available information and experience. Further research for improvements and finding answers to gaps in knowledge may be needed.

Summary and Conclusion

In the Aral Sea Basin the annual irrigated area has increased from 2.5 mha at the beginning of the century to 8 mha by the year 2000. The salinity of the water of the river Amu Darya and Syr Darya, which earlier used to be about 250 mg/l in the upper watershed has been progressively increasing. The salinity levels in the lower reaches of the rivers are much higher. The major contributory source to the progressively increasing salinity is the discharge of saline drainage water from irrigated lands into the rivers. The water salinity of return water in the upper reaches is 1500–3000 mg/l, in the middle reaches 3500–6000 mg/l and in the lower reaches 5000–7000 mg/l.

The waterlogging and salinity are on the increase in the irrigated land, causing loss in agriculture productivity. The problem is proposed to be overcome by cutting down on the volume of water that is excessively used for irrigation and by extension and reconstruction of the horizontal subsurface and open drainage systems.

There is no doubt that better water management to reduce the quantity of water being used for irrigation would help and is a step in the right direction. But it is doubtful whether horizontal subsurface and open drainage systems can be sustainable in the long term for the entire basin taken as a whole. The discharge of saline drainage water from irrigated lands into the rivers is the major cause of the problem. How then can it be a part of the solution? The right approach should be to reduce and ultimately stop altogether the discharge of saline drainage water into the rivers. This cannot be achieved by reconstructing and extending the horizontal subsurface and open drainage systems. This may further aggravate the situation.

The vertical drainage systems can be effective where the groundwater is of good quality and where assured electric power is available for pumping. But in this system the salts are merely recycled and not removed from the system.

Biodrainage, in which excess groundwater is removed from the soils by tree plantations through transpiration can be a feasible option. They can prevent the flow of saline drainage water into the river and restore the salinity levels in the water to the original natural levels of less than 250 mg/l. This in turn can help in achieving salt balance by removal of enough salts through the harvested crops and foliage.

THE INDUS BASIN - PAKISTAN

World Bank's Technical Paper No. 166 titled "Irrigation Planning with Environmental Consideration" presents a case study of Pakistan's Indus Basin by Masood Ahmad and Gary P Kutcher. The data on area, surface and groundwater flows and salinity concentration is shown in Table 6.3.

About 14 million ha area (CCA) is provided irrigation by using 122 bcm of canal fresh water and 52 bcm of groundwater. The salinity concentration in surface water is 200–250 mg/l. The groundwater salinity in Punjab in fresh areas is 900 mg/l and 3000 mg/l in saline areas. The groundwater salinity in Sind under fresh areas is 1000–1500 mg/l and 4000 mg/l under saline areas.

Table 6.3 *Water use data in Indus Basin - Pakistan*

Particulars	Punjab		Sind	
	Fresh areas	Saline areas	Fresh areas	Saline areas
Gross area (million ha)	7.55	2.00	1.72	4.32
Surface water canal flows (billion cubic metres bcm)	51.33	15.20	14.78	41.04
Groundwater use bcm	47.39	—	4.80	—
Salinity concentration of (ppm)				
Surface water	200	200	250	250
Groundwater	900	3000	1000–1500	4000

The degradation of soils has occurred due to waterlogging and salinity. Around 25 per cent of the irrigated lands are saline due to waterlogging, of which 60 per cent are unsuitable for cultivation and another 30 per cent suitable for selective cultivation only. A surface drainage network of nearly 15000 km has been constructed within the Indus system, but much of it is in ill-repair and not capable of handling run-off of heavy rains.

The major source of water in-flow to the groundwater is seepage from irrigation channels and infiltration from farms during surface irrigation. In the Sind saline zones, these seepages account for 92 per cent of total recharge, since rainfall is so low.

The major source of salt addition to the basin is through surface water used for irrigation. The total estimated annual addition for the basin is 24.27 million tonnes, excluding the tube well operations as these do not represent a net import or export of salt to the basin.

Comment

The simulation studies on salt balance and salt distribution ignore the quantity of salt contained in crops and biomass that are harvested and removed from the fields. Data on biomass yields and mineral content in the biomass are not presented in the report. The total water used for irrigation in the basin is 174 bcm (122 bcm from canals and 52 bcm from tube wells). With an overall average water use of 650 mm, this should enable irrigation over an area of 26.7 million ha in a year. Assuming a dry biomass yield (grain + fiolage) of 10 tonnes per ha with mineral (cation) content of 3.33 per cent the total salt evacuated from the system should be about $26.7 \times 10 \times (3.33 \times 3)/100 = 26.7$ million tonnes, indicating the position of an overall salt balance. This goes to show that if rise of groundwater table can be controlled to prevent waterlogging and secondary salinisation, the system can attain both water and salt balances and productive irrigation can be sustained indefinitely. Biodrainage with the help of tree plantations can help to achieve this.

The following conclusions can be drawn:

- The river water quality is quite good, containing 200 mg/l salts in upper reaches and 250 mg/l in lower reaches.
- The groundwater is relatively more saline but salinity level is less than 2.5 dS/m in fresh areas and less than 6.2 dS/m in saline areas. Tree plantations of several species easily grow at such salinity levels.
- The practised drainage methods of subsurface (horizontal), the vertical, the out-fall drain and evaporation ponds have been only partially successful. Around 25 percent of irrigated land is suffering from waterlogging and salinity.
- Biodrainage holds good potential of overcoming the problem, by achieving both water and salt balance, and enabling sustained productive irrigated agriculture on a long-term basis.

NORTH VICTORIA (AUSTRALIA) - PLANTATION STUDY

The concept of controlling waterlogging and salinity problems in irrigated lands, by biodrainage with the help of agro-forestry has received some attention in recent times. In North Victoria (Australia), eucalyptus species were planted over a 2.4 ha area in 1976.

The observed data and results are reported in the following paper:

"Salt and water dynamics beneath a tree plantation growing on – shallow water table" (May 1995) *A.F. Heuperman, State of Victoria, Australia*

General Description

In August 1976, a 2.4 ha site at the ISIA–Dairy Research Centre in Kyabram in the Goulburn Valley, North Victoria was planted with a range of eucalyptus species by the then Forest Commission of Victoria, to assess their growth potential under irrigated conditions. The species planted were southern mahogany (*Eucalyptus botryoides*), flooded gum (*E. grandis*), river red gum (*E. camaldulensis*), southern blue gum (*E. globulus*) and Sydney blue gum (*E. Saligna*). The planting layout is shown in Figure 6.1.

Rainfall at the site averages 480 mm/yr (1942–1988) and the mean annual evaporation (standard A_{pan}) is 1403 mm (1970–1988). Rainfall is spread over the year with an average monthly rainfall of 34 mm (maximum monthly rainfall 39 mm, minimum 29 mm) during the period November–April and 47 mm (maximum average 50 mm, minimum 42 mm) during the period May–October. Of the annual evaporation, 1076 mm or 77 per cent occur during the October–March period. The climatic conditions are not considered suitable for commercial tree growing without irrigation.

The soil at the plantation site is classified as 'lemons loam' (Skene and Poutsma, 1962), a red-brown duplex soil which is the principal soil type in the Shepparton Irrigation Region.

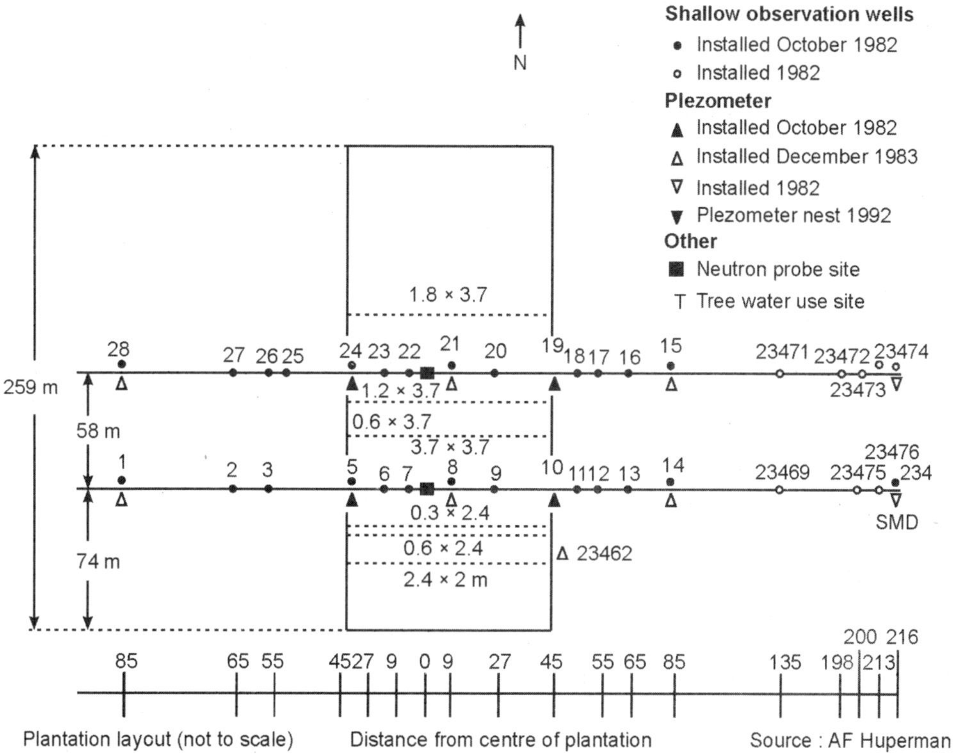

➲ **Figure 6.1** *Planting layout at Kyabram, North Victoria, Australia*

The trees were irrigated from 1976 until 1982 with height increments of up to 2.5 m/yr during that period for most species (Heuperman et al., 1984). In 1982, irrigation ceased because of water shortages during the drought. Since then, no irrigation has been applied with the exception of some occasional un-scheduled events caused by channel overflow before 1990.

Sub-Surface Hydrology

The water table level measurements commenced in October 1978 when three shallow observation bores (3 m deep) were installed in the plantation. Information on the location of the bores is not available but as the trees at that time were only three years old and planted at close spacing intervals (0.6 × 3.7 to 3.7 × 3.7 m), their location in relation to the trees is not important. The water table levels were measured at monthly intervals until June 1979. No salinity information was collected in these bores.

In November 1982, shallow observation bores (4 to 5 m deep) and piezometers (9 to 10 m below the surface with a slotted bottom section ranging from 0.5 to 1.5 m) were installed in two sections through the plantation block (Figure 6.1 section 1 south, section 2 north). The sections were spaced 58 m apart. In May 1984, additional observation bores (6 m deep) were installed in Section 2 inside the plantation at sites where the previously installed shallower bores were found to be permanently dry.

In May 1991, the bores transacts were extended eastwards into the irrigated pasture paddock, up to 170 m away from the edge of the plantation. Monthly waterable measurements were taken until February 1988 and resumed in February 1991. Water table salinity monitoring was less frequent during the 1982–1988 period (2 × per year).

Two nests of piezometers with screen lengths of 20 cm were installed in July 1992. Screens were positioned at 0.5 m depth intervals from 4.5 to 9 m below the surface. One nest was installed in the centre of the plantation while the other was located in an irrigated pasture, 40 m east of the plantation boundary. With water tables rising during the wet summer of 1993, two extra piezometers were installed in the plantation nest in February 1993. These were screened at 4 and 3.5 m below the surface respectively.

During the period October–December 1978, the average water table levels in the three bores installed in the plantation ranged between 139 and 185 cm below the surface with individual readings fluctuating between 73 and 207 cm below the surface. As the trees were only two years old at that stage and subjected to irrigation, these levels could be expected to be representative of the general water table status at KRI at that time. If anything, they appear to be slightly shallower than the levels measured at 3 other locations some distance away from the plantation which ranged from 98 to 215 cm (average 165 to 204 cm) below the surface during the same period. The average water table level measured in a series of observation bores at KRI in February 1977 was reported to be 194 cm below the surface (Mehanni and Repsys, 1978). No information is available on the number of observations on which this average level is based. However, its order of magnitude lends support to the measurements in the plantation.

At one section, the water table draw-down effect was observed up to a distance between 40 and 90 m away from the plantation edge. At another section, the draw-down effect extended only up to 5 m from the plantation edge.

The water table levels under the plantations were higher during winter than during summer. The water table depth variations are shown in Figure 6.2 (a) and (b).

The following conclusions can be drawn:

1. The draw-down under the plantations generally ranged between 3 m to 5 m, while the water table levels in the adjoining irrigated pasture lands were generally 1 m to 2 m below the surface.
2. The draw-down impact of plantations on the water table extended to a small distance (40 to 90 m at one section and 5 m at another section) beyond the edge of the plantation.
3. The water tables under the trees have risen significantly since monitoring commenced in 1982. This could be caused by either changes in irrigation intensities

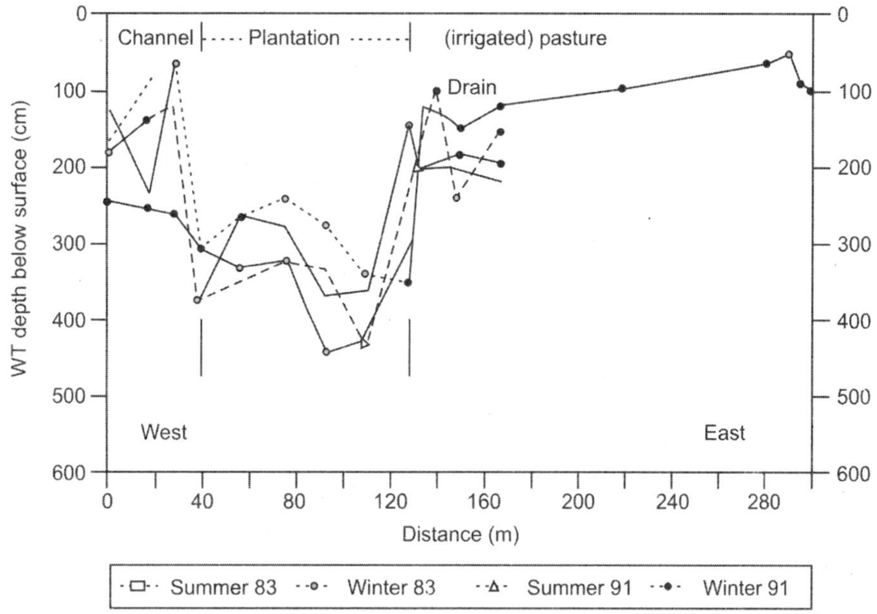

Figure 6.2 *(a) Summer and winter water tables in section 1 (South)*

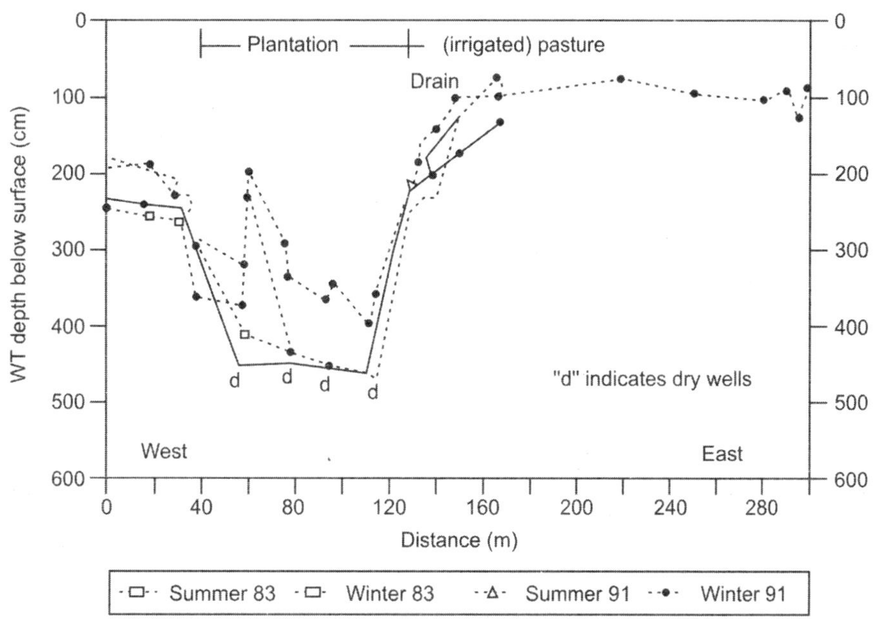

Figure 6.2 *(b) Summer and winter water tables in section 2 (North)*

in the surrounding land or by decreasing water use by the plantation caused by increasing water table salinities, or a combination of both.

4. Water table levels under the trees show a seasonal pattern, rising during winter and falling during summer. This trend is more marked under the trees than in the surrounding irrigated land.
5. The water table levels outside the plantation are land-use dependent. Rising water table levels in the paddocks surrounding the plantation result in increasing hydraulic gradients toward the plantation.

Rainfall and Evaporation Data

The rainfall and evaporation data near the project site is shown in Table 6.4.

Table 6.4 *Rainfall and evaporation data for Kyabram*

Year	Rainfall (mm)	Evaporation (class Apan) (mm)	Potential evaporation (Penman–Monteith) (mm)
1980	361	1700	—
1981	532	1710	—
1982	203	1910	—
1983	574	1660	—
1984	434	1563	—
1985	411	1608	—
1986	385	1399	—
1987	391	1385	—
1988	563	1343	—
1989	534	1283	—
1990	393	1322	—
1991	420	1440	—
1992	570	1336	1248
1993	699	1376	1036
Average	462	1503	

Of the annual evaporation, 1076 mm or 77 per cent occurs during the October–March period. The climatic conditions are not considered suitable for commercial tree growing without irrigation.

Quality of Irrigation Water

The trees were irrigated from 1976 until 1982 with height increments of up to 2.5 m/yr. In 1982 irrigation ceased because of water shortages during the drought. Since then, no

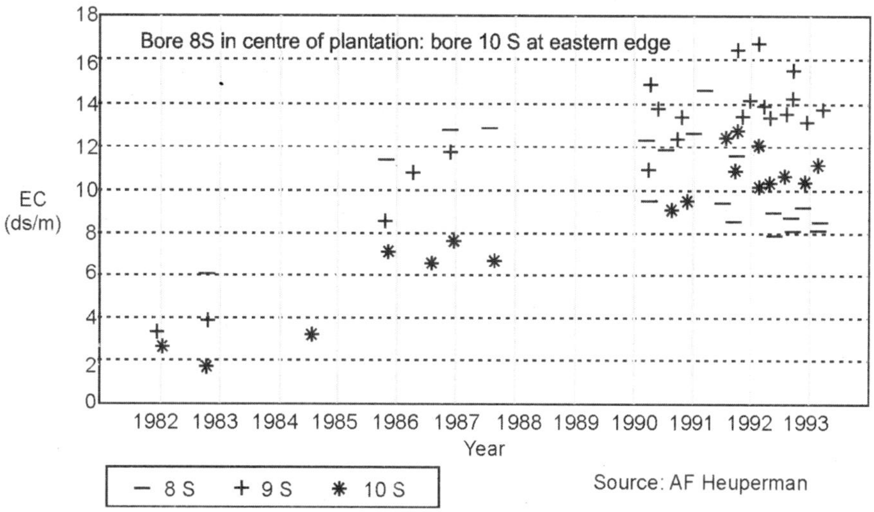

➲ **Figure 6.3** *Salinity [EC] of Bores Inside Plantation Section 1 South*

irrigation has been applied with the exception of some occasional unscheduled events caused by channel overflow before 1990.

The quality of irrigation water used in plantations (up to 1982) and in adjoining pasture lands is not reported.

Groundwater Quality

The water table *EC* values under the plantations, during the period 1982–1993, have shown a general rising trend with the passage of time, as shown in Figure 6.3.

The water table salinity under the trees has increased over time and is now 5 to 15 ds/m higher in the centre of the plantation than in the surrounding irrigated paddocks.

Afforestation

The plantation was established in a 2.4 ha site in 1976. The species planted were southern mahogany (*Eucalyptus botryoides*), flooded gum (*E.grandis*), river red gum (*E. camaldulensis*) southern blue gum (*E. globulus*) and Sydney blue gum (*E. saligna*). The spacing between the rows was generally 3.7 m and the distance between trees 1.8 m but spacing between rows of 2.4 m and between trees 0.6 m – 3.7 m were also used. The density of trees was 730 trees/ha.

Root Depth and Density

The root densities were highest in the 0–1 m depth gradually decreasing with depth of up to 5.5 m from the surface, though they were found to extend to a depth of 8.5 m. It appears that the tree roots were able to penetrate the saturated zone and extract part of their water requirements from below the phreatic surface. Greenwood et al. (1992) have reported that the roots of *E.globulus* and *E. camaldulensis* extend well into an unconfined aquifer below its lowest measured water table level.

It is stated that the apparent ability of the trees to extract water from the saturated soil profile would provide trees with a mechanism to avoid (or at least delay the onset of) growth reduction caused by salt accumulation. Some of the roots would "by-pass" the salt accumulation and explore the deeper, less saline, saturated layers.

Water Balance

The variations in summer and winter water tables during the year 1983 and 1991 are shown in Figure 6.2 (a) and (b).

The water tables underneath the plantations were between 2 and 4 m lower than in the adjacent paddocks. Water table levels under the plantations were higher during winter than in summer.

The draw-down impact of plantation on the water table did not extend beyond 90 m from the plantation edge. At some location it was limited to only 5 m.

Water tables, under the trees have risen significantly since monitoring commenced in 1982. This could be caused by either changes in irrigation intensities in the surrounding land or by decreasing water use by the plantation caused by increasing water table salinities, or a combination of both.

The trees were irrigated from 1976 until 1982, whereafter no irrigation was done. Since then (1982) the trees have been drawing their requirement of water almost wholly from the groundwater. This has caused a depression of the groundwater table (2 m to 4 m) under the plantations and of groundwater flow from the adjacent irrigated areas towards the plantations as shown in Figure 6.4.

Tree water use with a tree density of 730 trees/ha, measured over a short period during 1992, showed that the fraction "TWU/A_{pan}" ranged between 0.1 and 0.6. The published annual TWU values with the full canopy cover in western Australia (Marshall and Chester 1991) range between 0.6 A_{pan} for irrigated globulus to 1.9 A_{pan} for *E. camaldulensis* irrigated with sewage effluent (Morris and Wefner, 1987).

Tree water use (TWU) was estimated by Heuperman on the basis of the accumulated quantity of salts over a 1 m^2 area in 8 m deep soil profile over a period of nine years (11.9 kg/m^2) giving an average value of 1.3 kg/m^2/per year. This quantity of salt can be contained in a 430 mm column of water over 1 m^2 area with 3000 mg/1 salt salinity. On this basis, the tree water uptake from the water table was estimated to be 430 mm. This analysis does not account for:

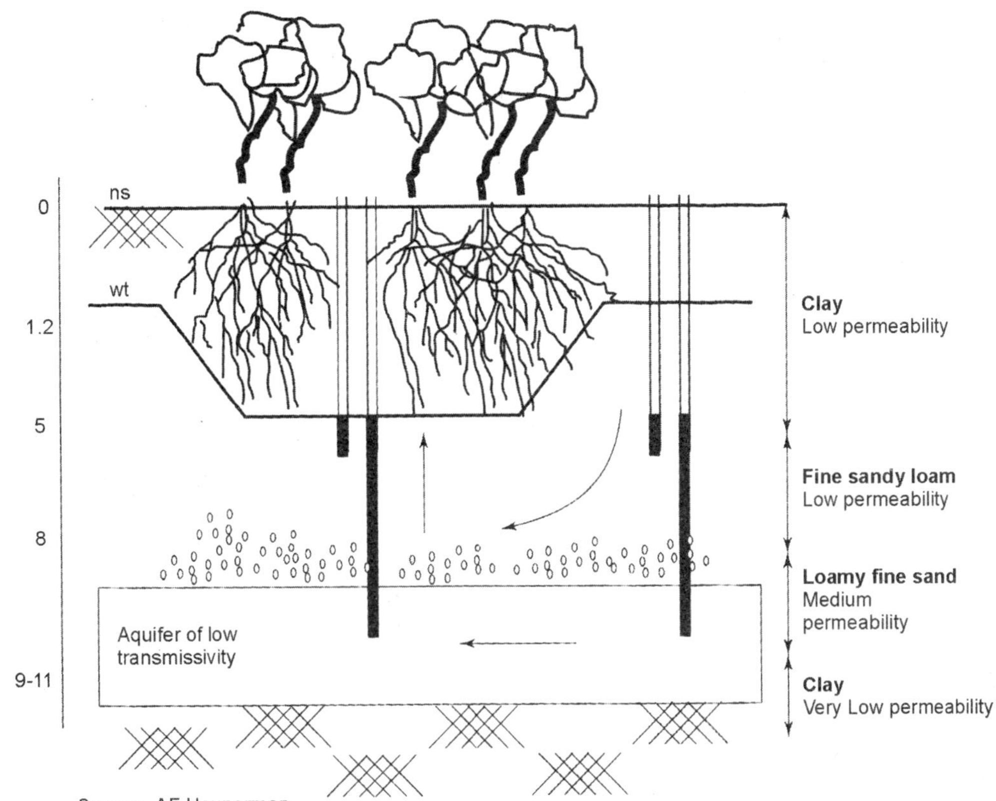

Source: AF Heuperman

Figure 6.4 *Water table scenario underneath the Kyabram plantation*

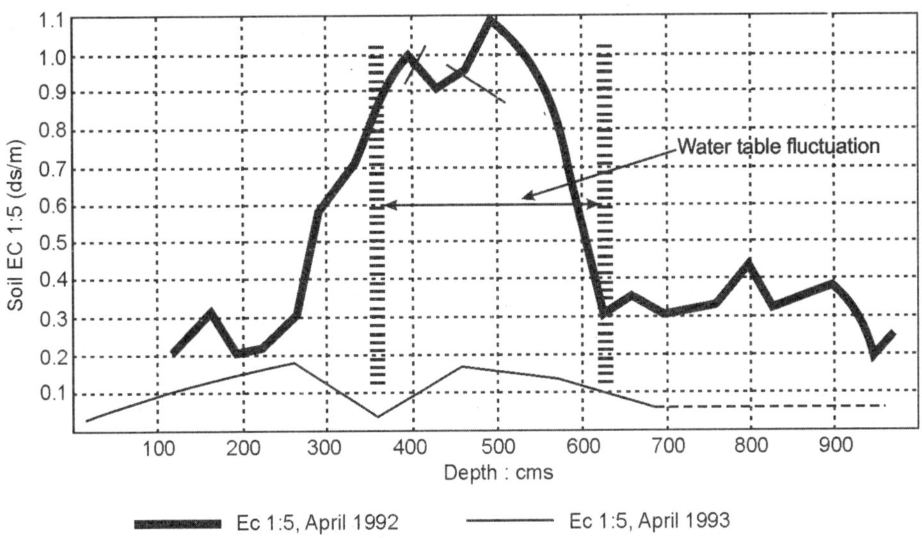

Source: AF Heuperman

Figure 6.5 *Ec 1:5 Profiles at site 21, section 2, North "Kyrbram Site, North Victoria Australia"*

- the quantity of accumulated salt that was subsequently dissolved and washed away during the rising phases of the groundwater table fluctuations during the period of nine years;
- the quantity of salt that was taken up by the plants alongwith water during transpiration;
- the quantity of water directly skimmed off by the tree roots from the surface of the groundwater table surface and used by the trees; and
- the average EC 1:1 values of 3.21 ds/m corresponds to salt content of 2150 mg/1 (not 3000 mg/1).

If all the above factors are properly accounted for, the total TWU is likely to be not less than 1.0 A_{pan}.

The water used by the trees was replenished by the groundwater flow from the adjacent areas. The important factors that would influence the rate of groundwater flow towards the plantations are the rate of recharge (from irrigation in adjoining pasture lands), the hydraulic conductivity of soil strata through which the flow is occurring, and the extent of draw-down of the water table underneath the plantations (which would depend upon the rate of transpiration).

The data on the rate of recharge from irrigation in the adjoining areas, and on the values of hydraulic conductivity of sub-strata are not reported. The size and extent of irrigated areas adjacent to and around the 2.4 ha plantation site are also not reported.

Salt Balance

The following observations are made about the water table salinity:

- The water table salinity under the trees has increased over time and is now 5 to 15 ds/m higher in the centre of the plantation than in the surrounding irrigated paddocks.
- There is evidence of salinity stratification taking place in the saturated zone of the profile underneath the plantation with the salinity increasing towards the water table.
- In 1992, elevated salt levels were found in the soil profile under the trees at Kyabram between about 2 and 5 m below surface; this salt accumulation was not evident in the 1983 sampling.

Figure 6.5 shows observed soil *EC* 1:5 values in 1983 and 1992 at various depths in the centre of the plantations.

It is seen that salts had accumulated between 2.5 m and 5.5 m depths from the surface with peak levels between 3.5 and 5 m. The salt bulge coincides with the water table fluctuation range.

The peak value of around 1 ds/m (*EC* 1:5) or approximately 3000 mg salt/kg dry soil is not very high. Some species such as *E. camaldulensis* are known to tolerate much higher salinity levels.

Heuperman has calculated and given the annual salt accumulation data for a range of assumed Tree Water Use (TWU) and water table salinity scenarios as shown in Table 6.5.

Table 6.5 *Annual salt accumulation increments (kg/m²), different TWU and water table salinity scenarios, assuming 300 mm effective annual rainfall*

Assumed total TWU	0.4 * A_{pan} (560 mm)	0.52 *A_{pan} (730 mm)	0.6* A_{pan} (840 mm)	1* A_{pan} (1400 mm)	1.2 * A_{pan} (1680 mm)
TWU from water table	260 mm	430 mm	540 mm	1100 mm	1380 mm
WT salinity					
2000 mg/l	.52	.86	1.08	2.20	2.76
3000 mg/l	.78	1.29	1.62	3.30	4.14
4000 mg/l	1.04	1.72	2.16	4.40	5.52

It has been shown that an average increase in *EC* 1:1 of 2.21 ds/m has occurred in a 6 m deep soil profile during a period of nine years. It has been concluded that soil salinisation underneath the non-irrigated plantations growing on a 5 ds/m water table can be expected to reach levels high enough to inhibit water uptake from the water table, during a period of about 20 years.

However, a reference to Figure 6.5 shows that whereas soil *EC* 1:1 in the zone between 3 to 5 m depth increased from about 1 dS/m to 5 dS/m during the nine years period, (1983–1992), that at depths deeper than 6 m increased to only about 2dS/m during the same period. Therefore, though soil salinities in the zone 3 m to 5 m deep may take about 20 years to increase from 5 ds/m to levels high enough (say + 12 ds/m) to inhibit water uptake, from the evidence available, the increase in the salinity of the water table at depths more than 6 m from 2 ds/m to 12 ds/m may take another 90 years. It follows that if tree roots have the ability to by-pass saline soil layers and find access to deeper, less saline groundwater in the saturated profile (conclusion 2, para 3.5 of the report), the plantations should be able to survive for long periods in spite of high salinity in the upper soil strata.

The plantations were put up in 1976. Their condition in the coming years and continued monitoring of salinity levels of groundwater table can throw very useful information.

Distance up to which Plantations can Provide Effective Biodrainage

The water tables underneath the plantations were between 2 and 4 m lower than in the adjacent paddocks but the draw-down impact on the surrounding paddocks was restricted to about 20 m from the edge of the plantation (1984). In 1991, the draw-down was measured at one location at least 40 m away from the plantation boundary. It was concluded that the extent of the protection effect in the Kyabram situation is not more than 40 m from the edge of the plantation.

This was possibly due to the reason that the tree plantation area was small in comparison to the large surrounding irrigated area that was contributing to the recharge.

▼ SAN JOAQUIN VALLEY CALIFORNIA (USA) – PLANTATION STUDY

This is described by F.F. Karajeh and K.K. Tanji (1991) 'Agroforestry Drainage Management Model'; published in the *Journal of Irrigation and Drainage Engineering,* ASCE Paper No. 6258.

General Description

The saline drainage waters from the irrigated land in San Joaquin valley were found to contain potentially toxic trace elements (selenium) and therefore, the San Luis Drain discharging into the Kesterson Reservoir which served as a terminal evaporation pond, had to be closed in 1986. Consequently the disposal of saline drainage waters became a major problem in one of the most productive agricultural regions in the world. On the west side of the San Joaquin Valley, about one-third of the 9,30,000 ha of irrigated land is waterlogged and salt-affected. Agroforestry systems have been recommended as one management option for lowering the shallow groundwater and using saline subsurface drainage water for irrigating salt-tolerant trees and halophytes. Irrigation management, shallow groundwater management, drainage water reuse and disposal, and other inter-related water quality, land productivity and environmental issues are required to be studied and integrated into the design and operation of on-farm and regional irrigation drainage plans in order to improve the current conditions in the valley.

Subsurface Hydrology

The groundwater table lies at a depth of approximately 2.0 m below the soil surface. A subsurface relief drain has been installed at a depth of 2.1 m. To isolate the site from the external lateral and salt flows, a perimeter drain was installed at a depth of 2.4 m around the plantation.

The soil profile is underlain by impermeable clay layers of low hydraulic conductivities at a depth of 3 m.

Figure 6.6 (a) describes the generalised water and salt flow pathway into and out of the eucalyptus root zone. Figure 6.6 (b) presents the cross-section of the soil profile with assumed boundary conditions based on experimental layout and monitoring data.

A finite element model for simulating the two-dimensional transient movement of water in variably saturated soil (Nour el-Din, 1986) was modified to consider the macroscopic root water uptake approach and the effect of the osmotic potential on the soil water movement into the trees roots and through the soil matrix. The model was applied to the agro-forestry demonstration site for sustainable agriculture at Mendota, California. The measured soil water pressures and volumetric soil moisture contents throughout the 1990 irrigation season compared well with the predicted data generated by the D-HYSAM model.

Rainfall and Evaporation Data

The cumulative amount of applied water (irrigation and precipitation) and tree evapotranspiration (*ET*) during the simulation period (220 days during April–November 1990) were 125.1 cm and 115 cm respectively. The total amount of precipitation during the simulation period was 4.1 cm. Irrigation water was applied by furrow irrigation for 24 hours every 10 days. The irrigation schedule and evapotranspiration for eucalyptus trees during the 1990 irrigation season are shown in Figure 6.7.

The irrigation application efficiency (ratio between required tree *ET* to applied amounts of irrigation water) was 75 per cent and the source of the irrigation water was saline subsurface drainage water (TDS of 8000 mg per litre) collected from the crop lands. Since enough data for the halophyte plantation was not available, the simulations were conducted for the eucalyptus plantation.

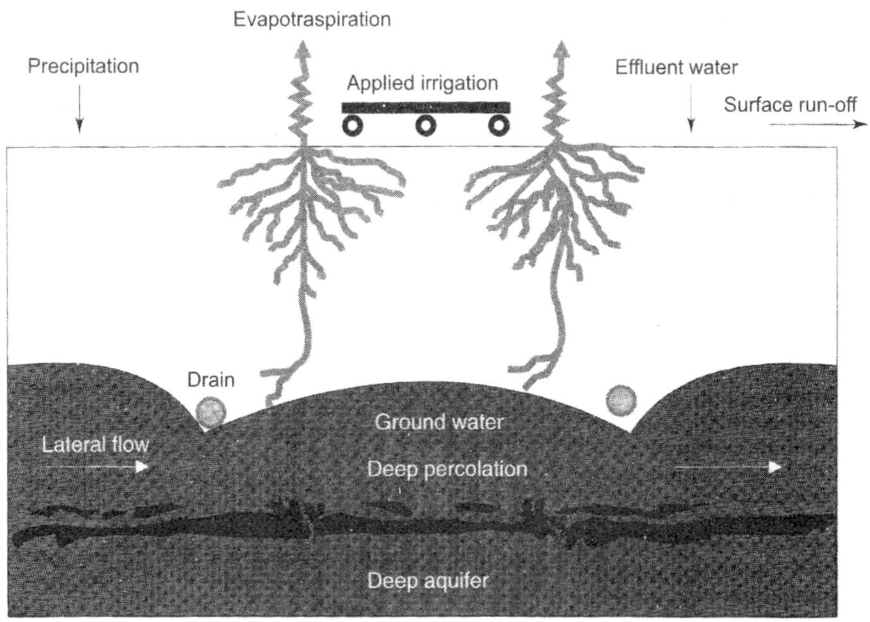

(a) Water and salt pathways into and out of crop root zone

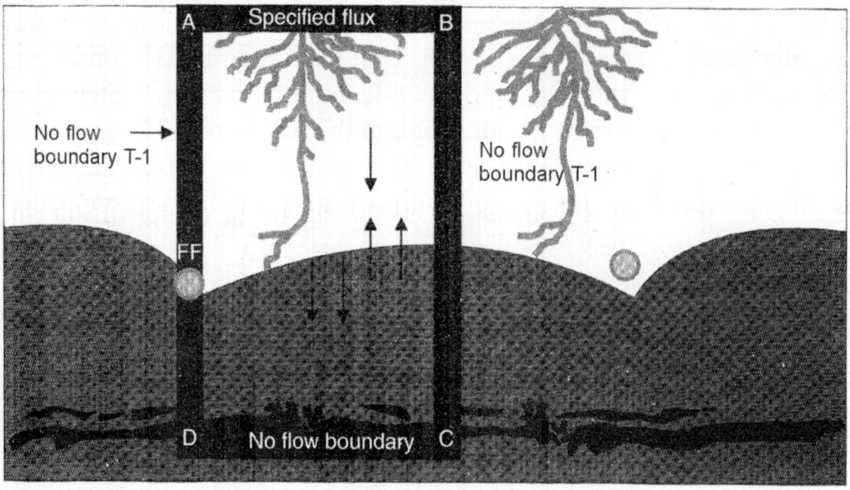

(b) Assumed water and salt boundary conditions

Figure 6.6 *Cross-section of a hypothetical tile-drained field with shallow groundwater*

Quality of Irrigation Water

The TDS of the irrigation water was 8000 mg per litre. The source of irrigation water was saline subsurface drainage water collected from the crop lands (Figure 6.7).

Groundwater Quality

The TDS of drainage water was measured as 14,000 mg/litre. The increase in salinity was due mainly to the concentrating effects of *ET*.

Figure 6.8 (a) plots the TDS of applied and drainage waters over the 220-day simulation period. Figure 6.8 (b) presents, for the study area, the mass of salt input and salt output for the root zone over 220 days. The cumulative salt mass entering the system was 6183 kg while the cumulative salt mass discharged from the system was 2355 kg or 38 per cent of the incoming salts. Figure 6.8 (c) gives the ratio between the input salt mass and the output salt mass as a function of the simulation time. Over the 1990 irrigation season, the ratio stabilised to a near steady state value of about 0.4 towards the end of the simulation time.

Afforestation

The plot under study was planted with 9.45 ha *Eucalyptus camaldulensis* (red gum) and 2.02 ha of *Atriplex canescens* (salt bush), a halophyte in 1985 and 1986. The six-year-old eucalyptus plantation had a maximum root penetration of 2.1 m (1991). The number and density of tree plantation is not reported.

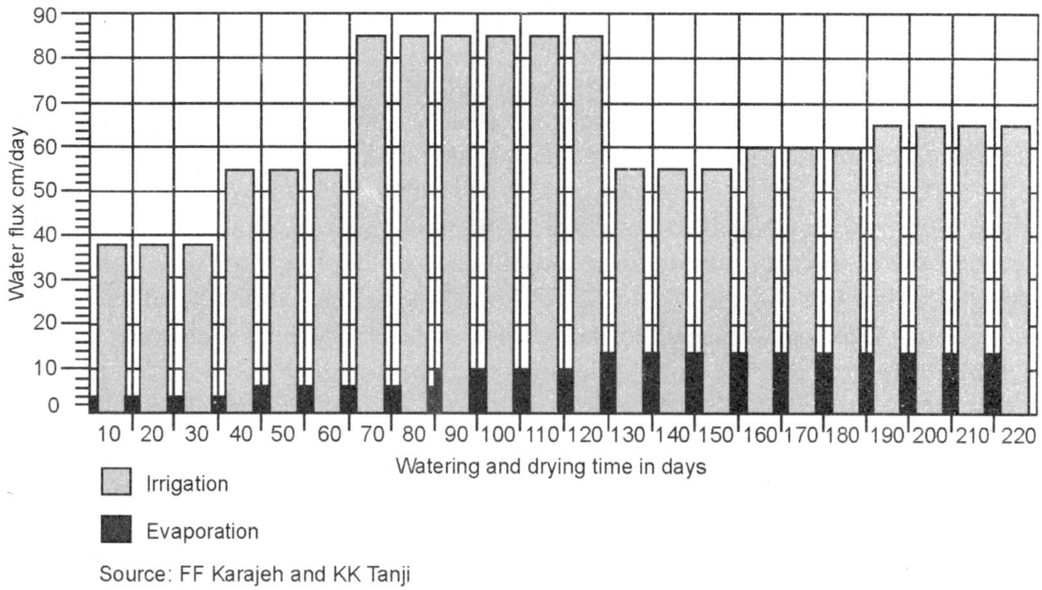

Figure 6.7 *Irrigation schedule and evapotranspiration for eucalyptus trees (During the 1990 irrigation season, San Joaquin Valley, California, USA)*

Root Depth and Density

Root-extraction Sub-Models

It is assumed that plant roots have full depth penetration equal to the root zone depth down to the surface of the water table. The assumed root density distribution is shown in Figures 6.9 (a) and 6.9 (b).

Water Balance

Here the plantation area (9.45 ha) was isolated from the external ground flow by installing a 2.4 m deep perimeter drain around the plantation. 1251 mm of water was applied (irrigation and precipitation) during a period 220 days. The tree evapotranspiration during the same period is reported to be 1150 mm, nearly equal to the applied water.

The groundwater table lies at a depth of approximately 2 m below the soil surface. No significant change in the water table depth is reported.

The objective in San Joaquin Valley (SJV) was to use crop land drainage water (8000 mg/l – TDS) for irrigating plantations and to reduce the volume of drainage water (whose measured TDS was 14000 mg/l). The water balance was achieved by transpiration from plants and by drainage of excess water. The control of the groundwater table, with the help of plantations, was not envisaged.

Salt Balance

The application of 8000 mg/l TDS water for irrigating the plantations, resulted in 14000 mg/l drainage water due mainly to the concentration effects of *ET*. For the study area, the mass of salt input was 6183 kg and salt output from the root zone over 220 days was 2355 kg or 38 per cent of the incoming salts. This means that 3828 kg of salt was left behind in 2.1 m deep soil strata. During the 1990 irrigation season, the ratio between the input salt mass and the output salt mass as a function of the simulation time stablised to a near steady state value of about 0.4 toward the end of the simulation time. The irrigation water salinity level was about 10 ds/m and that of the discharging drainwater about 17 ds/m. It is reported that the use of poor quality drain water for the irrigation of plantations, in the presence of subsurface drains, resulted in only a marginal deterioration of the shallow groundwaters.

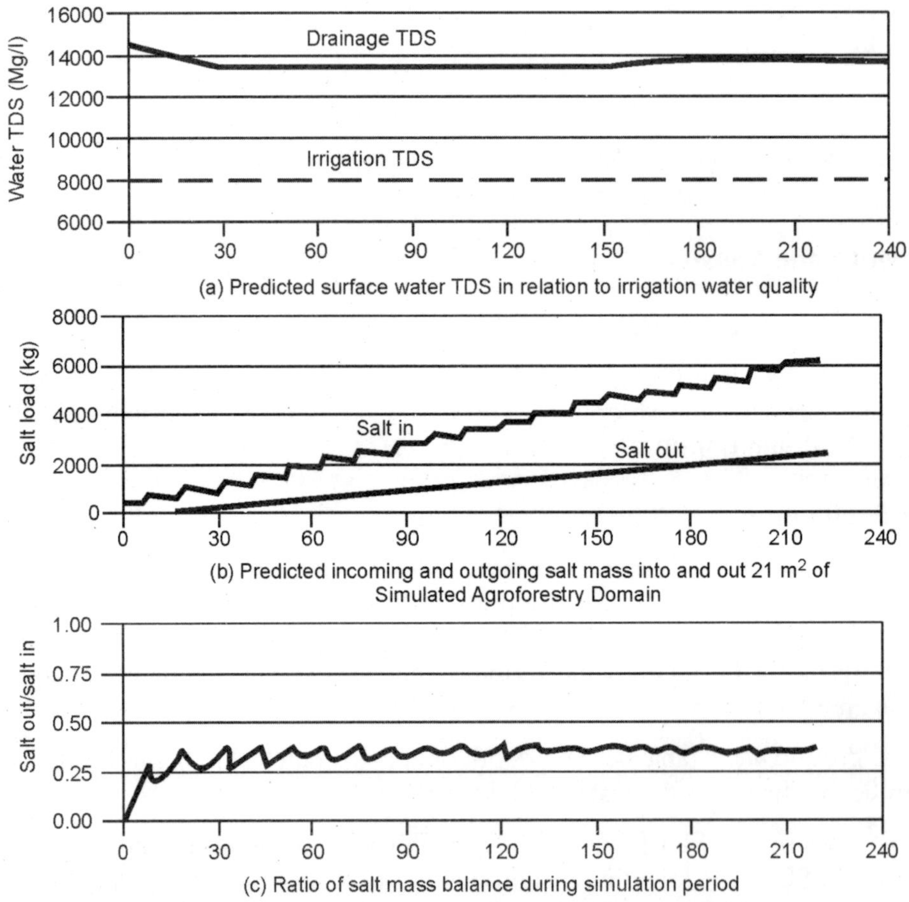

Source: FF Karajeh and KK Tanji

Figure 6.8 *TDS of applied and drainage waters over the 220-day simulation period*

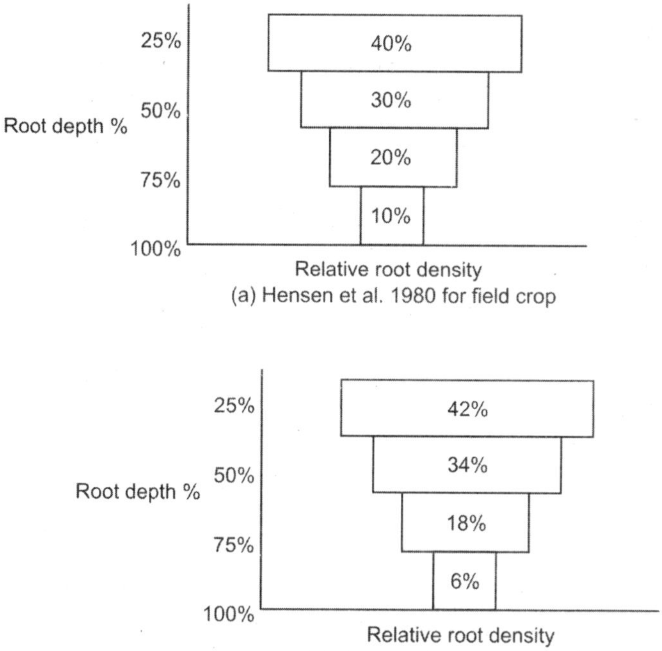

(a) Hensen et al. 1980 for field crop

(b) Six-year-old red gum eucalyptus

Source: FF Karajeh and KK Tanji

➲ **Figure 6.9** *Relative root density distribution in Vadose Zone*

It is not clear how despite a significant quantity of salts (almost 60%) being left behind from irrigation water, there was only a marginal deterioration of the shallow groundwater. The salts left behind should render the soil highly saline and totally unsuitable for supporting any vegetation. In case the salts are getting leached down to the groundwater table, then the salinity levels of the drainage water should increase quite rapidly and progressively from the reported figure of 14000 mg/l. A clarification is needed.

Distance up to which Plantation can Provide Effective Biodrainage

The groundwater table is located at a depth of 210 cm below the soil surface. Within the plot the tile drain and an interceptor drain around the plot are located at 210 cm and 240 cm depths respectively. To isolate the site from any external lateral salt and water flows, a perimeter drain was installed at a depth of 2.4 m around the plantation.

The total evapotranspiration by the Eucalyptus during the 220 day period of the 1990 irrigated season (April–November 1990), was 115.3 cm, while the cumulative applied irrigation during the same period was 121 cm. Therefore, there was no perceptible draw-down of the water table underneath the plantations or in the adjoining areas due to transpiration from the trees.

INTEGRATED SYSTEM FOR AGRICULTURAL DRAINAGE MANAGEMENT ON IRRIGATED FARMLAND, CALIFORNIA, USA

A Report by Cervinka et al. (1999) Describes the features of the system

The San Joaquin Valley Drainage Programme (SJVDP) report "A Management Plan for Agricultural Subsurface Drainage and Related Problems on the Westside San Joaquin Valley" (September 1990) used the following definition for agroforestry – "As used in this report, agroforestry is the practice of growing certain types of trees with drainage water. The trees act to dispose of applied drainage and shallow groundwater through foliar evaporation and at the same time produce a marketable commodity". This is the term established in 1985–86 at the beginning of the agroforestry programme. Since the programme involves growing crops and trees, this on-farm practice has been named 'agroforestry'. The research programme has since expanded to include additional components such as salt-tolerant crops and halophytes, source control, the sequential reuse of drainage water, semi-automatic control of water flows, use of trees as interceptors, a solar evaporator, salt recovery and utilisation, selenium management and wildlife safety. The basic premise of this expanded on-farm management system is to manage salt, selenium and drainage as resources rather than as wastes. The original 'agroforestry' concept has evolved into 'integrated on-farm drainage management'. The term 'integrated on-farm drainage management' (IFDM) describes the on-farm practices presently being demonstrated more precisely than agroforestry.

Project Location

The project site, Red Rock Ranch (RRR) is located in the WRCD area, in the Westlands Water District (WWS) in Fresno County. The location of the RRR demonstration project is about 4.8 km south of Mt. Whitney Avenue and 3.2 km west of Colusa Road, Fresno County. The chart in Figure 6.10 illustrates the project design.

Farm Characteristics

The project site lies within the alluvial fan and flood plain of Cantua Creek, east of the California Aqueduct. The elevation in the area varies from 35.8 m (117.5 feet) mean sea level (MSL) at the south-west to 26.5 m (88.37 feet) at the north-east with the ground surface sloping at 0.21 per cent in the south-north direction and at 0.38 per cent in the west-east direction.

The site soils consist of Ciervo clay, a saline sodic soil that is underlain by stiff clay averaging about 6.1 m (20 feet) in thickness. The soil has an average clay content of 40 to 60 per cent and the permeability of the soil is about 0.026 mm/sec. The deeper zones have alternating mixtures of clay, silt and sand, including Corcoran clay at a depth 157 m (515 feet) below the ground surface.

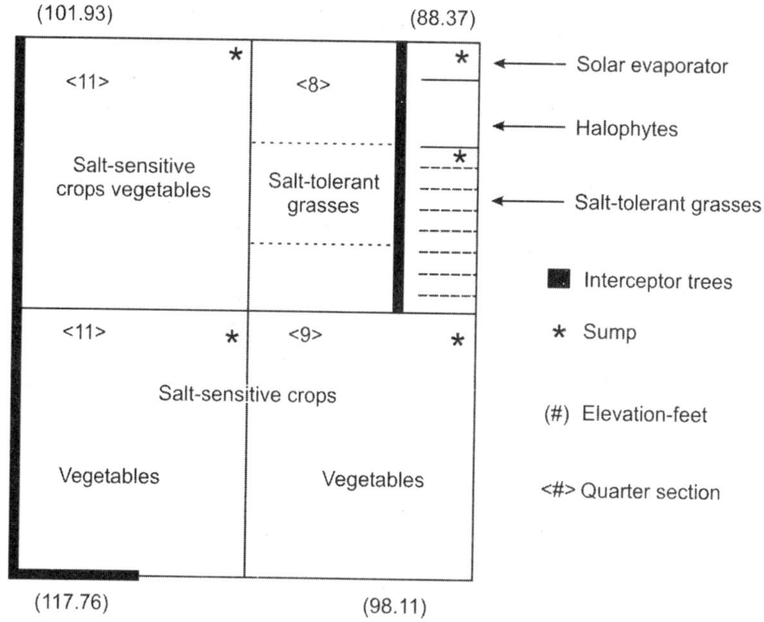

Figure 6.10 *Red Rock Ranch - Integrated on-farm drainage management*

Data obtained from the "Shallow Conditions" published by the Westlands Water District (WWD) shows that shallow groundwater conditions exist at the site. Shallow groundwater levels are typically highest in April after pre-irrigation and lowest following the cropping season in October. The depth of the shallow groundwater table is 1.5 m to 2 m (5 to 6.7 feet) below the ground level. This saturation zone extends to the depth of the Corcoran clay layer. "December 1993 Groundwater Conditions", published by the Westlands Water District, states that below the Corcoran clay layer, at depths ranging from 30 to 120 m (100 to 400 feet), is a confined water-bearing zone of fresh groundwater. The base of the fresh groundwater is at a depth of about 682 m (2275 feet) below the ground level.

Integrated on-Farm Drainage Management Concept

The Integrated on-Farm Drainage Management (IFDM) concept for salt management on irrigated farmland has these two objectives: (1) to utilise drainage water as a resource to produce marketable crops and (2) to remove salt from the drainage water directly on farms. It is a management system based on the principle that saline drainage can be sequentially reused to irrigate crops of progressively increasing salt tolerance, and salt can be recovered from the residual brine. The demonstration IFDM project at Red Rock Ranch in the Westlands sub-area indicates how this concept has been implemented on a farm in the San Joaquin Valley. The farm operation is divided into four salinity areas (zones). Relatively good quality irrigation water is applied to major commercial crops, such as vegetables grown in the non-saline zone, which represents 73.4 per cent of the farm area. Saline drainage is sequentially reused to irrigate salt-tolerant commercial crops, such as alfalfa or cotton, grown on a 20.3 per cent area of the farm in the low-saline zone. Salt-tolerant trees (such as Eucalyptus) and grasses (such as Wheatgrass or Perlagrass) are grown in the moderate saline zone (2%). Halophyte (such as Saltgrass, Iodine bush, Cordgrass or Salicornia) are grown in the high-saline zone (0.8%). This sequential water reuse process productively uses about 90 per cent of the drainage water. Salts concentrated in the remaining 10 per cent of the drainage water are collected in a solar evaporator for water evaporation and salt crystallisation. The solar evaporator is located on 0.3 per cent of the farm area. Saline drainage may also be used in a solar pond for heat/power generation, or can be processed by a solar/industrial still or by other methods, and refined into sodium sulphate, sodium chloride and other salts.

Project Design

Soil Conditions

Crop yields were declining due to a salinity build-up in the soil. Aerial photographs were available for the farm and for the NE 64 ha (160 acres) quarter section of the farm. This picture clearly indicates a severe salinity problem, typical for the soil in the entire farm area. The USDA–ARS–Salinity Laboratory surveyed salinity levels in two 64 ha blocks of the farm. The contour maps indicate salinity levels in the south-east (SE) quarter section and in the south-west (SW) quarter section. The measured data were analysed for the Electric Conductivity/Sodium Absorption Ratio (*ECe/SAR*) and for *ECe*/Boron relationships. The SE field had high salinity near its centre; the SW field had multiple high-salinity areas. Soil salinity was reaching the *ECe* 20 ds/m levels. There is mainly clay-loam texture in the SE field, and loam/clay-loam texture in the SW field. The relationship between *ECe*, SAR and Boron was well correlated.

Water Conditions

The water salinity fluctuated in various areas in correlation with the soil salinity. The levels were in a wide range from *EC* 5 to 25 dS/m. There was a subsurface flow of saline water from the south-west into the project site. Eucalyptus trees were planted along the western border of the farm to intercept the groundwater flow.

Cropping System

Cotton, alfalfa, sugar beets and grains were the major crops grown on the farm. J. Diener decided to convert his farm operation to high-yield, high-value vegetable crops requiring non-saline conditions. These would be established on three quarter sections (QS) of his farm. The drainage water from these three QS (about 188 ha/470 acres) would be sequentially reused to irrigate salt-tolerant crops on about 52 ha (130 acres). These salt-tolerant crops would include alfalfa, sugar beets, cotton, grains, and others.

Drainage water from these salt-tolerant crops would be again reused to irrigate salt-tolerant trees (5.2 ha/13 acres). A decision was made in 1997 to replace these trees with salt-tolerant grasses. Halophytes (2 ha/5acres) would be the last plants to receive the sequentially reused drainage water before its discharge into a solar evaporator.

Subsurface Drain Design

The design of the drains was based on the following investigations:

1. *Test Pits:* In the 256 ha (640 acre) area, 15 test pits were dug to evaluate the soil profile and to determine the static water table depth. This information was used to help determine the drain depth necessary to leach salts without collecting regional groundwater.
2. *Survey:* An extensive survey was completed to determine the overland slope of the field. This information was used to determine the grade at which the drains would be installed.
3. *Yield rate:* Two sumps were pumped to determine a water yield from the soil profile. This yield information was used to select the size of drains.
4. *Salinity maps:* Salinity maps were used to identify areas with high salt levels. The drains were placed to allow leaching of the salt in the root zone of the soil profile.
5. *Landowner input:* Discussions with the landowner were used to determine which areas had major drainage problems. This information influenced the decision about which areas would require drain tiles and which areas would need tree interceptors.
6. *Course of action:* The drains along the west edge of the SW quarter section had the highest priority since there was subsurface flows into this field from the west and south-west. Relief drains were installed at intervals of 126 m (420 feet) in the SW QS. A 1.6 km long tree interceptor was planted from south to north along the west end of the SW and NW quarter sections. A 400 m long tree interceptor was also planted east-west along the south end of the SW quarter section. The drains for the SE and NE quarter sections had the second highest priority. A drain system in the SE QS was installed at 120 m (400 feet). A drain system was installed for the NE 52 ha (130 acres) area consisting of three blocks of salt-sensitive crops. A separate drain system was installed in the moderate-saline area consisting of trees (5.2 ha/13 acres) with its own sump and pump in the SE corner of the halophyte field. Another separate drain system was installed under halophytes (2 ha/5 acres) and a solar evaporator (0.8 ha/2 acres) with a sump in the SE corner of the solar evaporator. Four 10.16 cm (4 inch) drain lines were installed at a depth of 1.98 m (6 feet). The sumps are 1.98 m in diameter and 2.9 m deep. The last drain installation, at 129 m (430 feet) and 78 m (260 m), was performed in the NW quarter section. All drains consisted of perforated polyethylene pipe placed in a gravel filter at least 7.6 cm thick on all sides of the pipe. The drain tiles were installed at depths from 1.65 to 2.13 m (5.5 to 7.0 feet).

Water Distribution Design

WWD canal water and pumped groundwater are used to irrigate salt-sensitive crops. Drain water is collected in a sump separately for each quarter section. Drain water from quarter sections NW (11) and SE (9) is transferred into the middle of the farm (central point of all quarter sections). This water together with the drain water from the quarter section SW (10) is applied to salt-tolerant crops (NE quarter section. 52 ha, 3 blocks). Drain water is then sequentially reused to irrigate salt-tolerant grasses; drainage water is collected in the sump and reused on halophytes. The halophyte drainage water is discharged into a solar evaporator. The tail water from salt-sensitive crops is collected in a pond located in the north-east corner of the quarter section NE. This tail water is then reused to irrigated salt-tolerant crops in the NE quarter sections.

Commercial Crops

The installed drains had a positive effect on the changing cropping system. The commercial crops grown at RRR since 1995 are shown in Table 6.6.

Table 6.6 *Commercial crops grown at Red Rock Ranch*

Area		*1995*	*1996*	*1997*	*1998*	*1999*
NE	block 1	**cotton**	corn, cotton	sugar beets	wheatgrass	wheatgrass
	block 2	**cotton**	canola, broccoli	sugar beets	alfalfa seed	alfalfa seed
	block 3	**cotton**	wheat	cotton	sugar beets	wheat
SE		cotton	**wheat, tomatoes**	corn, broccoli	safflower	onion
SW		**wheat**	alfalfa	alfalfa	alfalfa, broccoli	tomatoes
NW		cotton	cotton	**wheat**	alfalfa	alfalfa

The identification of crops in **bold letters** indicates the year of drain installation. It takes 2 (or 2.5) seasons to reclaim saline land and to make it suitable for growing salt-sensitive crops. The quarter section SW had drains installed in 1995 and broccoli was grown in the fall of 1998. The quarter section SE had drains installed in 1996 and onion was planted in the spring 1999. The quarter section NW had drains installed in 1997, and after two years of alfalfa, this field expected to be ready for a vegetable crop in 2000.

Salt-tolerant Grasses and Halophytes

The term 'salt-tolerant grasses' (STGs) is used for plants tolerating drainage water of *EC* ranging from 8 to 15 ds/m, and the term 'halophytes', for plants tolerating drainage water above *EC* 15 ds/m. Using water salinity of *EC* 15 as a separating limit is a rather artificial method, but it can be a basis to say that halophytes tolerate higher salinity than salt-tolerant grasses.

The drainage water from the salt-tolerant crops is reused to irrigate 5.2 ha (13 acres) of salt-tolerant grasses. This area is divided into 14 blocks to improve water distribution and to test various salt-tolerant grasses. Drainage water from this 5.2 ha area is sequentially reused to irrigate 2 ha (5 acres) of halophytes. The halophyte area is divided into 5 blocks for the same reasons as mentioned above.

STG's and halophytes should preferably be perennial plants to facilitate higher flows of drainage water during the winter/spring period. The other required characteristics include high water demand, tolerance to frequent flooding, frost tolerance, and marketability of the harvested biomass. STGs and halophytes are mainly used for the management of drainage water and salt. They are grown on a relatively small area of the farm (2 % and 0.8 %). The commercial value of vegetables and salt-tolerant crops (sugar beets, alfalfa, and others) is the most important economic consideration. However, economic value is also considered in the selection of STGs and halophytes.

The logical use of these grasses is livestock feed. Digestibility tests on these grasses were conducted by a Dairy Advisor, UCCE, Tulare County. Alfalfa was used as a control plant. The STGs and halophytes can be ranked in this order of their digestibility: (highest) Salicornia > Iodine bush (young plant) > Atriplex > Sesuvium > Five Hook Bassia > Saltgrass > Sacaton > Cordgrass > Morro Bay Saltgrass > Creeping Wildrye > Tall Wheatgrass > Utica Saltgrass.

It is apparent that some very promising grasses, such as Tall Wheatgrass or Creeping Wildrye may have a limited nutritional value. Both Wheatgrass and Wildrye produce a high volume of biomass. Future research projects will cover the conversion of this type of STGs into bio-energy and/or industrial feedstock.

An automatic irrigation system has been used to apply drainage water to both STGs and halophytes. Each of 14 STGs blocks is irrigated individually. Solenoid valves control the water flow. Two solenoid valves also control flows to the halophytes. An electronic timer is used for each system.

Trees

Trees use and evaporate drainage water. This is achieved through sequential reuse, by intercepting the flow of drainage water from up-slope, or through the uptake of shallow groundwater. Trees can be viewed as biological pumps.

Trees have been planted in these two project areas: (1) on the west side of SW and NW quarter sections and the south side of the SW quarter section, and (2) on the east side of three blocks of the fields of 'salt-tolerant crops' in the NE quarter section.

The role of the trees on the west/south-west borders is to intercept the flow of subsurface saline water from the up-slope surrounding areas. The trees planted included Eucalyptus 4573, 4543, 4544, and 4590 (these are identification numbers assigned to selected trees by the Eucalyptus Improvement Association). The trees are performing very well in the SW quarter section. The salinity levels were high in the NW quarter section, and this has affected tree growth. The salt is leached in this area, and some trees were expected to be replanted in 2000. The trees were planted in four and five rows. Two rows of trees on the west side of this interceptor were planted close to the PGE power line and the height of these trees was controlled.

Four rows of trees were planted in the field of 'salt-tolerant crops' in 1999. Pistachio trees (45 trees) were planted on the south of this area. Eucalyptus trees 4543, 4544, and 4573 were planted in the remaining area. Pistachio trees were planted as an experiment to evaluate their suitability for saline conditions and for the sequential reuse of drainage water; they could have a potentially high economic value. The role of eucalyptus trees is to lower the water tables and to occasionally receive reused drainage water, and thus, to assist in reducing the volume of drainage water to be managed.

At the beginning of the project, trees were planted in the STG area (5.2 ha), called a 'tree' area at that time. The idea was to sequentially reuse drainage water on trees, similarly as it has been done in the Mendota experimental project. These trees did not perform well and then suddenly died in the fall/winter of 1996. A senior plant pathologist from the California Department of Food and Agriculture inspected the trees and concluded that "There is a very high probability that the trees were damaged by stress caused by the combination of salt accumulation within the root system and inadequate irrigation". The cause of this tree failure has never been sufficiently explained. The possible causes may include saline soil conditions, inadequate salt leaching, accidental chemical damage or some other causes.

This experience resulted in the modification of the project design when blocks of salt-tolerant trees were planted in combination with fields of salt-tolerant crops and grasses. Four rows of trees were planted between the fields of salt-tolerant crop and salt-tolerant grasses. These trees are being established with good irrigation water (*EC* 0.6 dS/m) applied by drip irrigation. Both salt-tolerant plants and trees uses drainage water and reduce its volume. The trees take saline groundwater to lower water tables, intercept subsurface water flows, sequentially reuse drainage water, and create a biological barrier between low-saline and high-saline areas. Drainage water is mainly applied to salt-tolerant plants and only occasionally to the trees (e.g., during the high flows of drainage water).

Solar Evaporator

Design

The RRR solar evaporator is used for the final disposal of the drainage water, its evaporation, and for the crystallisation of salt. It was installed in July 1995. The solar evaporator consists of one cell. Its area and dimensions are:

Area	0.8 ha	2.0 acres
Size	117 ×76 m	370 ×250 ft

The solar evaporator is on a levelled area surrounded by borders on four sides. The levelled area and borders are covered with a (PVC 10 mill) liner to control water seepage. It consists of individual sheets (30 × 6 m, 100 × 20 ft), overlapped and glued together, and anchored along the elevated borders of the solar evaporator. Four drain tiles are installed underneath the solar evaporator to further protect the quality of groundwater. Any potentially collected drain water is returned into the halophyte sump and blends with the halophyte drainage water. This water is then pumped from the halophyte sump and 72 sprinklers distribute it in the solar evaporator. A programmed electronic timer controls the application of drain water on the evaporator surface. The operation of the solar evaporator is automated. The volume of drainage water discharged daily into the solar evaporator is equal to the daily evaporation rates (*ET*). The Westlands Water District had been providing the values of average evaporation rates since 1980. These data were initially used for the design and operation of the solar evaporation system.

The method of discharging water into the solar evaporator has been changed from flooding to a system of sprinklers. 72 sprinklers are installed at a spacing of 12 × 15 m (40 × 30 feet) and they discharge drainage water into the salor evaporator. These sprinklers operate in 9 (60 minute) cycles per day. The cycles are programmed according to the daily evaporation rates. During the low evaporation period, sprinklers may operate for 1 minute 'On' and 59 minutes 'Off' per each hour cycle. During the high evaporation period, sprinklers may operate 8 minutes 'On' and 52 minutes 'Off' per each hour cycle. The sprinklers have provided an even distribution of water in the solar evaporator and have eliminated water ponding. The sprinklers system is also reliable and smooth in operation.

The daily *ET* values, obtained from the California Irrigation Management Information System (CIMIS) system, are used to programme the water discharge system. The water is discharged at a pumping rate of 720 l/min. (190 *GPM*). The required layer of water is calculated, and the time of pumping is programmed as required. The programming of the solar evaporator can be illustrated through this example: Daily *ET* = 5.1 mm, required time of pumping = 55 minutes/day = 6 minutes per cycle. The solar evaporator is usually programmed for slightly lower discharge time values to reduce the chance of water ponding. During the late summer-fall season, the time periods are further reduced due to the declining levels of evapotranspiration.

Discharge of Drainage Water

The annual volumes of drainage water and salt discharged into the solar evaporator are shown in Table 6.7.

Table 6.7 *Annual volumes of water and salt discharged into the solar evaporator*

	Operational days	Volume of drainage water		Est. Volume of salt
		ml/yr	AF/yr	m t/yr(*)
1995	135	5.6	4.6	74
1996	164	4.8	3.9	51
1997	200	4.3	3.5	42
1998	109	3.1	2.5	29
Total	608	17.8	14.5	196

(*) Calculation: ml × 11 m t/ml (equiv. 11,000 mg/l)
Note: Salt concentrations fluctuated from 9.3 to 13.2 m t/ml for individual years.

The solar evaporator operates for about 240–270 days per year. There is only limited evaporation during the winter season. The solar operator was not operational for some days due to installation of the system, repairing the damaged plastic liner, and testing of an *ET* sensor.

The calculated *water depth per day* in the solar evaporator was typically within the following ranges:

	mm/day
1995	0.2 to 18.1
1996	0.1 to 7.7
1997	0.5 to 5.6
1998	0.7 to 5.4

The levels of water depth are correlated with evaporation rates. Higher water levels in 1995 were needed to cover the plastic liner after its installation to protect it from wind damage. Testing the operation of the solar evaporator could also have caused these high water levels. The 1995 season was a 'learning' period.

Quality of Drainage Water Discharged

The electrical conductivity (ds/m) of drainage water discharged into the solar evaporator fell within the following ranges:

Year	EC (ds/m)	No. of samples
1995	15.0 to 18.0	8
1996	12.0 to 14.7	18
1997	10.7 to 18.6	19
1998	11.8 to 14.2	20

The water *EC* values fluctuated between 10.7 and 18.6 ds/m from 1995 to 1998. The values will increase, and their variations may be reduced, as the sequential reuse system of the drainage water will be completed.

The selenium levels have decreased after the initial leaching of the soils containing selenium that had accumulated for many years:

Date	Se ppb
8/30/95	1386
11/21/95	1653
4/23/96	1214
8/22/96	1041
4/24/97	920
7/25/97	811
8/20/97	792
9/30/97	839
3/3/98	1030
9/21/98	870
12/1/98	900

The laboratory results of these 10 samples (11 for *Se*) were within the following ranges:

	From	To	
pH	7.2	7.4	ds/m
EC	11.8	16.3	ppm
HCO_3	256	305	ppm
NO_3–N	70	180	ppm
Cl	2226	3182	ppm
SO_4	2686	3720	ppm
Na	2207	3018	ppm
Ca	629	847	ppm
Mg	162	258	ppm
B	17	21	ppm
Se	0.792	1.653	ppm

Evaluation

Solar evaporators compare favourably with evaporation ponds. A 256 ha (640 acre) farm would require about 25 ha (62 acres) of land; this represents about 10 per cent of the farm area. An IFDM system based on the sequential reuse of drainage water requires only 0.8 ha (2 acres) of solar evaporator; this represents about 0.3 per cent of the farm area. This small area, with crystallised salt and minimum water ponding, is not attractive to wildlife, thereby minimising its access to the sources of selenium. Drainage water is used to produce salt-tolerant crops, trees, salt-tolerant grasses and halophytes. Operational data indicate that this system has the potential to save about 20 per cent of the irrigation water and reduces drainage volume by about 199 acre-feet per a 640 acre farm area (0.3 acre-foot per acre). This system also eliminates the discharge of drainage water, salt and selenium into ponds, rivers, and the delta, thereby improving the water quality in these systems.

Water Management and Conservation

Irrigation water from the Westlands Water District (WWD) is applied to salt-sensitive crops in the area of 470 acres at an average rate of 2.5 acre-feet/acre. The estimated volume of irrigation water used is about 1175 acre-feet. If the WWD irrigation water is not available, groundwater is pumped.

Drainage water from the area of salt-sensitive crops (area 'A') consists of leached water and regional groundwater. The outflow of drainage water is 123 acre-feet. The 'A' area also generates about 188 acre-feet of tailwater that is collected in a 'tailwater' sump located in the NE corner of the farm.

About 351 acre-feet of water are required to irrigate 130 acre of salt-tolerant crops (area 'B'). This irrigation water consists of 36 per cent drainage water, 53 per cent tailwater, and 11 per cent WWD irrigation (new) water. The outflow of drainage water from the 'B' area is 53 acre-feet.

This drainage water is reused to irrigate 13 acres of salt-tolerant grasses (area C), at the rate of about 4.1 acre-feet/acre. WWD canal water and tailwater have been used to establish salt-tolerant grasses. The outflow of drainage water is 22 acre-feet.

Drainage water from the 'C' area is again sequentially reused on halophytes (area D, 5 acres). The irrigation rate is about 4.4 acre-feet/acre. The drainage water from halophytes about 11 acre feet in volume, is then discharged into a solar evaporator.

The following amount of irrigation water was applied to salt-sensitive crops in 1998 :

Quarter section	Crop	Irrigation rate	Water applied
SE (9)	safflower	2.42 acre-feet/acre	375 acre-feet
SW (10)	alfalfa, broccoli	2.86	443
NW (11)	alfalfa	5.37	832
	Total		1650

The following amount of irrigation water was applied to salt-tolerant crops in 1998 :

Quarter section	Crop	Irrigation rate	Water applied
NE (8) block 1	wheatgrass	2.58 acre-feet/acre	129 acre-feet
block 2	alfalfa seed	2.03	80
block 3	sugar beet	3.5	140
	Total		350

In 1998 the NW quarter section was being reclaimed while growing alfalfa. Since the salt was leached from the NW quarter section in 1998, a higher volume of irrigation water had to be applied. When leaching is accomplished, the data from the SE and SW quarter sections indicates that the volume of irrigation water used would be about 1175 acre feet as estimated when the system is designed. The amount of water applied to salt-tolerant crops was 350 acre-feet. This amount includes about 11 per cent (40 acre feet) of WWD water; the remaining is drainage water and tailwater from the 'A' area of salt-sensitive crops.

The area of salt-tolerant grasses and halophytes was still under development in 1998. The recorded amounts of drainage water were 25.9 acre feet from salt-tolerant crops (B) to salt-tolerant grasses (C), 6.6 acre feet from salt-tolerant grasses to halophytes, and the recorded amount of drainage water discharged into the solar evaporator was 2.5 acre-feet. These lower values for drainage water sequential reuse indicate the use of WWD water and tailwater for establishing salt-tolerant grasses and halophytes. The solar evaporator was operated for only 109 days in 1998.

It is estimated that the sequential reuse of drainage water through a system of salt sensitive crops > salt-tolerant crops > salt-tolerant grasses > halophytes results in saving 22 per cent of irrigation water that would be applied to 620 acres under conventional practices on other farms in the San Joaquin Valley. Drainage water us used to produce salt-tolerant plants and trees, and its volume is thus reduced by about 199 acre-feet on this farm. This water would have otherwise contributed to regional groundwater problems.

Economic Benefits

The management of salt and drainage water as resources must offer economic benefits to growers for this technology to be implemented on many farms in the San Joaquin Valley.

The results of this demonstration project indicate that Integrated on-Farm Drainage Management not only offers environmental advantages, but also makes good economic sense.

The total RRR project investment was $376,000; which is equal to $606 per acre of land. This investment included the installation of drains in the 620 acre field, six sumps, a water distribution system, materials for a solar evaporator on two acres, automatic irrigation system, the purchase of salt-tolerant grasses and trees, and other expenses.

The land value was between $500 and $1,000 per acre before the project started. The value of the reclaimed land increased to about $2,500 to $3,000 per acre. The present value of the reclaimed land is between $1,39,000 and $1,700,000. The difference in the land value before and after reclamation is about $1,060,000. The increase in the land value is $1,656 per acre.

The main purpose of this project was to control salinity, sustain high crop yields, and to change from the production of low value crops (cotton, alfalfa, and others) to the production of high-value vegetable crops.

The net return from growing low value crops was between $150 to $200 per acre; which is between $93,000 and $124,000 for this 640 acre farm. A net return from the production of vegetable crops grown on 470 acres is between $400 to $700 per acre. It is assumed that the net return from salt-tolerant crops, grown on 120 acres, will remain between $100 and $200 per acre. The total net return will be between $2,05,000 and $3,62,000 from the production of vegetables and salt-tolerant crops. This represents an increase of the net return between $1,16,000 and $2,42,000 per year or between $181 and $378 per acre/year.

The investment of $606 per acre has increased the land value by $1,656 per acre and the crop net return by $181–378 per acre year. The salt will also be marketed in the future. Finally, the project allows the farmer to manage salt and drainage water on his farm, thus avoiding environmental expenses for the taxpayers of California.

Land and Water Use

The land and water use on the farm can be summarised as shown in Table 6.8.

The total canal and drainage water used for irrigation is 1175 + 40 = 1215 acre feet and 123 + 53 + 22 = 198 acre feet respectively.

The *EC* values of canal irrigation water is not reported but on the basis of the reported content of 400 tonnes of salt in 1175 acre feet of water, it is estimated that the *EC* value is less than 0.5 ds/m and the salt content is 0.028 per cent or 280 mg/l. 93.7 per cent of the farm area is used for commercial and salt-tolerant crops, 2.8 per cent for salt-tolerant trees and grasses and only 0.3 per cent for solar evaporator. About 90 per cent of the drainage water is used for irrigation within the farm and the requirement of canal water is reduced by about 22 per cent.

Table 6.8 *Land and water use on the farm*

Block No.	Area ha (acre)	Percentage area	Crops	Irrigation water			
				Inflow (acre feet)		Outflow (acre feet)	
A	188 (470)	73.4	Commercial	1175	(canal water)	188	(tail water)
			Crops – Vegetables			123	(drainage water)
B	52 (130)	20.3	Salt-tolerant commercial crops	188	(Tail water from 'A')	53	(drainage water)
				123	(Drainage water from A)		
				40	(Canal water)		
				Total	– 351		
C	5.2 (13)	2.0	Salt-tolerant tree and grasses	53	(drainage water from 'B')	22	(drainage water)
D	2 (5)	0.8	Halophytes	22	(drainage water from 'C')	11	(drainage water)
Solar	0.8 (2)	0.3	—	11	(drainage water)	–	
evaporator	0.8 (2)	0.3	—	11	(drainage water from 'D')	—	

1 hectare metre = 8.2 acre feet

Salt Management

The salt inflow-outflow and management can be summarised as shown in Table 6.9

Table 6.9 *Salt inflow and outflow on the farm*

Block No.	Crops	Water source	Salt addition (Tonnes)	Cumulative salt flow (Tonnes)
A	Salt-sensitive crops	Irrigation water	400–72	328
		Groundwater	146	474
B	Salt-tolerant crops	Irrigation water	14	488
		Tail water	72	560
		Groundwater	66	626
C	Salt-tolerant grasses	Groundwater	36	662
D	Halophytes	Groundwater	18	680

The position of salt actually received in the solar evaporator is shown in Table 6.10

Table 6.10 *Salt received in the solar evaporator*

Year	*Operating days (Nos)*	*Salt received in solar evaporator (tonnes)*
1996	164	51
1997	200	42
1998	109	29
Total	473	122

The lower number of operational days was due to installation problems and it is expected that the solar operator will operate for about 240–270 days in a year. On a proportionate basis, the quantity of salt expected to be received would range between 60 to 70 tonnes per year.

Salt Imbalance in the IFDM System

An important point to be noted is that the total annual input of salt in the farm from irrigation water (1215 acre feet) is 414 tonnes while salt received in the solar evaporator in a year is 60–70 tonnes, which is only about 16 per cent or one-sixth of the input quantity. What happens to the balance 84 per cent of the salt quantity that is brought in annually ? The outflow of the drainage water from the farm is successfully prevented in the IFDM system but only a small part of the incoming salt is effectively removed through the solar evaporator. The major portion of incoming salts is left behind in the farm area. Groundwater flow would also tend to increase groundwater salinity in the adjoining areas. This appears to be a weakness in the IFDM system.

Salt Removed By Biomass Not Accounted For

In reality, the situation may not be as alarming as it seems from the preceding paragraph. All plants, grasses and trees contain minerals which are acquired from the soil water where they grow. If we know the quantity of annual biomass produce that is harvested and removed from the farm area, and the percentage mineral content in the biomass, then the quantity of salts removed annually from the area can be estimated. This quantity is likely to be much more than the quantity of salts removed through the solar evaporator. It is not clear why this aspect has been ignored in the IFDM study.

Biodrainage As An Alternative To IFDM System

There can be alternative model to the IFDM. The net quantity of canal water applied for irrigation in block A is 1175–188 = 987 acre feet providing drainage water quantity of 123 acre feet (12.5%) which is used to irrigate block 'B'. In case 90 per cent of the farm area (instead of 73.4%) is used for growing commercial crops (vegetables) requiring

1395 acre feet (90% of 1550 acre feet), the resulting quantity of drainage water would be about 175 acre feet. Trees planted over 10 per cent of the farm area (about 60 acre) should be able to transpire this whole quantity from the groundwater. Properly selected and planted trees can transpire as much as evaporation from a standard pan (A_{pan}). The trees may need irrigation by good quality canal water only in the initial stages. Once established, they would draw their whole water requirement from the groundwater. The option of irrigating the trees by pumped groundwater would also be available. If this is feasible, then the trees can fully take care of the quantity of drainage water and the groundwater table can be maintained at a safe depth without the need for a subsurface drainage system of perforated pipes.

Trees can be planted in blocks of suitable sizes or preferably in parallel strips to sufficiently depress the groundwater table underneath so that the groundwater table underneath the crop area does not rise above critical levels. There are methods to estimate and plan this.

In a biodrainage system, trees provide the required drainage and water balance. The harvested crops and utilisable biomass produce from tree plantations together can also provide an acceptable salt balance, where the canal irrigation water is of good quality.

The following suggestions on the IFDM system are worth consideration and trial:

(i) The net annual import of canal water in the IFDM system is 1215 acre feet, while the quantity of drainage water received in the solar evaporator is 11 acre feet, i.e., almost 99 per cent of the imported water quantity is consumed within the farm system. Similarly out, of 414 tonnes of salt imported along with irrigation water, only 60–70 tonnes reaches the solar evaporator, i.e., 84 per cent quantity is consumed or left behind in the system. With minor modifications, it should be possible to consume or use all the quantity of imported water and salt within the system so that no quantity of drainage water and salt are left to be discharged into the solar evaporator. This would meet the basic objective of the IFDM system of 'no drainage' out of the farm area without the need of a solar evaporator.

(ii) The drainage water (123 acre feet) from commercial crop area (Block 'A') is used to irrigate salt tolerant commercial crops (Block 'B'); 53 acre feet of drainage water from Block B is used to irrigate salt tolerant trees and grasses (Block 'C'); 22 acre feet of drainage water from Block C is used on the halophytes (Block D). The water is sequentially reused with the help of sumps and pumps. The total quantity of drainage water used in this manner is 198 acre feet. It should be possible to alternatively use an equal quantity from groundwater in a different manner. Trees planted in suitable blocks or strips over 64 acre area (10% of the farm area), should be able to biopump and consume more than this quantity of water by transpiring it into the atmosphere. This would enable to do away with the need of subsurface drainage system, sumps and pumps.

(iii) Use of saline drainage water (of salinity 10 ds/m or more) to surface irrigate tree plantations causes progressive build up of soil salinity in the root zone, which in the long run may reach such high levels so as to jeopardise the sustenance of the tree plantations. On the other hand when trees use water by biopumping, the concentration of salts would be confined to the capillary zone. The trees would be able to draw water continuously from the groundwater table without threat to their survival.

(iv) The quantity of minerals evacuated from the farm along with the harvested vegetable crops and utilisable biomass produce from the trees and grasses should equal or exceed the quantity of salt imported annually with the 1215 acre feet of canal water. An exercise needs to be done to make a check on this. However, if there is an imbalance, the rate of increase in groundwater salinity can be estimated which is likely to be too slow to be a cause of concern.

Economic Benefits

Under the IFDM system, the net return from the produce of vegetable crops is between $ 400 to $ 700 per acre and from salt-tolerant crops, between $ 100 to $ 200 per acre. Commercial crops are grown over a 73.4 per cent farm area and salt-tolerant crops over a 20.3 per cent area.

In the alternative biodrainage system commercial vegetable crops can be grown over 90 per cent of the farm area and the net return per acre would increase by about 15 per cent as follows:

IFDM System

0.73 acre @ $ 550/acre (mean value)	=	$ 401.5
0.21 acre @ $ 150/acre	=	$ 31.5
0.94 acre	=	$ 432.0

Biodrainage System

0.90 acre @ $ 550/acre	=	$ 495

Besides higher net return, the costs involved in installing and operating a subsurface drainage system and solar evaporator would be saved in the biodrainage system.

AGROFORESTRY - IFDM EXPERIMENTS, MENDOTA, CALIFORNIA, USA

Cervinka et al. (2001) in the report "Drain Water, Salt and Salenium Management utilising IFDM/Agroforestry System" describe the design and operation of an experimental Integrated On-Farm Drainage Management (agroforestry) project located near Mendota, California USA during the period from 1994 to 1997. The project goals were

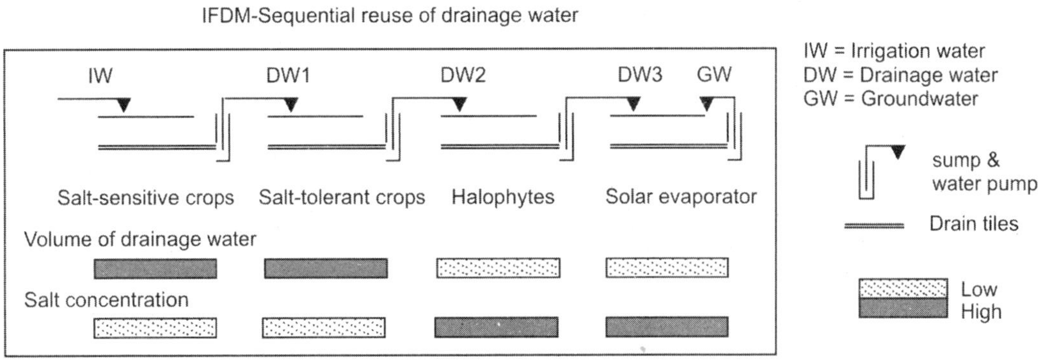

Figure 6.11 *Experiment at Mendata, California, USA*

to manage drainage salt and selenium within farm boundaries and thus, to eliminate any need to discharge salt and selenium into State water bodies. The IFDM method moves salt and selenium through a sequential reuse of drainage water to irrigate salt tolerant trees, plants and halophytes. The separation of salt and selenium occurs at the last reuse stage through the evaporation of the remaining drainage water. The Figure 6.11 illustrates this method.

The main components of the experimental IFDM system were (1) trees tolerating drainage water of about EC 8 to 12 dS/m; (2) halophytes tolerating drainage water above 15 dS/m and; (3) solar evaporator consisting of a levelled area where the separation of salts from drainage water occurs.

The project area was 9.6 ha. The sub-area of trees was 5.9 ha, halophytes 2.4 ha and the solar evaporator 1.3 ha. The solar evaporator had a plastic liner to prevent the seepage of accumulated drainage water and salt.

The trees received drainage water from the regional collector drains. The drainage water from the tree sump provided irrigated water for halophytes. Drainage water from the halophytes was pumped and evenly distributed into each of the three sections of the solar evaporator.

The tree area included Eucalyptus and athel species. The best performing trees were Eucalyptus 4543, 4544 and 4573. The trees did not tolerate extended water ponding. Halophytes included salicornia, salt grass, cord grass, iodine bush and native plants (such as five-hook basia). Salicornia, salt grass are promising plants deserving further development.

The maximum quantities of water applied were in 1996. The irrigation rate was 13.5 ML/ha/yr for the trees and 13.5 ML/ha/yr for halophytes. The water evaporation rate in solar evaporator was 11.9 ML/ha/yr during this season. About 40 per cent of water applied to the trees was leached and sequentially reused for the irrigation of halophytes.

The trees were receiving drainage water from farm crops at an average EC of 9.4 dS/m. The drainage water from trees that was applied to halophytes had an average EC of 31 dS/m. The drainage water from halophytes, discharged into the solar evaporator had an average EC of 28.2 dS/m. The decrease of salt concentration in the drainage water discharged into the solar evaporator was caused by water leaking from the collector line located in the proximity of the project area.

Water tables had typical seasonal characteristics from about 0.8 m in April/may to about 1.8 m in October. Available soil data indicated the effect of the drainage system on sustaining soil quality. The drainage water from the farm crops had an EC about 9 dS/m. The initial value of soil ECe was 10 dS/m in 1987. That value increased to 29 dS/m in three years. After installing drainage tiles the ECe value was lowered to 11 dS/m in four years.

It is claimed that salt balance is achieved through the sequential reuse of drainage water. As initial salt in the tree area was high before the installation of a drainage system, the estimated salt flow to halophytes exceeded the salt inflow into the trees. The drainage water was diluted (due to the leakage of a collector line) at the final stage of discharge and consequently not all salt was transferred into the solar evaporator. The salt balance would have been achieved if the interference of the leakage in the collector line had not occurred.

Comment

The data presented in the report for the three years 1995 to 1997 indicates that the net salt-out (1033 tonnes) and salt-in (1683 tonnes) ratio relating to the project area was 0.61, which means that 39 per cent salts brought into the tree area along with irrigation water were left behind in the project area.

An assumption is made in the report that dilution of drainage water (from halophyte area to solar evaporator) occurred due to water seepage from the nearly regional collector line, and if this had not happened, the EC of drainage water would have been 44 dS/m resulting in salt balance condition. The validity of this assumption is doubtful. In case the leakage from collector line (carrying drainage water of salinity 10 dS/m) diluted the salinity of drainage water from the project area to 28 dS/m from 44 dS/m, then contribution to the volume of drainage water into the solar reservoir from the project area would be 53 per cent and the balance 47 per cent would come from the leaking collector line. This would correspondingly reduce the estimated quantity of salt-out from the project area as well as the salt-out: salt-in ratio. It appears that the salt-out from the groundwater in the project area are, somehow, escaping into the adjoining areas (to an extent of about 40%) and the project objective to manage drainage water, salt and salenium within farm boundaries is being achieved only partially.

THE UNDERA SERIAL BIOLOGICAL CONCENTRATION (SBC) EXPERIMENT IN NORTHERN VICTORIA, AUSTRALIA

Heuperman (2000) has described the experiment at the Undera SBC site in north Victoria.

The Undera SBC Site in Northern Victoria

In 1994, a farm-scale pilot project commenced on a dairy farm at Undera in north-central Victoria. The property was severely affected by salinity and parts of the farm were retired from productive use with bare patches, dead trees and barley grass dominating. The site has a flat topography, typical of the flood plains of the Goulburn Valley and a shallow water table (< 1 m from the surface). The area has a semi-arid climate with annual evaporation and rainfall of 1400 and 480 mm respectively and monthly evaporation exceeding monthly rainfall during nine months of the year. Irrigation is applied from August to May. The project consists of a groundwater pump protecting about 20 ha of the property ('Stage 1'), a 3 ha tile-drained SBC area ('Stage 2') and a 1.5 ha evaporation pond system ('Stage 3'). The site layout is presented in Figure 6.12

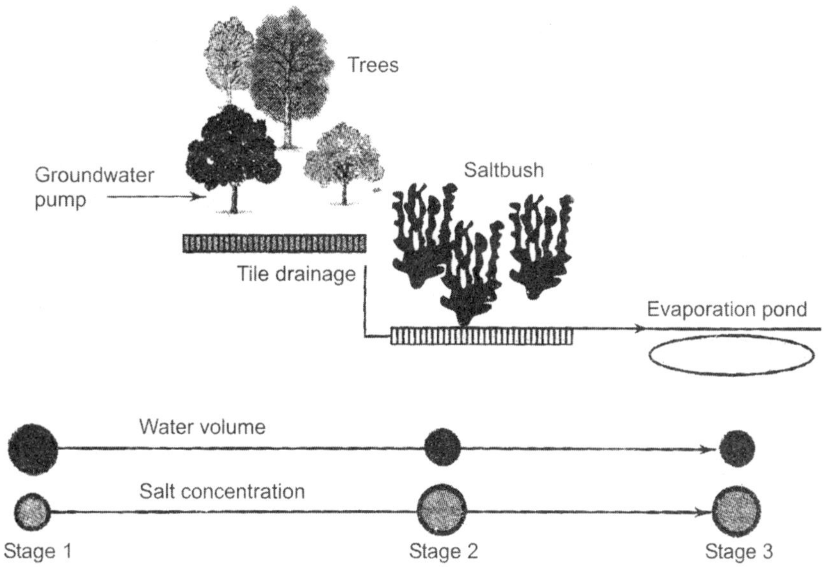

Figure 6.12 *Serial biological concentration*

The Groundwater Pump

Exploratory drilling in the southern part of the property identified a shallow aquifer at 6–8 m and a deeper aquifer at 15–17 m, both with groundwater salinities ranging between 9 –11 dS/m. A pump test in the shallow aquifer indicated that a yield of 30–35 ml (mega litres) could be achieved over the irrigation season, providing salinity control for an estimated 20–25 hectares on the property. The groundwater pump is located 600 m south of the SBC site.

Crop Selection

The groundwater is used to irrigate River Red Gums (*Eucalyptus camaldulensis*), a range of other tree species and a combination of saltbushes (*Atriples spp*). The Stage 2 tile-drained area is located outside the area of influence of the pump. The tree/saltbush crops are grown on raised beds, spaced three metres apart with intra-row spacings of 1 m for the saltbush and 1.5 m for the trees. Salt-tolerant Tall Wheatgrass (*Thinopyrum elongatum*) and Puccinellia (*Puccinellia cilata*) are grown between the tree rows to increase water use.

Saltbush was selected because it is a perennial plant with species that cover a wide range of salinity tolerances, has a high biomass production potential and requires few inputs after establishment. A combination of the erect species, Old Man Saltbush (*A.nummularia*) and the prostrate species, River Saltbush (*A. amnicola*) has been planted at the Undera site. Trees were established in the first season with fresh water (1995–96) and have received straight groundwater thereafter. During 1997–98, approximately 16 megalitre were applied to the 3 ha wood lot. At an *EC* of 10 ds/m, this amounts to the equivalent of about 100 tonnes of salt.

Leaching Capacity of the Soils

The sustainability of the system will be primarily dependent on the ability of the soil to prevent the accumulation of the salts in the root zone beyond tolerable levels for tree crop production. Using Rhoades' predictive leaching requirement model (Rhoades, 1989; equation 1), it has been predicted that the average soil salinity (*ECe*) under the trees and saltbush would not exceed 16 dS/m.

Value-adding Components

To make the best possible use of the SBC site, a number of tree species, not commonly used under saline conditions, have been tried at the pilot site. These include tree oil, timber and firewood species. In addition, the saline evaporation basins have been used to test a range of fish and mollusc species (mariculture).

Alternative Tree Species

A number of potentially high value tree species were planted in November 1995 to field-test their viability and production potential in this saline environment. The spp. were selected on the basis of their survival in laboratory pot trails by the CSIRO Division of Forestry in Canberra. They include *Eucalyptus grandis, E. horistes, E. largiflorens, E. occidentalis, E. polybractea, E. spathulata, E. suggrandis, E. trabouti, Melaleuca alternitifolia, M. uncinata,* and the *Poplar alba, P. bolleana, P. euphratgea, P. nigra and P. euaphratea x P. deltoides.*

Both the Eucalyptus and melaleuca species, selected for their oil-producing potential, produced commercial volumes of oil (at a 'per tree' basis), three years after establishment. The quality of the melaleuca oil was not good enough to qualify as commercially acceptable 'tea tree oil'; the quality of the eucalypt oil as measured in bulk samples from 20 trees was commercially acceptable. The potential to produce commercial yields of tea tree (provided appropriate varieties are selected) and eucalpyt oil by using saline irrigation water has been shown. Gross annual returns measured at the site were about \$500/ha and \$1200/ha for eucalyptus and melaleuca respectively. The frost sensitivity of certain melaleuca species has also been identified.

Mariculture (Saline Aquaculture)

The objective of the Undera mariculture trial was to identify species which have the potential to grow at commercially acceptable rates in evaporation basins filled with saline groundwater. The first two of the four basins were used during this study. The first basin (basin 1) was supplied with water from the tile drains, while the second basin (basin 2) was fed by the overflow water from basin 1. Cage trials were conducted on two oyster species, two prawn species and nine fish species, including freshwater, estuarine and marine species. More detailed information on the trials is presented by Ingram et al. (1996). Atlantic salmon (*Salmo salar*), silver perch (*Bidyanus bidyanus*), Australian bass (*Macquaria novemaculata*) and rainbow trout (*Oncorhynchus mykiss*) were identified as the most promising species.

Comment

The reported salinity of groundwater which is used to irrigate the trees ranges between 9–11dS/m. It is predicted that the average soil salinity (EC_e) under the trees and salt bush would not exceed 16 dS/m. This can be possible only if the leaching fraction equals or exceeds 0.625 (10 ÷16). It is not explained whether this is feasible. If not, the soil salinity under the trees would increase progressively putting their long-term survival under doubt.

BHAWALNAGAR, PAKISTAN STUDY ON BIOLOGICAL CONTROL OF WATERLOGGING

Choudhary et al. (2000) describe a study over 18.2 ha canal irrigated area in which about 4 ha area was planted with six year old eucalyptus plants near Bhawalnagar, Pakistan. There were 1340 plants per hectare. The surrounding area was canal irrigated with major crops of cotton, rice and sugarcane. The reported findings of the study were:

- The water table was deeper in the area under plantation as compared to the area under crops. The water table depth ranged between 135 cm to 258 cm.
- The water table fluctuated during different periods depending upon the irrigation applied and rainfall received.
- The soil salinity remained below critical limits (< 4 ds/m^{-1}) in the area under tree plantation.
- The annual water use by the plantation ranged between 300 to 2100 mm.
- Due to eucalyptus plantation, the groundwater movement showed as a front towards the plantation area from the north to south direction.
- The environment in the planted area was improved.

The methodology used to determine annual tree water use is not reported. The plantation area was surrounded by irrigated crops like rice, sugarcane, etc., which generally cause high groundwater recharge.

▼ WATER TABLE CONTROL BY BIODRAINAGE – ISRAEL

Gafni and Zohar (2000) describe an Israeli experiment in their paper. Lands in the 200 km^2 western Yizre'el valley were dry cultivated since the early 1920s. In the mid-1960s, when cotton became the dominant crop, irrigation was introduced, and many water harvesting and storing installations – reservoirs were constructed in the valley to secure irrigation waters for the long summer season. This caused disturbance in the existing hydrological balance: the shallow groundwater level rose and the inevitable salinisation plagued the valley.

The biodrainage approach was tested at five different, waterlogged and salinity affected plots representing a broad range of environmental conditions, including rainfall gradient, subsurface hydrological regimne, soil salinity, sodicity status etc. The 0.62 ha sites had been abandoned by the farmers because of waterlogging and/or salinity built-up. Typical soils were clayey and measurements prior to planting showed SAR values to be mostly within the range of 5 to 10 with some extreme values of 23 to 26 being measured. Six provenances (seed sources) of *Eucalyptus Camaldulensis* and two provenances of *E. occidentalis* were selected and planted in four replications at each of the five sites.

Shallow groundwater tables were routinely monitored over the period of the experiment in six observation wells per site, both inside and outside (control) the plots. Growth rates of trees were measured periodically using Diameter at Breast Height (DBH) and tree height as variables. After five years, trees with average DBH were cut in order to estimate the average animal biomass yield per hectare. The transpiration rate of *Eucalyptus Camaldulensis* drawn from the two most promising seed sources (Hadera,

Israel; and Lake Albacutya, Victoria, Australia), growing at the Nahalal site (Western Yizre'el Valley), were determined by using the calibrated heat pulse technique.

The trials proved to be very successful. Groundwater tables declined constantly during the 5-year monitoring span and dropped to more than 3 m below the soil surface at each of the sites in summer of 1999. At the tree age of 3.5 year, transpiration on summer days was between 4.5 and 5.1 mm per day. Zohar and Schiller (1998) calculated the cumulative depth of water uptake by the trees at Nahalal site to be 1882 mm in 708 days. Excluding rainfall, the trees consumed 932 mm directly from underlying groundwater, thus depressing it by 1.2 m.

Growth rates were no less impressive. At the end of the fifth year of the trial, in the best plot in that year, the biomass yield reached 30 tonnes per ha per year for the Broken Hill ecotype (NSW, Australia) and 57 tonnes per ha per year for the Hadua (Israel) ecotype.

A related study, recently conducted both at the Nahalal site under controlled conditions, examined the tolerance of four *E. Camaldulensis* ecotypes to salinity and flooding stresses (Lev–Hari, 1999). The study found that the high tolerance of *E. Camaldulensis* to salinity is based on avoiding the uptake of sodium and chloride ions at the root level. Even though the Na^- concentration in the soil and irrigation solutions was 200–400 times higher than the K^- concentration, the molar ratio between these ions in the leaves was close to 1. Chloride ions, too, did not accumulate in these leaves. The conclusion drawn from this was that *E. Camaldulensis* roots are highly selective and are able to inhibit salt uptake. In the light of these findings, and of the concerns expressed by others, regarding the risk of eventual salt accumulation in the root zone, monitoring of the salinity dynamics will continue for some period of time.

The economic viability of growing eucalyptus on abandoned lands in Israel was also studied as part of the biodrainage research effort. It was concluded that, without irrigation, under current conditions, communical forestry of elite *E. Camaldulensis* for both Medium Density Fibre board and other timber products is a viable agricultural alternative that could bring higher returns than wheat, corn or cotton (Hadas, pus. comm). At present commercial forestry is readily expanding in Israel with more than 600 ha being offered by farmer for this perspective venture.

YINEBI IRRIGATED PLAINS, STUDY - CHINA

The Yinebi alluvial plain in China spread over an area of 3700 km^2 is irrigated (gravity system) by using 2.5 billion m^3 water annually from the Yellow River. Xie et al. (1998) report that inspite of comprehensive drainage network about 40 per cent of the irrigated land suffers from salinity and waterlogging and crop yields are depressed by about fifty per cent. Some of the alluvial plain has been irrigated for as long as 2000 years. The groundwater level is near to the ground surface. The salt content in irrigation water is about 400 mg/l with *EC* value around 0.6 ds/m. SAR values are not high.

The groundwater salinity generally ranges between 2–6 ds/m with a few high values of up to 50 ds/m. The SAR values do not exceed 15, whenever *EC* values are not more than 7.5 ds/m.

The water balance study shows that in the total input of 3.3 billion m^3, major contributions were from irrigation water (75%) and precipitation (20%). The major outputs were evapotranspiration (63%) and discharge by drains (35%). The annual inflow exceeded the outflow by 4.7 million m^3.

According to the salt balance study, out of the total input of 1.5 million tonnes, major source of salt input was irrigation water (75%) followed by pumped water (22%). The salt removal was mostly by drainage water (98%). It was estimated that 210 million kg of salt (14%) was left behind.

Comment

It is observed that the salt balance study does not account for the salts evacuated from the soil by the harvested crops and vegetative biomass. The quantity of salts evacuated in this manner should exceed the estimated balance quantity of 210 million kg. Therefore, the root cause of the problem appears to be waterlogging and not the adverse salt balance. Salinisation of soils is the consequence of waterlogging.

The practised drainage measures have not fully resolved the waterlogging problem. Forty per cent of the irrigated land is producing only 50 per cent of potential yields. Biodrainage should be an effective answer. It would depress the groundwater tables sufficiently to overcome the waterlogging problem.

CONCLUSIONS

Aral Sea and Indus Basins

In the Aral Sea Basin a significant cause of progressive increase in salinity of river waters is discharge of saline drainage water from irrigated fields into the natural streams. The problem would be further aggravated on implementing the programme of constructing more and more subsurface and open drainage systems. A rethink is needed to plan an appropriate strategy in which biodrainage can be an important component.

In the Indus Basin (Pakistan), the practised drainage methods of subsurface, the out-fall drain and evaporation ponds have not given the desired results. Biodrainage, by itself or in conjunction with other measures holds good potential of overcoming the problem of waterlogging and salinity.

Evapotranspiration from Trees

The studies show that trees can transpire significant quantities from groundwater.

In the Victoria, Australia study, with tree density of 730 trees/ha, tree water use during February–March 1992, ranged between 0.1 to 0.6 A_{pan} (Class A evaporation pan). The published annual TWU values range from 0.6 A_{pan} to 1.9 A_{pan} for eucalyptus species in Western Australia. The annual tree water use was estimated as about 1400 mm.

In the California USA study, the total evapotranspiration from tree plantations during 220 days (April–November 1990) was estimated as 1153 mm.

Depth to which Tree Roots can Extend

The depth to which tree roots would extend would largely depend upon the depth of the groundwater table and other local conditions.

In the Victoria, Australia study, where the groundwater table fluctuated between 3 m to 5.5 m, the root densities were maximum up to about a depth of 3 m and very low below a 5.5 depth. High root density was also observed at 8.5 m depth, perhaps on account of the textural change in the soil strata.

In the California, USA study, it was assumed that plant roots have full depth penetration equal to the root zone depth down to the surface of the water table, which was 2.1 m.

Extent of Depression of GWT Underneath Plantation and its Effect on Adjoining Lands

The tree plantations depress the groundwater table directly underneath due to evapotranspiration. The extent of depression of the groundwater table underneath the plantation and its effect on groundwater table levels in the adjoining areas, would depend upon the rate of recharge (from precipitation and irrigation), the hydraulic conductivity of soil strata, depth to barrier layer and the distance between plantation blocks.

In the Victoria, Australia study the groundwater table was depressed about 5.5 m depth underneath the plantation and the draw-down effect was observed up to about 40 m distance from the plantation boundary. The data on rate of recharge, hydraulic conductivity, depth to barrier layer and distance between plantation blocks is not available.

In the California, USA study, there was no perceptible depression of the groundwater table underneath plantations and in the adjoining area, because the plantations were irrigated with water supply exceeding the evapotranspiration capability.

Uptake of Salts by Plants and Soil Salinity Underneath Plantations

As a result of tree water use, there occurs some concentration of salts underneath the plantations. A reanalysis of the available data, updating of information and further research and study are needed.

In the Victoria, Australia study, the salinity rose to about 5 ds/m (from 1 ds/m) in the soil zone at a depth of 3 m to 5.5 m during a period of nine years (1983–92). However, salinity at greater depths (below 6 m) rose to a level of only about 2 ds/m during the same period. It is assumed that most plants (including trees) use water, not salts; they exclude salts at their root surface. But no experimental or observed data is given to support this assumption.

In the California, USA study, irrigation water was continuously applied in the plantation area. The salinity level of irrigation water was about 10 ds/m (8000 g/l) and that of drainage water, about 17 ds/m (14000 gm/l). About 40 per cent of the imported salt was removed by drainage water. What happened to the remaining 60 per cent salt is not explained. Was any salt taken up by the plantations? Was there an increase in the salinity level of the soil at a 0–2.1 m depth? Is the salinity level of drainage water increasing from year to year? There is no mention of any damage to plantations because of the high salinity.

Biodrainage by Plantations

The Victoria, Australia study suggests that non-irrigated plantations growing on a 5 ds/m water table can be sustained for a period of more than 20 years, and the period can be extended by either a selection of more salt-tolerant species or irrigation with water of a better quality than the underlying groundwater. Since the salinity of groundwater at a depth of more than 6 m has risen to only 2 ds/m (from 1 ds/m) in a period of 9 years (1983–92) it is possible that it may take a much longer period (about 90 years) to reach the threshold limit that would inhibit trees to draw water from the groundwater table because of high salinity. A suggested potential management option is to rotate plantation sites every 20 to 30 years.

Another option can be to leach down the accumulated salts underneath the plantations with good quality irrigation water at suitable periodical intervals, which interval may be 10 years or 20 years as may be practical. If leaching operations can be carried out in a period of 1–2 years, these can be practised by rotation in different blocks of plantations by turns.

The California, USA study confirms the feasibility of growing plantations under quite saline conditions. The IFDM study proves the capability of trees and grasses to transpire significant quantities of water. But the solar evaporator receives only about 16 per cent of the total quantity of salt imported alongwith the irrigation water. What happens to the balance quantity of salt? The electrical conductivity (ds/m) of drainage water received in the solar evaporator during the four-year period, 1995–1998, does not show any increasing trend. This suggests that vegetable crops, soil-tolerant crops, trees, grasses and halophytes are consuming a bulk of the salt quantity contained in the irrigation water.

In the Undera SBC study the groundwater is pumped from a sump in the farm area to depress the groundwater table underneath and the pumped water is used to irrigate a wood lot 600 m away. The same objective can be achieved by locating the wood lot adjacent to the farm area. Trees would act as biopumps and draw water supply directly from the water table. If need be, the groundwater can be pumped from a sump located in the wood lot area to provide supplemental irrigation to the trees. There would then be no need to provide subsurface drainage under the wood lot. In case the quantity of salt received in the salt concentration ponds is a small part of the total quantity of salt imported alongwith the irrigation water, their usefulness would be very limited.

The Bhawalnagar, Pakistan study shows that tree plantations can control the rise of the groundwater table to facilitate irrigated agriculture in the surrounding areas.

The Israel experience confirms the feasibility of biodrainage in reclaiming water-logged and saline lands and their use for productive agriculture.

The Yinebi Plains (China) experience shows that problems of waterlogging and salinisation can persist despite conventional drainage measures. The pollution of river water by discharge of saline drainage water into the river and use of such river water for irrigation in down stream reaches can accentuate the problem. If seems that biodrainage, which has not yet been tried, can provide the needed relief.

Kijne et al. (SWIM Paper 2, IIMI 1998) note that the use of salt-tolerant trees, especially eucalyptus varieties, in reclaiming saline areas by lowering the groundwater level is important. Bari and Schofield (1992) have reported the control of human induced salinity through reforestation with eucalyptus species in West Australia, where the groundwater levels decreased in 10 years by 1.5 m under the forest and increased by 1.8 m under near by pasture land. Interestingly, groundwater salinity levels under the forest were reported to have decreased by 11 per cent.

REFERENCES

Cervinka, V., C. Finch, M. Martin, F. Menezes, D. Peters and K. Bucknoff (2001). *Drainwater, Salt and Selenium Management Utilizing IFDM/Agroforestry Systems*, Report prepared by USBR, p. 21.

Cervinka, V., J. Diener, J. Erickson, C. Finch, M. Martin, F. Menezes, D. Peters and J. Shelton (1999). *Integrated System for Agricultural Drainage Management on Irrigated Farmland*, Report of the Westside Resource Conservation District to the US Bureau of Reclamation, Sacramento, CA.

Chaudhry, M.R., M.A. Choudhary and K.M. Subhami (2000). 'Biological Control of Waterlogging and Impact on Soil and Environment', Paper presented at the 8th ICID International Drainage Workshop, February 4, New Delhi, India.

Dukhovny, V.A. and K.H. Yakubov (2000). 'Return Waters of the Aral Sea Basin and Possibilities of

Their Utilisation on the Base of Green Desert Programme', Paper Presented at the 8th ICID International Drainage Workshop, New Delhi, Vol. III, pp. 329–31.

Dukhovny, V.A. and P.D. Umarov (2000). 'The Land Salinization and Drainage Problem in Central Asia', Paper presented at the 8th ICID International Drainage Workshop, New Delhi, Vol. I, p. 71 and pp. 79–84.

Gafni, A. and Y. Zohar (2001). 'Sodicity Conventional Drainage and Biodrainage in Israel', *Aust. J. Soil Science*, 39: 1269–78.

Heuperman, A.F. (1995), *Salt and Water Dynamics Beneath a Tree Plantation Growing on a Shallow Water Table*, Department of Agriculture, Energy and Minerals, Victoria, Australia.

— (2000). 'Biodrainage: An Australian Overview and Two Victorian Case Studies', Paper presented at the 8th ICID International Drainage Workshop, February, New Delhi.

Kijne, J.W., S.A. Prathapar, M.C.S. Wopereis and K.L. Sahrawat (1998). 'How to Manage Salinity in Irrigated Lands: A Selective Review with Particular Reference to Irrigation in Developing Countries', SWIM Paper 2, International Irrigation Management Institute, Colombo, Sri Lanka.

Kapoor, A.S. (2000). 'Biodrainage Feasibility and Principles of Planning and Design', Paper presented at the 8th ICID Workshop, New Delhi, Vol. IV, pp. 23–25.

— (2001). *Biodrainage: A Biological Option for Controlling Waterlogging and Salinity*, Tata McGraw Hill, New Delhi, pp. 71–74.

Karajeh, F.F. and K.K. Tanji (1991). 'Agroforestry Drainage Management Model', *Journal of Irrigation and Drainage Engineering*, ASCE Paper No. 6258, March/April.

Kitamura, Y., T. Yano and S. Yasuda (2000). Proceedings of the 8th ICID International Drainage Workshop, February, New Delhi, Vol. I, p. 540.

Masood, Ahmed and Gary P. Kutcher (1992). 'Irrigation Planning with Environmental Consideration in Indus Basin, Pakistan', World Bank Technical Paper No. 166.

Serial Biological Concentration (SBC) (1999). Murray Darling Basin Commission, Victorian State Salinity Programme, ISIA, Tatura, Pamphlet.

Takano Yoshihiro, GIF Japan (1994). 'Our Actions Towards the Rehabilitation of the Aral Sea', Paper presented at the Special Sessions Programme on Aral Sea Basin, Bulgaria (19 May 1954), ICID Report, p. 35.

Usman, U.K. et al. (2011). Improving irrigation water operation in the lower reaches of the Amu Darya river – Current status and suggestions. Irrigation & Drainage, 60(5): 600–612.

Xie, G., J. Breburda, A. Kollender-Szych and A. Battenfeld (1998). 'Relation between Soil Salinity and Water Quality as well as Water Balance in Yinebi Plain PR China', in *Towards Sustainable Land Use, Furthering Cooperation between People and Institutions,* Vol. I, Advances in Geolocology 31, International Society of Social Sciences (ISSS), pp. 485–93.

7

Research Studies in India on Biodrainage during 2004-2012

ICID AND IPTRID RECOMMENDATIONS

Following the recommendations of the 8th ICID International Drainage Workshop held at New Delhi in the year 2000 and the suggestions contained in the International Programme for Technology and Research (TPTRID–FAO) knowledge Synthesis Report (2002) on the subject of 'Biodrainage, principles Experience and Applications' several research programmes were initiated in India. Studies were carried out in different parts of the country with different climatic and physiographic conditions. Agricultural universities, forest institutes and research stations of good standing were entrusted with the task of undertaking research work. Experimental methods, their results and findings in brief are described in this chapter.

STUDY BY THE HARYANA AGRICULTURAL UNIVERSITY, HISSAR

Lysimeter Study

The Haryana Agricultural University, Hissar (India) undertook a study on ecophysiology of tree species to evaluate biodrainage potential for waterlogged soils. 1.45 m diameter and 1.25 m deep lysimeters were set up in the field as shown in Figures 7.1 and 7.2.

One and half year old saplings established in nurseries were transplanted into PVC lysimeters in July 2006. In all experimental measurements water table was maintained at a depth between 80–90 cm.

The main stem of the plant was pruned to have a height of 45 cm at the time of transplantation. The height of the trees and girth at breast height were measured monthly.

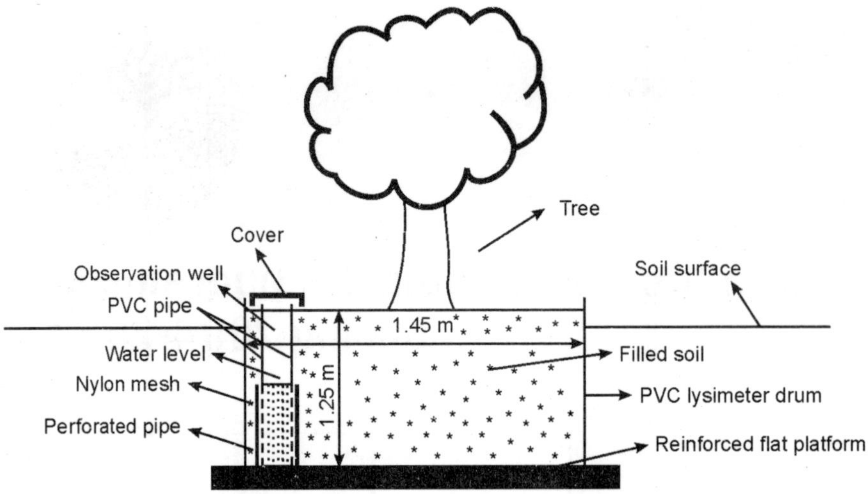

➲ **Figure 7.1** *A schematic diagram of the lysimeter setup (see text for details)*

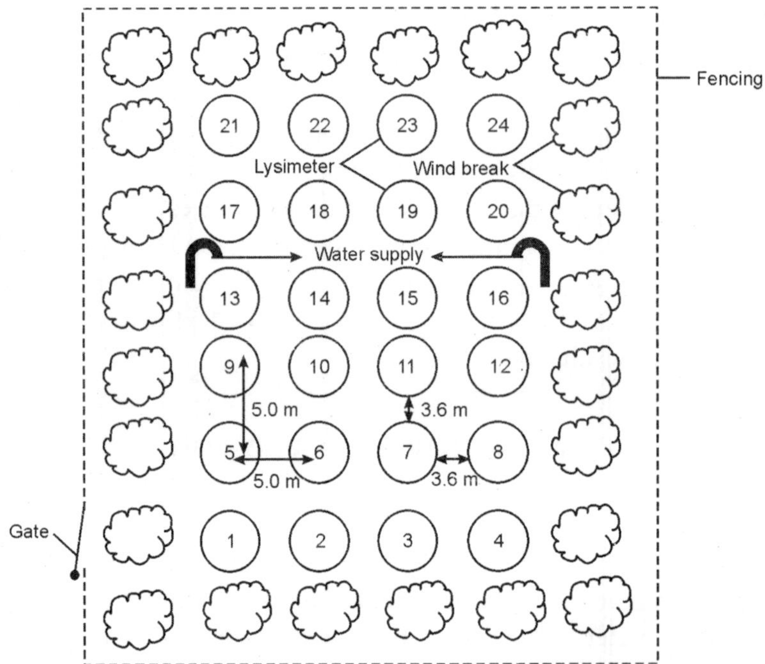

➲ **Figure 7.2** *Field layout of the lysimetric set up. Figures within the parenthesis denote lysimeter number of the species in a completely randomised design. The twenty four lysimeters were planted with C. equisetifolia (2, 8, 11, 13), E. tereticornis C-10 (4, 15, 18, 21), M. azedarach (7, 10, 16, 23), P. juliflora (3, 5, 12, 14), T. arjuna (6, 9, 19, 24) and control (1, 17, 20, 22). The lysimeters are surrounded by a wind break of Tamarix aphylla trees on all the four sides. A barbed wire fencing protected the plot.*

In general tree height showed a gradual increase during the period of two years (August 2007 to August 2009). In *M.azzdarach, P. Juliflora* and *T. arjuna* there was no increase in height during winter. Modest increase was observed during the period April–September in *C. equisetifolia* and *E. terecticornis C – 10*, maximum growth occurring in the months of August and September. 3 years after transplantation *E tereticrones C – 10* attained a height of 824 cm (in August 2009) which was 690 cm for *E. equisatifolia*, 374 cm for *P. Juliflora* 291 cm for *T. arjuna* and 226 cm for *M. azedarach*.

The maximum girth at breast height was attained by *P. Juliflora* (39 cm) followed by *E. tereticornis* (35 cm), *C. equisetifolia* (30 cm), *M. adedarach* (29 cm) and *T. arjuna* (23 cm).

The leaf area of the five tree species in lysimeters during the period (2007–2009) was relatively low in winter and high in summer. The range was 40–100 m^2 for *C. equisetifolia*, 20–80m^2 for *e.tereticornis*, 20–25 m^2 for *T. arjuna*, 0–40 m^2 for *M. azedarach*, 10–45 m^2 for *P. juliflora*.

Meteorological data of temperature, relative humidity, wind speed, sun shine hours and pan evaporation etc. was collected from the observatory located at a distance of 800 m from the lysimeter site.

The Potential Evaporation (E_p) was low during winter, 1–2 mmd^{-1} reaching a peak of 10 mmd^{-1} in the summer month of May.

The observed values of water biodrained (W_b) in litres per day by the tree species were as follows:

Tree species	W_b during winter	Peak W_b value in May	Average W_b per day
C. equisetifolia	10 Ld^{-1}	90–119 Ld^{-1}	47 Ld^{-1}
E. tereticornus	10–16 Ld^{-1}	89–93 Ld^{-1}	46 Ld^{-1}
M. azedarach	4 Ld^{-1}	33–43 Ld^{-1}	18 Ld^{-1}
P. Juliflora	8–15 Ld^{-1}	63–84 Ld^{-1}	38 Ld^{-1}
T. arjuna	5–12 Ld^{-1}	17–31 Ld^{-1} (Peak value of 31 Ld^{-1} was attained in the month of September)	16 Ld^{-1}

The overall co-relation between E_p and W_b in respect of the five tree species was as follows:

Tree species	E_p: W_b ratio	W_b: E_p ratio
C. equisetfolia	0.809	1.236
E. tereticornis	0.906	1.103
M. azedarach	0.869	1.150
P. Juliflora	0.869	1.150
T. arjuna	0.557	1.795

Comments

The trees in the experiment were grown in lysimeters of 1.45 m diameter which put a constraint on their growth. Trees growing in the field without such constraints would show much better growth. The experiment was done on trees of 5-year age. They had not yet attained their full potential growth. In the field experiment *E. tereticornis* trees attained a height of about 10–15 m and girth at breast height of 50–60 cm at the age of 6 years. The capacity of trees to biodrain water is dependent on their growth, therefore, the experimental results underestimate their ultimate biodrainage potential. Even so, the finding that W_b exceeds E_p or A_{pan} for all species of trees and average observed value of 46 Ld^{-1} for *C. equisetifolia* and *E. tereticornis* of 5-year age, support the feasibility of biodrainage as a remedial measure of waterlogging.

Field Experiment in Waterlogged Land

The experiment was conducted in 12 ha waterlogged land which had been abandoned from agriculture use.

In January 2005, twelve parallel embankments were made each 2.90 m broad at base 2.60 m broad at top and 0.45 m in height as shown in Figures 7.3 and 7.4. Each embankment was about 150 m long and 60 m apart from each other.

In August–September 2005 saplings of 11 tree species were planted on the embankments. These were: *M. azaderach, E. teriticornis c–10, T. arjuna, P. perinata* on bunds 1 to 4 in that sequential order. There was no planting on embankment No. 5 (control). *T. aphylla, T. lanceolatus, E.tetriticornis c–130, E.tetriticornis c–3, E. hybrid, P. juliflora* and *D. strictus* species of trees were planted on embankments 6 to 12 in that sequential order. Trees were planted in two rows on each embankment while row to row distance was 1.3 m in all cases, plant to plant distance was 1.5 m in all four Eucalyptus clones and 3 m in all other species.

Observation wells were installed to measure water table fluctuations at 0, 5, 10, 15 and 20 m (eastwards) from the middle of each bund as shown in Figure 7.4.

In November 2009, *Eucalyptus* species attained a height of 14–15 m with *Eucalyptus* c–130 being the tallest. *P. juliflora, T. aphylla and M. azedarach* were in the 6–7 m range and others were low showing slow mode of growth.

In November 2009, *Eucalyptus* species attained a girth at Breast Height (GBH) in the range 49–61 cm with *Eucalyptus* c –130 showing the best results. *Prosopis Juliflora* attained GBH of 47 cm, *Tamarix aphylla* of 45 cm, *M. azadirach* of 38 cm, *Pongmia pinnata* of 33 cm, *T. arjuna* of 31 cm, and *Callistemon lanceolatus* of 22 cm.

All four *Eucalyptus* species, *C. lanceolatus*, and *T. aphylla* were 'evergreen' and showed minimal decline in leaf area in winter months. *M.azedarach* was completely deciduous in winter. *T. arjuna* showed low leaf area values during the month of April. *P.*

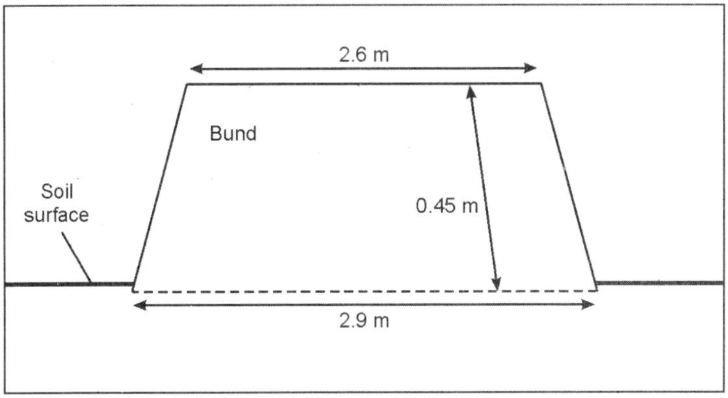

➲ **Figure 7.3** *Diagramatic cross-section representation of a field bund.*

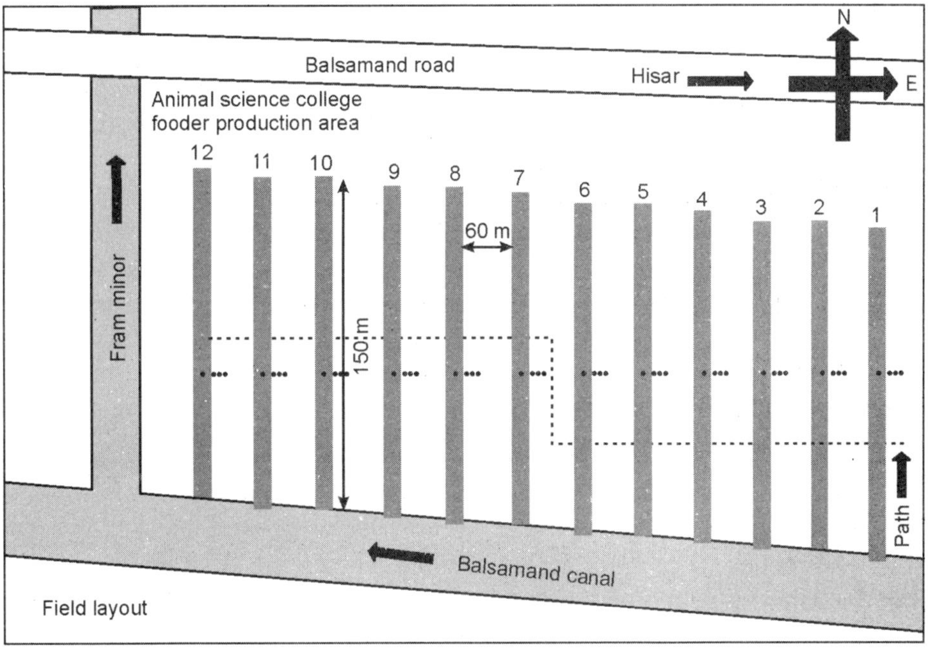

➲ **Figure 7.4** *Field layout of biodrainage plantations. Bars denoted with numbers are bunds 60 m apart on which strip plantations of 1. M. azaderach, 2. E. teriticornis C-10 3. T. arjuna, 4. P. pinnata, 5. Control (blank bund), 6. T. aphylla, 7. C. lanceolatus, 8. E. teriticornis C-130, 9. E. teriticornis C-3, 10. E. hybrid, 11. P. juliflora and 12. D. strictus and C. equisatifolia (failed plantations) were raised. Dark black indicate position of the observation wells. Figure not to scale.*

juliflora exhibited high leaf area from April to August and very low in winter months of December and January.

The four *Eucalyptus* clones and *T. aphylla* had very low (below 5%) mortality. *M. azedarach, P. pinnata, P.juliflora, T. arjuna* had mortality rates varying from 11–20 per cent. Bamboo species *D. strictus* showed near total failure (95%) because of soil salinity (2–8 dSm^{-1})

C. lanceolatus, the four *Eucalyptus* clones, *P. juliflora* and *T. aphylla* exhibited remarkable tolerance to waterlogging. Tree species most susceptible to waterlogging were *M. azedarach* and *T. arjuna* with survival rates of 40 and 68 per cent respectively.

Up to the 3rd year there was no significant change in the overall water table, which started to decline from the 4th year. In the 6th year the water table was lowered to an extent of about 25 cm. There was a mixture of ten tree species and the decline could have been more if a single fast biodraining tree like *E. hybrid* had been planted in all the strips.

Comment

Although strip plantation of trees had shown some lowering of groundwater table, it was not enough to reclaim the land for agriculture by overcoming the prevailing waterlogging condition. The reasons are obvious. The strip plantations cover only about 5 per cent of the field area. Many of the strips were planted with tree species that are poor biodrainers. With growth of trees beyond 6th year, better results are expected. The experimental area is flanked on two sides by irrigation canals. It seems leakage/seepage from them is the cause of waterlogging. Reducing recharge to the groundwater table due to leakage/seepage from canals should also help in quicker reclamation of waterlogged fields.

Field Experiments on Farm Land

Eucalyptus tereticornis trees were planted in year 2001 in two rowed strips (Figure 7.5 and 7.6) about 130 m long on five embankments with tree to tree distance of 1.50 m and row to row distance of 1.60 m. Seventeen observation wells were installed in east (P 17) to west (P 1) direction to monitor the water table beneath the biodrainage plantations as well as in the adjacent field both towards east and west where there were no tree plantations.

In November, 2007 the *E. tereticornis* strip plantations attained tree like proportions with tree height about 10–15 m and girth at breast height (GBH) in the range of 50–60 cm. The canopy had considerable expanse.

The mean groundwater level beneath the biodrainage tree plantation area was about 40–50 cm lower than that in the surrounding fields on yearly average. It was observed from profiles of salt distribution in underneath soils that there was no salt accumulation in soils beneath the six year old plantations.

The impact on yield of wheat and paddy crops in the adjacent fields due to shading impact of *E. tereticornis* tree plantations was assessed. The paddy and wheat yields were

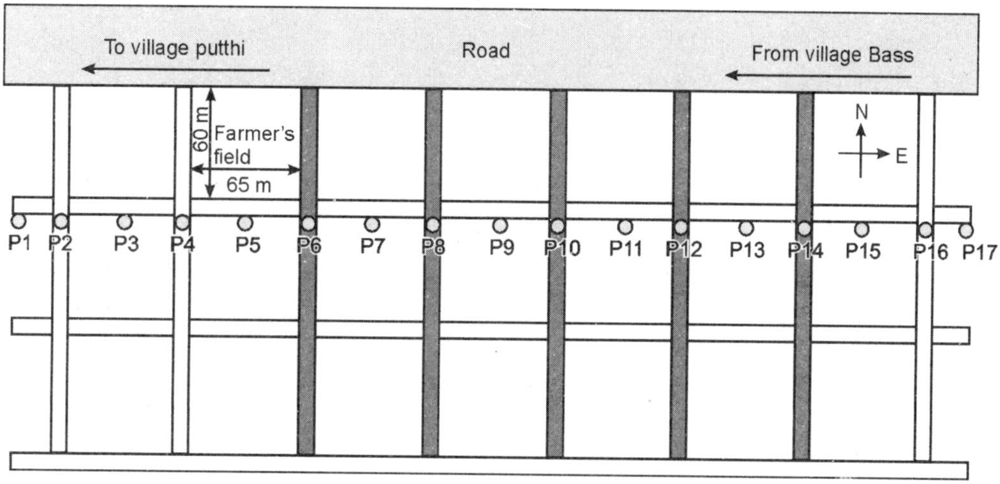

Figure 7.5 *Diagrammatic layout of the farmer's field experiment. Grey strips indicate Eucalyptus tereticornis biodrainage plantations on field bunds. Blank strips are fallow bunds. P1 to P17 denote location of observation wells in the west to east direction.*

adversely affected up to 7–8 m from the tree strips. In case of paddy the yield declined by 55 per cent at 0–1 m from the tree strip. There was no adverse impact on yields beyond 8 m distance.

The mean ECe of soil profile up to 180 cm depth below the plantation bunds was 3.99 as compared to 3.82 underneath bunds without tree plantations suggesting there was little accumulation of salts in the soil underneath the tree plantations.

STUDY BY THE WATER TECHNOLOGY CENTRE, BHUBANESHWAR (ORISSA)

Waterlogging Problem in Deltaic Regions

The Water Technology Centre for Eastern Region, Bhubaneshwar, India conducted a study during the period 2004–2009 on use of biodrainage vegetation to control water-logged situation caused by (a) accumulation of water in topographic depressions and (b) sea water intrusion in deltaic areas. Two sites in coastal deltaic Orissa with problem of waterlogging due to sea water intrusion and two other sites with waterlogging problem due to topographical depression were selected for the experiments.

Casuaina equisetifolia was selected for cultivation in waterlogged areas under influence of sea water intrusion. Seedlings of *casuaina equisetifolia* were raised on nursery beds. Seeds sown in the month of November germinated within ten days and seedlings attained a height of 10–15 cm within 45 days. At this stage seedlings were taken out and placed in polythene bags. After another 30 days they attained a height of about 40 cm and were ready for transplantation.

Figure 7.6 *(a) A view of the Eucalyptus tereticornis strip plantation in north to south direction raised at the farmer's field near village Putthi. Note the wheat crop on east and west sides of the plantations. (b) A close up view in between the strip plantation. The two rows of Eucalyptus tereticornis trees were seen. Also note the harvesting of the wheat crop on east and west of the strip.*

Superior clones of Eucalyptus were developed through survey and collection. Eight to ten leaved seedlings 25–30 cm height were raised in moulded polythene trays.

The texture of soil at experimental sites varied between silty loam to loamy sand. The water use by the *Eucalyptus* and *Casuarina* plantation was determined with the help of a sap flow system based on the principle of thermal dissipation. The probe consisted of two needles, one corresponding a (–) Reference T-type thermo-couple another one incorporating a (+) T-type thermocouple and a heating unit. Heated upper needle temperature is compared to lower ambient temperature needle (dT). Maximum dT occurs when the needle is hottest representing situation of no flow and correspondingly minimum dT occurs when the needle is coolest suggesting high flow. Measurement of apparent sap velocity through tree trunk was recorded using four 30 and 50 mm TDP thermocouple sensors each attached to a heater. The sap flow rate was recorded at 30 minutes interval with the help of a multi channel data logger. The sap flow velocity was multiplied with sap wood area of respective tree to arrive at the total quantum of flow per tree on daily basis.

(a) Water Stagnating Areas

Eucalyptus were planted at Alisibindha and Ambapada at spacing of 4 m × 4 m (625 trees per hectare) where above ground stagnation of water condition continued for more than 3–4 months in a year. Even through the *Eucalyptus plants* could survive waterlogged environment, the extended period of waterlogged conditions retarded the growth of the plantation. In 26–27 months old vegetation at Alishibindha, rate of transpiration recorded was 4.90 m mol $m^{-2}s^{-1}$ and in 39 month old plantation it was 6.24 m mol $m^{-2}s^{-1}$. The sap flow measurements showed average sap flow of 22.19 litres day^{-1} $plant^{-1}$. However, due to prolonged period of waterlogging at Ambapada, the average sap flow was only 7.62 litres day^{-1} $plant^{-1}$.Maximum total annual evapotranspiration of the *Eucalyptus* plantation including that of intercrop, at Alishibindha was 2146 mm during the year 2007–08 and 2306 mm during 2008–09.

The progressive drawdown of water level at phreatic surface monitored through shallow observation bores showed significant lowering in successive years. The mean of first two years (2004–05 and 2005–06) lowest water level at Alishibindha was – 168.5 cm which declined to – 185.33 cm (mean of last three years, i.e., 2006–07, 2007–08 and 2008–09). Similarly at Ambapada decline was from –150 cm to 200 cm. Therefore, there was a clear draw down in level of water table underneath biodrainage vegetation. When the drawdown in water level in observation bore wells underneath biodrainage vegetation was compared with that of observation wells in non vegetated area 300 m away from the experimental site, the effect of draw down was more pronounced at all the four sites.

(b) Sea Water Intrusion Areas

In the Casuarina vegetation (tree spacing 4 m × 4 m) the average rate of transpiration measured with a steady state porometer (M/S PP System USA) after 27 months of planting at Baghadi and Patna was 5.20 and 5.10 m mol $m^{-2}s^{-1}$ respectively. In the study at Patna the soil EC (1:2) ranged between 1.18 to 2.15 ds m^{-1} and EC of bore well water was up to 2.5 dsm^{-1}. The rate of transpiration after 39 months of planting was 4.95 m mol m^{-2} sec^{-1} which was less than that at Baghadhi (5.68 m mol m^{-2} sec^{-1}). The impact of salinity on *Casuarina* vegetation was increasingly manifested at later stage of plant growth. Leaf area per plant also showed concomitant decline at higher salinity level. The total leaf area per plant was 30–34 per cent less at Patna as compared to Baghadi. This was reflected in quantum of total transpiration per plant. On an average the *Casuarina* plant at Baghadi showed total transpiration of 16.5 litres day^{-1} $plant^{-1}$ whereas at Patna it was 7.89 litres day^{-1} $plant^{-1}$. The highest total annual evapotranspiration of the *Casuarina* plantation and intercrops calculated from pan evaporation data at Baghadi was 1280 mm during 2007–08 and 1507 mm during 2008–09.

In the deltaic Orissa at the experimental sites the annual rainfall is more than 1400 mm. As a result large areas remain inundated with water standing above ground for a period of 10–12 weeks. Under such conditions, primary objective of biodrainge would be to bring the fields to cultivable condition as soon as possible after the rainy season, to enable growing of post monsoon crops ('rabi' crop). At all the experimental sites the accelerated draw down of water table due to biodrainge by tree plantation enabled advancing of winter cultivation by about 2–3 weeks earlier than in the fields without biodrainage vegetation. This established that in real time scale, quicker establishment and better growth of vegetation is a critical determinant for providing biodrainage.

Rice was taken as intercrop along with *Casuarina* at Baghadi and Patna. The final yield in (t/ha) of the paddy obtained was 1.73 t/ha in 2005–06, 2.20 t/ha in 2006–07 (var. kanchan); 1.5 t/ha in 2007–08 and 1.6 t/ha in 2008–09 (var. Pauli). At Patna, where level of salinity was more, the summer paddy yield improved from 0.89 in 2004–05 to 1.53 t/ha in 2005–06.

Similarly watermelon crops were successfully raised as intercrop inside *casuarina* plantation.

In *Eucalpytus* plantation fields at Alisibindha, rice was taken as intercrop. The yield of paddy declined progressively during the period 2005–06 to 2008–09 from 3.51 t/ha to 2.12 t/ha due to adverse impact of shading and release of allelochemicals. However, intercrop of groundnut in winter was satisfactory as compared to that in nearby farmers fields. Green grain cultivation as intercrop under *Eucalyptus* plantations exhibited declining trend.

Overall, under both *Casuarina* and *Eucalyptus* plantations, most intercrops gave a benefit cost ratio of 2.0 or more to the farmers.

A clear-cut draw down of groundwater up to 109 cm underneath *casuarina* and up to 115 cm beneath *Eucalyptus* vegetation compared to non-vegetated area (without biodrainage vegetation) at the end of the season underneath four year old vegetation shows the success of biodrainage. Farmers in the area acknowledge the positive impact of biodrainage.

STUDY BY THE TAMIL NADU AGRICULTURAL UNIVERSITY

The Research Study was done during the period 2005–2009 on genotypes of *Eucalyptus tereticornis* and *E. Camaldulennis* and *E. saligna* (obtained from Australia).

Selection of Candidate and Candidate Plus Trees

Sample trees were selected from nine different sources. A sample plot size of 10 m × 8 m was placed in the plantation, likewise three sample plots were placed randomly in a population. One plot accommodated 20 trees and in all the trees height, diameter at breast height (DBH), volume up to DBH and total volume were estimated. The physiological characters of transpiration, photosynthesis, stomatal conductance and inter cellular Co_2 concentration (CINT) were measured for all trees in each sample plot.

Most plantation sources subjected for selection are in coastal areas of Tamil Nadu, India in very high water table area. In the first phase candidate trees (CTs) were selected for which the entire population was perambulated to study the status of tree growth in relation to the variation in topographic and edaphic characteristics. Based on Gardne (1961) recommendations, the entire population was divided into grids to check the soil heterogeneity. In each grid, phenotypically superior trees were selected.

Physiological parameters of transpiration, stomatal conductance, photosynthesis and internal CO_2 concentration (CINT) were measured using portable photosynthesis system CID, Inc (made in USA). The growth parameters of height, DBH, volume up to DBH and total volume were measured and estimated.

Selection index criteria was established and from the candidate trees (CT), those showing best results were designated as candidate plus trees (CPTs).

Multiplication and Field Plantation

The nodal cuttings of candidate plus trees (CPTs) were mass multiplied in propagation chambers. The rooted cuttings were transplanted to polybags and hardened in three phases, first inside the hardening chamber under double layer shade net and then under single layer shade and finally the clones were moved outside the hardening chamber to expose them to full sunlight. After three months of hardening in full sunlight, the clones were planted in the clonal bank. 32 clones were successfully multiplied through vegetative propagation. All the 32 clones were raised in three replications with 6 ramets in each replication in the clone bank. All clones were planted in the clonal bank with a spacing of 1 m × 1 m.

The ramets of all the clones were planted in field during August 2006. Two fields were laid out in completely Randomised Block Design (CRBD) with three replications and a spacing of 3 m × 2 m. The biometrical and ecophysiological observations were recorded at three monthly intervals to evaluate the clones for biodrainage efficiency.

Statistical Analysis

The data gathered from the field experiments were subjected to statistical analysis. Estimates of mean, coefficient of variation and standard error were worked out using the method described by Panse and Sukhatme (1978). The significance test was carried out by referring to the standard 'F' table described by Snedcor (1961).

Out of 35 clones tested under waterlogged conditions some clones showed very good performance, some moderate and some very poor performance. This highlights the importance of identifying appropriate clones that would be suitable for biodrainage under prevailing conditions at each project site.

The Study recommends that biodrainage technique offers greater scope to alleviate waterlogging problems at cheaper costs as against engineering drainage systems.

Transpiration Rates

The reported mean transpiration rates at five locations of Methupalayam, Nager coil, Tiruchandur, ITC and Ariyar ranged from 16.45 to 19.62 m mol $m^{-2}s^{-1}$ while those at locations of Ramanathapuram, Sivagangai and Pudu Kothai ranged between 9.03 to 10.74 m.mol $m^{-2}s^{-1}$. (Conversion factor: 15.42 m.mol H_2O $m^{-2}s^{-1}$ = 1 kg H_2O $m^{-2}hr^{-1}$)

▼ STUDY BY THE ARID FOREST RESEARCH INSTITUTE, JODHPUR

Arid Forest Research Institute, Jodhpur (India) carried a study of characteristic features pertaining to biodrainage potential of some selected tree species (2004–2010). Three experiments were conducted.

Lysimeter Study

In the first experiment lysimeters (2 m × 2 m × 2 m depth) were used to create waterlogging at 50 cm and 100 cm soil depth. Seedlings of *A.nilotica*, *E. Camaldulensis* and *T. aphylla* were planted in the lysimeters in September 2007. Water with different salinity level (S_1: no salinity, S_2: 7–12 dsm^{-1}, S_3: 12–24 dsm^{-1}) was used. At the age of 19 months salinity levels were increased to 12 and 24 dsm^{-1} in S_2 and S_3 respectively.

Observed growth of 19 month old plants under different levels of waterlogging and salinity was as shown in Table 7.1

Table 7.1 *Growth (cms) of 19 month old plants as influenced by different levels of waterlogging and salinity*

Treatments	*E. camaldulensis*			*A. nilotica*			*T. aphylla*		
	S_1	S_2	S_3	S_1	S_2	S_3	S_1	S_2	S_3
Height									
W1	393	422	336	339	310	309	294	306	218
W2	605	447	407	395	311	347	407	363	332
W3	620	561	411	375	262	267	406	387	316
Crown diameter									
W1	149	145	147	253	246	242	189	165	180
W2	216	177	146	439	265	267	263	170	223
W3	243	168	160	435	260	235	206	153	198
Collar growth									
W1	6.5	5.7	4.3	5.6	5.9	5.1	5.4	4.6	4.5
W2	10.2	7.5	6.8	9.4	5.6	6.2	6.2	5.8	5.4
W3	10.3	8.1	6.5	8.1	6.4	4.5	6.8	5.8	5.5

1. Salinity level: S_1: Normal water; S_2: 7dSm^{-1};S_3: 12 dSm^{-1}
2. Waterlogging: W_1: watering at field capacity once in a week; W_2: at 100 cm soil depth; W_3: at 50 cm soil depth.

Average water use (in litres per day) by different tree species was as shown in Table 7.2

Table 7.2 *Average water use (litre/day) by different tree species plants under the influence of varied levels of waterlogging and salinity in lysimeters*

Month/treatment	W2S1	W2S2	W2S3	W3S1	W3S2	W3S3
Ecalyptus camaldulensis						
May–09	33	23	12	38	23	37
June–09	37	13	11	33	23	31
July–09	32	10	11	29	17	24
August–09	31	12	10	29	25	20
September–09	29	11	9	37	18	19
October–09	44	10	9	29	18	19
November–09	31	7	6	27	15	18
December–09	39	7	6	26	13	17
January–10	40	6	5	33	14	17
February–10	51	6	6	31	19	20
March–10	61	6	5	32	21	20

Month/treatment	*W2S1*	*W2S2*	*W2S3*	*W3S1*	*W3S2*	*W3S3*
Acacia nilotica						
May–09	30	8	14	46	28	18
June–09	24	12	10	50	26	17
July–09	20	14	10	53	21	15
August–09	30	18	11	61	22	13
September–09	36	18	11	65	21	14
October–09	29	10	8	56	21	14
November–09	32	9	6	63	25	13
December–09	31	17	10	51	19	16
January–10	35	16	14	49	19	19
February–10	38	20	13	62	22	23
March–10	38	16	13	56	20	23
Tamarix aphylla						
May–09	35	28	12	25	37	16
June–09	36	25	12	20	30	13
July–09	30	21	11	17	26	11
August–09	23	20	10	14	27	12
September–09	25	17	9	16	26	11
October–09	25	18	8	14	26	11
November–09	20	21	6	11	23	8
December–09	17	20	7	15	20	6
January–10	12	16	6	18	19	5
February–10	21	17	6	9	19	6
March–10	25	17	6	19	20	5

Plant age: From the age of 20 month to 30 month

Waterlogging: W_1 watering at field capacity once in a week

W_2 at 100 cm. soil depth

W_3 at 50 cm. soil depth

Salinity level: S_1, Normal water

S_2: 12 dSm^{-1}

S_3: 24 dSm^{-1}

E.Camaldulensis exhibited 39 per cent, 26 per cent and 43.5 per cent reduction in plant height, diameter and collar girth respectively under W_3S_3 conditions as compared to that under W_1S_1 conditions. *Eucalyptus camaldulensis* plants in W_3S_3 treatment started wilting permanently at the age of 2 years.

Field Experiment at Anupgarh Branch Canal IGNP

Eucalyptus of different species had been planted along the canal. Transpiration and photosynthesis rates were recorded using ADC make and Lci portable photosynthesis system. There were ten sample plots, 6 in plantation raised in the year 2004 and 4 in plantations raised in the year 2003. Water table at the site of tree plantations was at 3 to 4 m depth. The growth of *Eucalyptus Camaldulensis* trees was as showing in Table 7.3.

Table 7.3 *Growth of Eucalyptus camaldulensis at Anupgarh shakha, IGNP*

Plot number	Height (m)	Clear bole (m)	GBH (cm)	Crown dia. (m)
Three and half year old				
1	13.2	4.0	49.1	3.5
2	12.0	3.8	50.7	3.3
3	12.5	4.8	49.3	3.2
4	11.0	4.4	46.7	3.7
5	13.0	5.4	40.1	2.6
6	16.1	5.1	42.8	2.8
Mean	13.0	4.6	46.5	3.2
Four and half year old				
1	16.7	5.0	53.6	5.6
2	17.6	6.3	56.7	5.0
3	19.3	5.9	56.2	4.8
4	17.0	4.8	47.9	4.9
Mean	17.7	5.5	53.6	5.1

Field Experiment at RD 1357, IGNP Main Canal

The growth parameters of four and half year old plants, measured in February 2008 were as shown in Table 7.4.

Table 7.4 *Growth of four and half year old plants under waterlogged condition at 1357 RD, IGNP*

Species	Height (cm)	Crown dia (cm)	GBH (cm)
Eucalyptis rudis	1386	412	56
E. Camaldulennis	1250	232	27
E. fastigata	1201	214	29
Corymbia tessellaris	1060	237	31

The rate of transpiration was high in October and low in January for all species. The observed values of transpiration rates were as shown in Table 7.5

Table 7.5 *Temporal variation in transpiration recorded in different species*

Species	*April*	*July*	*October*	*January*
E. Camaldulennis	2.04	5.03	6.36	1.58
E. fastigata	1.94	4.13	7.69	1.56
E. rudis	4.47	4.39	4.81	3.11
C. tesselaris	3.35	4.56	5.27	3.03

Transpiration in m mol $H_2O\ m^{-2}s^{-1}$

Higher biomass in *E. rudis* and proportionately higher allocation to leaf and branches as compared to *E. camaldulensis, E. fastigata* and *C. tessellaries* indicated relatively high adaptability of *E. rudis* under perennial waterlogging condition. Higher water use in *E. rudis* is also supported by the fact that the species exhibited more steady rate of transpiration throughout the year in comparison to other species.

Seepage water from the canal as well as runoff water from the nearby fields had inundated an area of 3 ha. with average water depth of 15–25 cm. Raised embankments (60 cm. high, 60 cm wide and 2 m apart) were prepared in waterlogged area to provide comfortable root zone for young seedlings, groundwater level receded by 145 cm (from stagnant water of 20 cm to a depth of 125 cm.) in *E. rudis* plot. In *C. tessellaris, E. Camaldulensis*, and *E. fastigata* water level receded by 90 cm., 70 cm and 60 cm. respectively.

The soil at the experimental site is deep sandy in nature with pH ranging from 8.4 to 9.1, electric conductivity (EC) ranging between 0.25 to 2.30 dsm^{-1}.

STUDY BY THE TROPICAL FOREST RESEARCH INSTITUTE, JABALPUR

The Tropical Forest Research Institute, Jabalpur conducted a study (2012) to identify suitable tree species and other vegetation for biodrainage in Bargi command area. One experiment was on the field over 10 ha. area and the other by using forty leak proof lysimetric tanks of 2m × 2 m × 2m size in which water levels were maintained at 0 to 0.25 m, 0.25–0.5 m and 0.5 to 0.75 m depth. The mean annual rainfall at experimental site was 1345 mm and average measured annual evaporation (A_{pan}) was 1660 mm. The water quality was very good with total dissolved solids 200 mg/l and salinity 0.37 dSm^{-1}.

Lysimeter Study

One year old seedling of *Eucalyptus hybrid, Pongamia pinnata, Dalbergia sisoo, Acacia Nilotica, Albizia procera, Albizia lebbeck* and *Terminatia arjuna* were planted in lysimetric tanks in July 2008. The increase in height of *Eucalyptus* was in the order: 0.25 m – 0.50 > 0.50 – 0.75 m > 0 – 0.25m > control tank while growth of breast height (GBH) was in the order: 0.50 m – 0.75 m > 0 – 0.25 m > 0.25 m > 0.50 m > control. It was concluded that the trees had greater rate of growth in waterlogged conditions in comparison to controlled conditions and consumed more water than any other tree species and agricultural crop.

Transpiration Rates

Transpiration rates were measured using CID make CO_2 gas analyser (Model 340). Species vise rate of transpiration was found maximum in *E.hybrid*, followed by *P.pinnata* and *T.arjuna*. Minimum transpiration rate was found in *A.nilotica*. Peak transpiration rates were observed when water was under 0–0.25 m water regime. Average water use (litres per day) by tree water species was as showing in Table 7.6.

Table 7.6 *Average water use (litre/day) by tree species under different water regimes in lysimeters*

Month	Water level (m)		
	0–0.25	0.25–0.50	0.50–0.75
Eucalyptus hybrid			
April 2010	54	51	47
May 2010	64	61	58
June 2010	61	56	56
July 2010	31	31	29
August 2010	32	34	27
September 2010	29	30	30
October 2010	36	32	31
November 2010	39	35	32
December 2010	31	35	40
January 2011	48	48	40
February 2011	55	43	39
March 2011	57	55	49
Pongamia pinnata			
April 2010	51	47	42
May 2010	60	56	48
June 2010	56	52	54
July 2010	34	30	29
August 2010	31	31	23
September 2010	28	29	22
October 2010	31	29	28
November 2010	32	33	31
December 2010	35	33	27
January 2011	44	40	41

Month	Water level (m)		
	0–0.25	0.25–0.50	0.50–0.75
February 2011	42	46	37
March 2011	50	52	48
Dalbergia sissoo			
April 2010	44	46	40
May 2010	46	42	41
June 2010	44	42	36
July 2010	21	25	22
August 2010	23	23	21
September 2010	25	23	20
October 2010	26	26	19
November 2010	30	25	22
December 2010	32	32	31
January 2011	32	25	32
February 2011	35	35	34
March 2011	31	29	24
Acacia nilotica			
April 2010	40	37	33
May 2010	46	46	41
June 2010	46	38	37
July 2010	22	17	20
August 2010	22	17	21
September 2010	25	19	17
October 2010	31	20	21
November 2010	27	28	25
December 2010	25	33	35
January 2011	29	33	25
February 2011	39	35	31
March 2011	31	27	30
Albizia procera			
April 2010	48	42	36
May 2010	52	48	33
June 2010	46	44	36
July 2010	23	23	29
August 2010	26	22	20
September 2010	29	25	25

Month	Water level (m)		
	0–0.25	0.25–0.50	0.50–0.75
October 2010	36	41	31
November 2010	29	27	31
December 2010	36	30	32
January 2011	41	42	41
February 2011	39	42	37
March 2011	42	40	40
Albizia lebbek			
April 2010	52	48	44
May 2010	51	46	41
June 2010	46	46	39
July 2010	27	23	27
August 2010	24	20	23
September 2010	20	17	21
October 2010	39	35	35
November 2010	37	31	31
December 2010	36	36	37
January 2011	38	40	35
February 2011	40	37	35
March 2011	45	44	37
Terminalia arjuna			
April 2010	45	37	41
May 2010	46	42	40
June 2010	40	37	35
July 2010	23	20	21
August 2010	25	19	19
September 2010	21	17	17
October 2010	37	32	29
November 2010	32	31	30
December 2010	31	32	30
January 2011	35	39	37
February 2011	38	41	40
March 2011	38	35	36

Notes: (1) 1 year old seedlings planted in lysimeter in July 2008.
(2) Canal water EC 0.368 dSm^{-1} (200 mg/l)
(3) Average (annual) A_{pan} = 4.55 mm/day
(4) Mean annual rainfall = 1345 mm

For *E. hybrid* the average annual tree water use of 3–4 year old plants was 39 ld^{-1}, for *Pongamia pinnata* 36ld^{-1} and for *Dalbergia sisoo* 28 ld^{-1}, for *Acacia nilotica* 28 ld^{-1}, for *Albizia procera*. 33 ld^{-1}, for *Albizia lebbek* 34 ld^{-1}, for *Terminata arjuna* 31 ld^{-1}.

Field Experiment

In the field experiment plantations of *Pongamania pinnata, Acacia Nilotica, Albizia lebbek, Albizia procera, ailanthus excelsa, Eucalyptus hybrid (Jabalpur), Eucalyptus hybrid* (FRI 4 and FRI 5) were raised along a canal in year 2007. All species except *Eucalyptus hybrid* (FRI 4 and 5) were planted at 2 m × 2 m spacing. *Eucalyptus FRI* 4 and 5 were planted at both 2 m × 2 m and 1 m × 1 m spacing. The growth characteristics of seedlings of *E. hybrid* are shown in Table 7.7 and Figures 7.7 and 7.8

Table 7.7 *Growth characteristics of seedlings of Eucalyptus hybrid planted along left bank canal of Bargi command area*

S.No.	Month	Height (m)	Girth (cm)
1.	Sept. 2007	1.05	1.98
2.	Dec. 2007	1.19	2.59
3.	March 2008	1.30	4.31
4.	June 2008	1.40	5.03
5.	Sept. 2008	1.63	7.02
6.	Dec. 2008	1.75	8.85
7.	March 2009	1.92	9.14
8.	June 2009	2.21	9.79
9.	Sept. 2009	2.26	10.92
10.	Dec. 2009	2.49	12.15
11.	March 2010	2.56	13.97
12.	June 2010	2.62	15.35
13.	Sept. 2010	3.19	17.55
14.	Dec. 2010	3.92	19.18
15.	March 2011	4.01	19.56
16.	June 2011	4.53	22.48

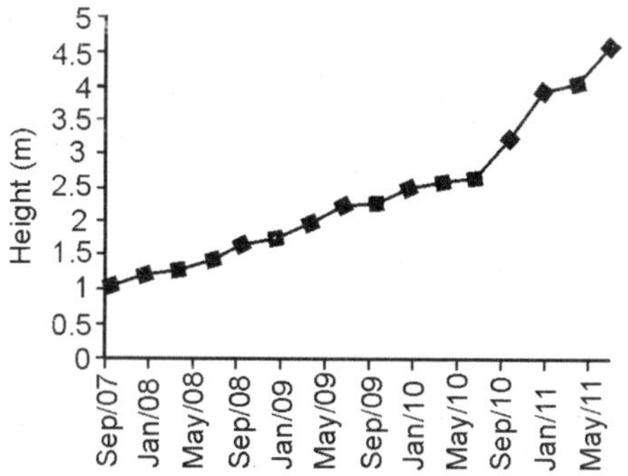

➲ **Figure 7.7** *Height of E. hybrid seedlings*

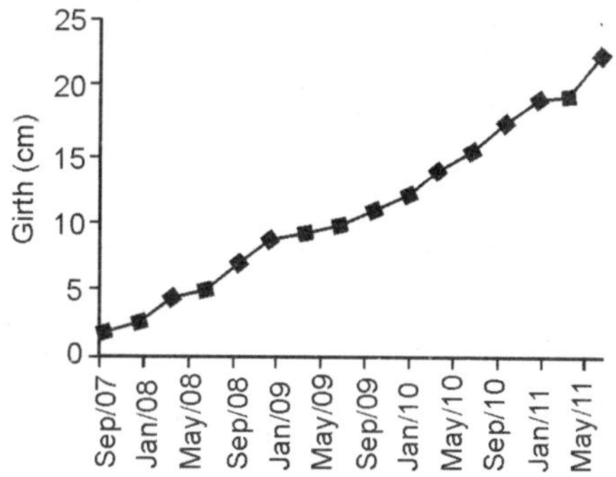

➲ **Figure 7.8** *Girth of E. hybrid seedlings*

Biomass

The above ground dry biomass (kg/plant) of four and half year old plants was 44 kg for *Eucalyptus*, 16 kg for *Pongamia pinnata*, 13 kg for Dalbergia sissoo , 9 kg for *Acacia Nilotica*, 11 kg for *Albiza procera*, 10 kg for *Albizia lebbek* and 12 kg for *Ternimalia arjuna*. The dry biomass of 44 kg *Eucalyptus* per tree in four and a half year translates to an annual produce of about 25 tonnes/ha with 2500 trees on a hectare.

Groundwater Table Movement: Observation wells were installed in the field to measure water table fluctuations. The water table depression was maximum under *E.hybrid* trees. This ranged between 9–29 cm between the months of January to June. During the rainy season (July–October) the water table came up to ground surface. Excessive recharge from canal seepage and irrigation influenced the groundwater table during most months while in the rainy season this coupled with the impact of heavy rainfall totally neutralised the uptake of water by trees and brought the water table to ground surface.

STUDY BY THE AGRICULTURAL RESEARCH STATION OF MAHARANA PRATAP UNIVERSITY OF AGRICULTURE AND TECHNOLOGY, UDAIPUR

The Agricultural Research Station of the Maharana Pratap University of Agricultural and Technology, Udaipur, India conducted a study (2006–2010) on identification and evaluation of potential vegetation for biodrainage in semi-arid environment of Chambal command area of Rajasthan. The soils in the area are clay loams with low water intake rate of about 0.25 cm/hr. on surface and are almost impermeable at 1.20 m depth. The average annual rainfall was 800 mm and evaporation 1800 mm. The range of maximum temperature ranged between 20° (winter) to 45° (summer) and minimum temperature between 5° (winter) to 30° (summer). The groundwater salinity was less than 1.0 dSm^{-1}.

Two experiments were conducted, one to identify local specific effective plant species for biodrainage and other to evaluate effectiveness of different drainage systems.

The plant species tested were *Eucalyptus camaldulensis, E.tereticornis, Delbergia sissoo, Pongamia pinnata* and *Zizipus marautiana*. Maximum plant height (522 cm) and diameter (6.25 cm) of stem was observed for plant species *Eucalyptus tereticornis* followed by *Eucalyptus Camaldulensis* and minimum plant height and stem diameter were recorded for plant species *Zizipus marautiana*. The plant spacing did not produce significant effect on plant height and diameter. Tables 7.8 on and Table 7.9 show the effect of plant spacing on plant height and diameter under shallow water table conditions.

Table 7.8 *Effect of plant species and spacing cn plant height under shallow water table condition.*

Treatments	Mean plant height (cm)								
	Jan. 08	April 08	July 08	Oct. 08	Jan. 09	April 09	July 09	Jan. 10	April 10
A. Plant species									
Eucalyptus camaldulensis	137.92	238.38	313.50	361.00	400.17	435.67	466.33	490.33	509.33
Eucalyptus tereticornis	167.54	246.29	318.50	366.83	411.50	441.50	489.16	501.50	521.83
Delbergia sissoo	86.25	163.54	210.83	252.83	287.17	323.67	357.50	367.33	385.16
Pongamia pinnata	47.13	99.49	126.33	160.50	187.83	195.17	225.00	228.66	238.33
Zizipus marautiana	92.79	96.00	121.67	148.67	174.50	182.83	199.00	207.83	223.33
CD @ 5 %	7.61	7.66	2.60	2.40	8.16	5.86	6.47	2.75	3.31
B. Plant spacing									
(2 × 2) m	109.2	164.00	213.9	255.5	289.2	309.7	342.33	257.06	372.06
(3 × 4) m	103.4	173.00	222.4	260.4	295.3	321.8	352.46	361.20	379.13
CD @ 5 %	NS	4.85	1.64	1.52	5.16	3.70	4.09	1.74	2.09

Table 7.9 *Effect of plant species and spacing on mean plant diameter under shallow water table condition.*

Treatments	Mean diameter (cm)								
	Jan. 08	April 08	July 08	Oct. 08	Jan. 09	April 09	July 09	Jan 10	April 10
A. Plant species									
Eucalyptus camaldulensis	1.55	2.06	3.33	3.75	4.07	4.62	5.19	5.42	6.04
Eucalyptus tereticornis	1.61	2.60	3.62	4.01	4.27	4.86	5.83	5.99	6.25
Delbergia sissoo	1.18	1.55	2.80	3.45	3.88	4.08	4.49	4.64	4.77

Treatments	Mean diameter (cm)								
	Jan. 08	April 08	July 08	Oct. 08	Jan. 09	April 09	July 09	Jan 10	April 10
Pongamia pinnata	1.10	1.40	2.85	3.06	3.10	3.52	4.71	4.88	5.03
Zizus marautiana	1.52	1.80	2.47	2.79	2.87	2.94	3.43	3.57	3.70
CD @ 5 %	0.42	0.09	0.15	0.25	0.35	0.10	0.08	0.06	0.23
B. Plant spacing									
(2 × 2) m	1.5	1.9	2.8	3.3	3.5	3.8	4.67	4.83	5.03
(3 × 4) m	1.3	1.9	3.2	3.6	3.8	4.2	4.80	4.97	5.28
CD @ 5 %	NS	NS	0.10	0.16	0.22	0.06	0.05	0.04	0.15

The depression of groundwater table was greatly influenced by rainfall and seepage from canals. The mean water table was about 20 cm lower under *Eucalyptus* trees as compared to that under *zizipus marautiana* (April, 2010).

In the field study to evaluate effectiveness of different drainage systems, the experimental area was divided into following sub units:

Unit 1 (biodrainage + surface drainage): Consisted of four rows of *Eucalyptus tereticornis* and surface drainage system (field drain, seepage drain and carrier drain) in remaining area.

Unit 2 (surface drainage): Consisted of surface drainage system.

Unit 3 (biodrainage): Comprised of plantation of *Eucalyptus tereticornis* in both recharge and discharge zones. Plantation of *tereticornis* was done in an area of 1.0 ha in recharge zone and 0.2 ha in discharge zone without providing any surface drainage.

Unit 4 (Control): The area under this unit was treated as control plot without any drainage system.

The groundwater table movement was largely controlled by rainfall and seepage from canals and irrigation. In January 2010, at a distance of 300 m away from the canal, the water table depth (cm) for different drainage systems was as follows:

Biodrainage + surface	92.6 cm
Surface drainage	89.7 cm
Biodrainage	68.5 cm
Control	45.6 cm

The results show that in high rainfall areas surface drainage is most important and that surface drainage coupled with biodrainage gives the best results.

STUDY BY THE FOREST DEPARTMENT, HARYANA

The Chief conservator of Forests, Haryana, India undertook a study on the subject of control of waterlogging and salinity through agroforestry interventions. Location and size of the research plot are shown in Figures 7.9 and 7.10.

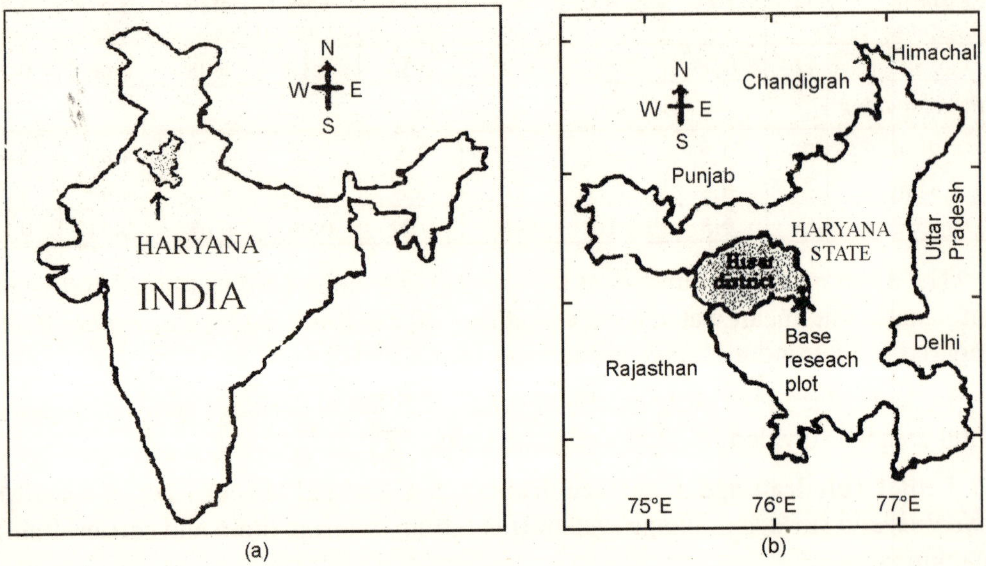

Figure 7.9 *Location of bass research plot*

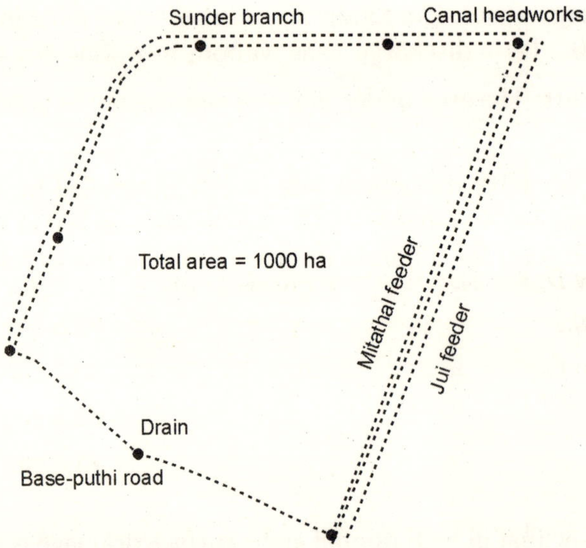

Figure 7.10 *Land-locked area of Bass research plot*

Layout of Experiment

The area of the plot is about 1000 ha, with irrigation canals on three sides and a road on the south side. The area in the North and West sides along the irrigation canal is highly affected by salinity. The experiment was initiated in the year 2008. The soil in the area is sand loam.

The layout of experiment is shown in Figure 7.11.

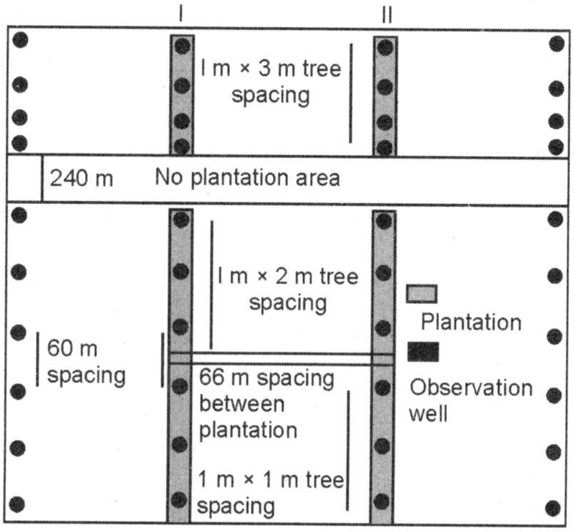

➲ **Figure 7.11** *Layout of the experiment 1*

Growth of Trees

Growth performance of two years old clonal *Eucalyptus* strip plantation was as follows:

Tree Spacing					
1 m × 1 m		*1 m × 2 m*		*1 m × 3 m*	
Height (m)	DBH (m)	Height (m)	DBH (m)	Height (m)	DBH (m)
7.21 ± 1.72	0.72 ± 0.108	8.13 ± 0.98	0.79 ± 0.085	8.60 ± 0.76	0.84 1 ± 0.076

Water Table Draw Down Due to Tree Plantations

There were two strips of tree plantations. Tree spacing of 1 m × 1 m, 1 m × 2 m and 1 m × 3 m represent density of tree plantation of 300 trees ha^{-1}, 150 trees ha^{-1} and 100 trees ha^{-1} respectively. The impact of lowering down of groundwater table due to tree plantations in the months of January, April and September in the year 2009 is shown in Figure 7.12.

The 2 year old trees had significant impact on lowering the groundwater table. It is expected that the tree plantations would have greater impact when they mature fully.

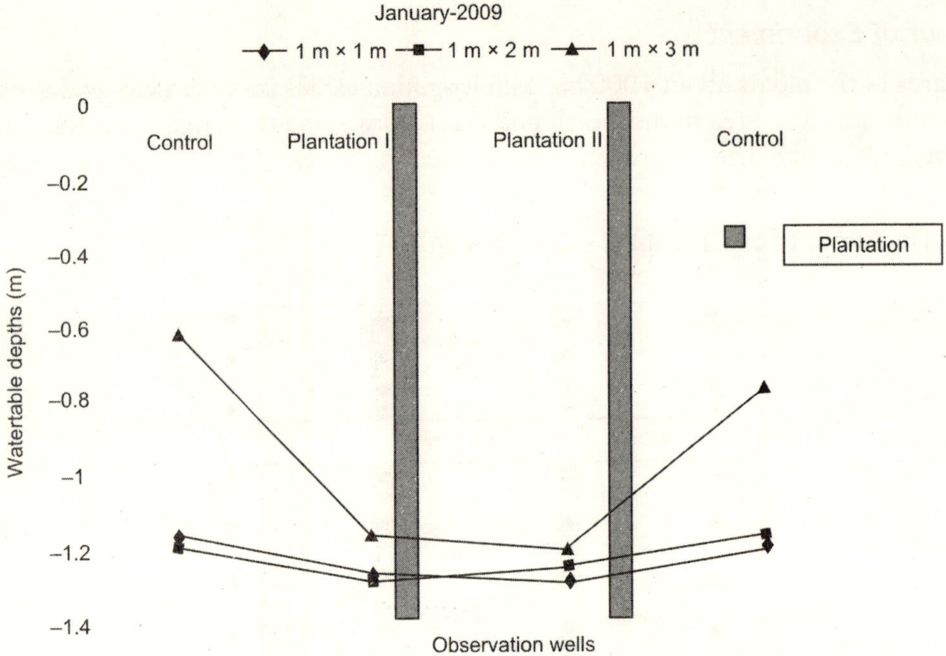

Figure 7.12 *Effect of plantations on water table drawdown*

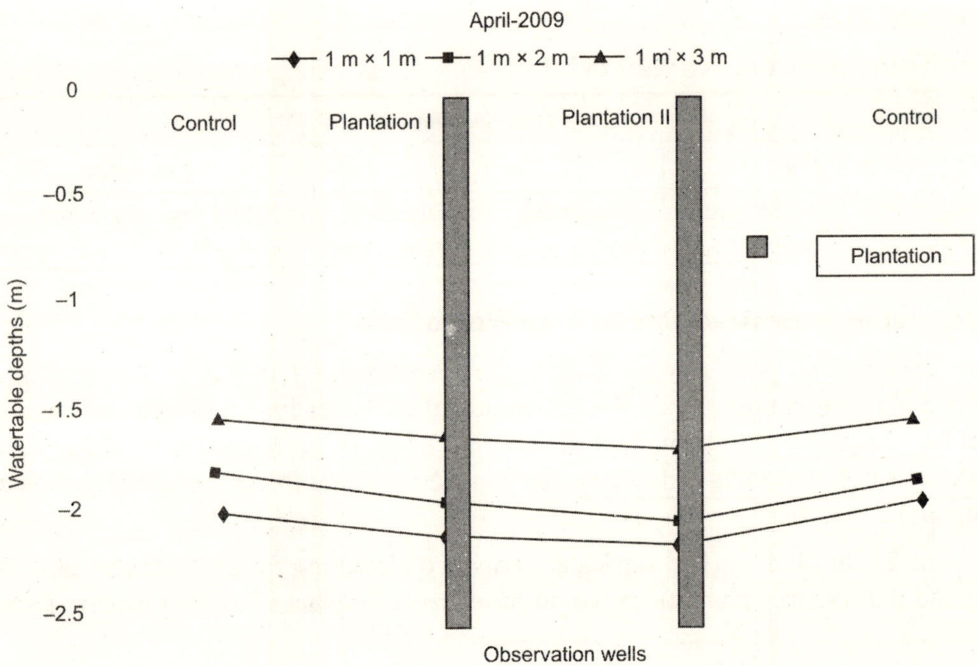

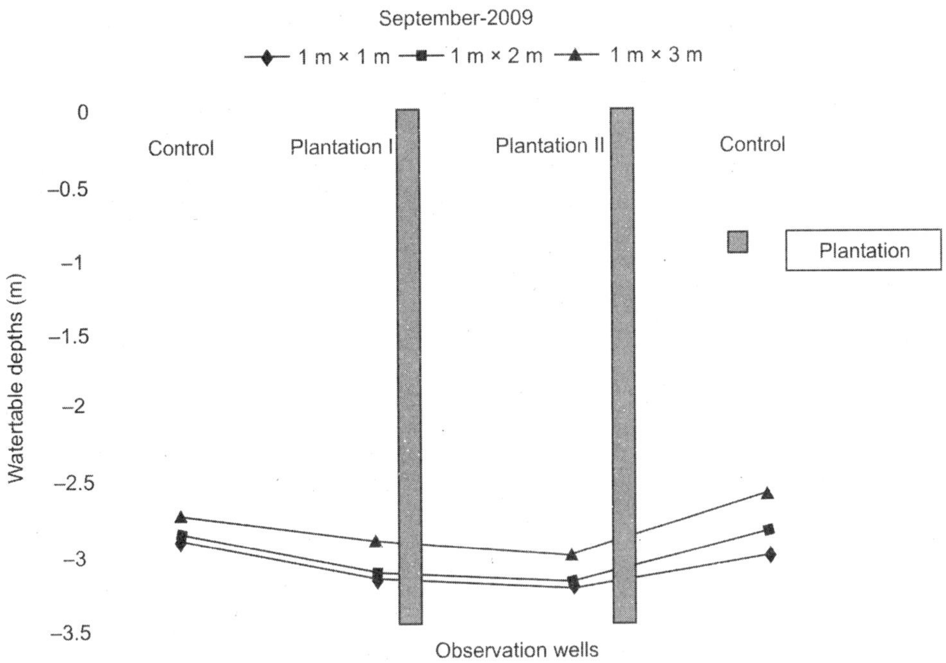

Figure 7.12 *Effect of plantation in different spacing on water table draw down*

REFERENCES

Agricultural Research Station of the Maharana Pratap University of Agriculture and Technology, Udaipur, India (2010). *Identification and Evaluation of Potential Vegetation for Biodrainage in Semiarid Environment of Chambal Command Area of Rajasthan*, Report Submitted to the Indian National Committee on Irrigation and Drainage.

Agricultural University, Hissar, Haryana, India (2010). *Ecophysiology of Tree Species to Evaluate Biodrainage Potential for Waterlogged Soils of Haryana*, Report Submitted to the Indian National Committee on Irrigation and Drainage.

Arid Forest Research Institute, Jodhpur, India (2010). *Study of Characteristic Features Pertaining to Biodrainage Potential of Some Selected Tree Species*, Draft Report Submitted to the Indian National Committee on Irrigation and Drainage, New Delhi.

Forest Department of the State of Haryana (2010). *Control of Waterlogging and Salinity Through Agroforestry Interventions*, Progress Report (up to March 2010), Submitted by the Chief Conservator of Forests to the Indian National Committee on Irrigation and Drainage.

Tamil Nadu Agricultural University, Mettupalayam, India (2012). *Screening Superior Genotypes of Eucalyptus for Biodrainage Through Ecophysiological Approaches*, Draft Completion Report Submitted to the Indian National Committee on Irrigation and Drainage.

Tropical Forest Research Institute, Jabalpur, India (2012). *Identification of Suitable Tree Species and Other Vegetation for Biodrainage in Bargi Command Area*, Report Submitted to the Indian National Committee on Irrigation and Drainage.

Water Technology Centre for Eastern Region (2009). *Fitting in and Evaluation of Biodrainage in Different Agroecological Situations of Deltaic Region of Orissa*, Draft Final, Report Submitted to the Indian National Committee on Irrigation and Drainage.

8

Principles of Biodrainage Planning and Design

The agronomic options for remedial management aim to prevent or mitigate the degradation of soil and water resources. Kijne et al. (1998) report that one of the key agronomic options is crop diversification. Different crops have different evapotranspiration patterns and lead to different rates of recharge of the groundwater. In areas with rising groundwater tables, planting sufficient trees can reduce or even reverse the trend. Varieties and species with higher salt-tolerance can be planted to replace those that yield less when the salt-content of the available irrigation water increases, as for example, when more irrigation supplies are drawn from pumped groundwater.

One of the agronomic options could be to deliberately reduce the area that is cropped each season, to match the demand for good quality water with the available supplies. This could be done by reducing crop intensities while maintaining the total command area, through the introduction of periodic fallow. This can be achieved by the rotation of canal supplies and restriction of the use of pumped groundwater.

METHODS OF DRAINAGE

Drainage enables sustainable agriculture land use. The following methods of drainage are generally practised at present:

(a) Surface drainage;

(b) Subsurface drainage, by means of

  (i) Horizontal drainage; or

  (ii) Vertical drainage; and

(c) Biodrainage.

SURFACE DRAINAGE

There are many regions with rolling topography and steep land gradients, in humid and sub-humid climates, where excess water from the land surface flowing through streams and rivulets finds its way into the natural rivers. Such areas generally occur in the upper regions of river basins. The excess drainage water from the irrigation fields in such areas generally finds its way into the nearby stream. Such lands need very little intervention for drainage and do not face the threat of waterlogging.

Another method of surface drainage, generally practised in paddy cultivation, in lands with sufficient slope, is to drain water from field to field. This method also helps to wash down the pollutants from the land surface.

Some regions in the flood plains are prone to submergence during floods in rivers. When the floods subside, the surface water drains away, but temporary submergence causes waterlogging in some areas for some time. The construction of embankments for rails, roads, and canals sometimes obstructs natural drainage pathways and results in the accumulation of water on the ground surface, which can result in waterlogging. This type of waterlogging is generally temporary and is a direct result of inadequate surface drainage. It can be taken care of by providing suitable surface drainage.

SUBSURFACE DRAINAGE

Horizontal Drainage

In agriculture, drainage is the process of removal of excess water from the soil and wherever it is managed by flow through the soil, it is termed as subsurface drainage. The primary objective of the drainage system is not to permit the rise of the groundwater table above a pre-determined level, which, in general, may be about 2 m below ground level.

The most widely practised method of subsurface drainage is to lay perforated pipes or tile drains in trenches that are sufficiently deep to draw both surplus irrigation water as well as water from the rising water table. The field drains are connected to sub-collector and collector drains which ultimately discharge drainage water into a natural drain or stream or into a reservoir or evaporation pond if the former is not available. The size and spacing of drains depends upon various factors including the rate of recharge, hydraulic conductivity of sub-strata and depth of barrier layers, among others.

Once properly installed and maintained, these drains function quite well in keeping the rise of the groundwater table under control and in sustaining agriculture on the drained fields. But the construction and operation of a separate drainage system to transport salt to the sea is an expensive proposition; Kijne et al. (1998) report that the construction cost of the 'left bank outfall drain' in Sindh, Pakistan, is approximately one billion US dollars, to provide drainage for approximately 0.5 million ha area.

Evaporation basins have been installed in recent years in many countries. There are over 200 evaporation basins ranging from 2 to 2000 hectares in size in the Murray Darling Basin, Australia, alone. They include constructed evaporation basins as well as natural depressions and salt lakes. Some of them are intended to hold saline water temporarily until it is released during high river flows, while others are expected to have operating lives of 50 to 150 years.

The Wakool/Tulakool subsurface drainage scheme in New South Wales, Australia, protects an area of 30,000 hectares, and the evaporation basin covers 2000 hectares, the evaporation basin to total area ratio works out 1:15.

Vertical Drainage

Whenever and wherever surface irrigation is introduced, some water inevitably seeps down into the groundwater reservoir. The quantity of water seeping down depends upon the efficiency of water conveyance and its use, and the method used for practising surface irrigation. The quantity of water going down to the groundwater reservoir as a percentage of diverted water in case of surface irrigation may range between 20–40 per cent.

In general, the available irrigation water is in short supply as compared to the total requirement of irrigation water for the available cultivable land, and it therefore has to be supplied in some regulated manner as a rationed supply. With the passage of time, as the groundwater table rises and if it is of usable quality, the farmers are able to draw groundwater through wells for supplemental irrigation. The use of groundwater in this manner augments the supply of water from canals. It also provides drainage to arrest or slow down the rise of groundwater table. The pumping of water from wells is sometimes called 'vertical drainage'.

Both horizontal and vertical subsurface drainage methods, have some demerits. In horizontal drainage, the major demerit is the high cost of installation, and the pollution of natural stream waters by saline affluent from drainage. In vertical drainage, the groundwater is recycled with no removal of salts from the system and therefore its long-term sustainability is suspect.

▼ BIODRAINAGE

All plants transpire water. The rate of transpiration depends primarily upon the climatic condition, the type and species of plantation, and the availability of soil moisture in the root zone. Agricultural crops consume a major part of the irrigation water by transpiration but the water lost in percolation during field application and that lost through seepage in the conveyance system, goes down into the groundwater reservoir. As and when the water table surface comes up to a sufficient height, and is within the reach of the roots of trees in plantations, the trees start drawing water from the groundwater reservoir through the process of transpiration. This process of withdrawal of groundwater by plantations is termed '***biodrainage***'.

Since plantations, particularly in the dry arid regions, can transpire a large quantity of groundwater, they can be used to control the rise of groundwater table. Plantations also draw salts and minerals from the soil to some extent. Where the irrigation water is of good quality, biodrainage through plantations can help achieve both the water balance as well as the salt balance in the groundwater regime.

For biodrainage to be effectively adaptable, the following requirements must be met:

(a) Water Balance: The quantity of water removed from the groundwater annually should equal the quantity of recharge.

(b) Salt Balance: The quantity of minerals removed annually should be nearly equal to the quantity of mineral import.

(c) Area under Plantation: Irrigation is practised primarily to promote agriculture, horticulture, dairy farming, etc. Therefore in terms of economic returns, afforestation or agro-forestry should be comparable with that from other alternatives uses of land. If not, afforestation may still be justified, on considerations of its environmental and drainage benefits.

(d) Water for Plantations: Under an ideal situation, trees in the afforestation area, on full development, should be able to draw most of their requirement of water from the groundwater table, so that the surface irrigation water can be put to other productive uses. If, this is not possible, plantation trees would need some irrigation water. They may also need some water periodically to leach down salts from the root zone, if and when the salinity levels approach threshold limits.

(e) Groundwater Quality: The quality of groundwater, when the water table approaches the root zone of trees, should be such as can be tolerated by the plant species, otherwise the trees would need to be supplied irrigation water.

(f) Effect on Lowering the Groundwater Table: Trees can lower the groundwater table directly underneath the plantation area, to a depth up to which the tree roots can extend. This can be up to 15 m from the ground surface or even more. To be effective as a drainage measure, the groundwater table must be lowered in the irrigated area to a minimum critical depth (say 2 m below ground level), at the farthest point from the edge of plantation area.

WATER BALANCE

Disturbance in Water Balance on Introduction of Irrigation

Before the introduction of irrigation, the groundwater system is in a state of equilibrium.

The inflows, mostly from natural precipitation, seepage from water bodies and groundwater inflow match the outflow on account of withdrawal of water for agriculture and other uses and the groundwater outflow. Apart from some fluctuations in the water table level from season to season and from dry year to wet year, the groundwater system over a period of time, reaches a state of equilibrium and remains fairly stable. With the advent of irrigation when a large quantity of water is brought from an outside area, the state of equilibrium is disturbed and the groundwater table no longer remains stable. Depending upon the quantity of the net incremental recharge, the groundwater table starts rising and continues to do so until a new balance is reached. As long as the balance is not reached, the water table continues to rise and may come up to the ground surface or rise even higher, thus causing waterlogging. Ultimately, evaporation from the ground surface in the waterlogged area and from the surface of the formed water pools alongwith other withdrawals, strikes a balance with the quantity of recharge. But by this time, large areas may be lost from agricultural use, on account of waterlogging.

To overcome the above problem, any drainage scheme aims to achieve a water balance before the groundwater table rises up to the critical depth, which, in general, may be taken as 2 m below ground level. This would be possible if the annual rate of withdrawal from the groundwater equals or exceeds the rate of recharge, when or before the water table creeps up to the critical depth.

Recharge of Groundwater

The major items contributing to the recharge of the groundwater in an irrigated area are :

(a) infiltration from natural precipitation;

(b) percolation from water bodies;

(c) groundwater inflow from the adjacent area;

(d) seepage from water conveyance system; and

(e) deep percolation loss from irrigation.

Out of the above, infiltration from natural precipitation and percolation from water bodies other than from those that are newly formed, was seen to occur even before the introduction of irrigation and their impact is accounted for in the groundwater balance under pre-irrigation conditions. The rise of the water table under the irrigated area generally leads to a greater ground outflow from such an area towards the un-irrigated adjacent areas, and not an increase in the groundwater inflow. The major factors disturbing the pre-established groundwater balance that cause the rise of the groundwater table are primarily seepage from the water conveyance system and deep percolation losses from irrigation.

The water conveyance loss would depend upon the method of conveyance. In piped system, this can be negligible. In open unlined canals, this would depend upon the permeability of strata through which the canals run. In lined canals, the percolation losses are considerably reduced. While data on the approximate estimates of water losses in different types of conveyance systems are available, experiments and trails under field conditions can generate more dependable data.

Deep percolation losses from field irrigation depend upon the method and efficiency of water use. These are minimal in the sprinkler and drip methods of irrigation. When surface irrigation is practised, these depend upon factors like the infiltration rate of soils, stream size, border strip dimensions, field grade and application efficiency. It is possible to make reasonable estimates of likely losses, which can be confirmed or corrected on the basis of the actual field study and observations.

If R_c be the water recharge due to water conveyance loss and R_p, that due to deep percolation loss from field irrigation in a year, then the total additional annual recharge to the groundwater system would be $R_c + R_p$.

Withdrawal from Groundwater

The withdrawal from a groundwater reservoir can be by:

(a) pumping out water

(b) groundwater outflow, or

(c) drainage measures like biodrainage.

For the present, it is assumed that there is no pumping out of groundwater. In case there is pumping, it can be estimated and accounted for. There would be some increase in the groundwater outflow from areas under irrigation as compared to the situation under pre-irrigation conditions, but this too is ignored for the present.

Transpiration

The withdrawal of groundwater by transpiration from trees depends upon:

(i) climatic conditions;

(ii) plant species;

(iii) density of plantations;

(iv) area under plantations;

(v) water supply; and

(vi) root system;

The effect of these factors on tree water use is described in Chapter 3.

Water Balance

If P be the plantation area, A_{pan}, the annual transpiration from trees, then the total tree water use would be $P \times A_{pan}$ in a year, and if all the water requirement of trees is met from the groundwater, then the annual withdrawal of water by plantations from the groundwater W_b

would be equal to $P \times A_{pan}$. For a stable water balance, the annual withdrawal by plantations should equal the net recharge due to irrigation, that is $W_b = R_c + R_p$.

SALT BALANCE

Source of Salts

The original and, to some extent the direct, source of all salts in the soil are the primary minerals found in soils and in the exposed rocks of the earth's crust as explained in Chapter 2.

Quantity of Salts in Irrigation Water

The amount of dissolved salts in irrigation water is generally expressed as total dissolved solids (*TDS*) in milligrams per litre (mg/l) and as electrical specific conductance, i.e., the conductivity per unit volume (1 cm^3) of saturated solution in siemens(s) per cm. (ds/m = ms/cm = m mhos/cm).

If V (in cubic metres) be the volume of the imported water used for irrigation containing m_w (in mg/l or g/m^3) minerals constituents, the total quantity of minerals brought in by irrigation water would $m_w V \times 10^{-6}$ tonnes.

Minerals in Plant Body Weight

According to Palladin, the bulk of dry weight of a plant is made up of:

Carbon	45.0 per cent
Oxygen	42.0 per cent
Hydrogen	6.5 per cent
Nitrogen	1.5 per cent
Mineral constituents	5.0 per cent
Total:	100.0 per cent

Pandey and Sinha give average chemical composition of plant body by weight as follows:

Carbon	20.0 per cent
Oxygen	62.0 per cent
Hydrogen	10.0 per cent
Nitrogen	3.0 per cent
Mineral constituents	5.0 per cent
Total:	100.0 per cent

It would be reasonable to assume that the mineral constituents in a plant, which are all derived from the soil water, form about 5 per cent of the dry body weight of the plant. In case more dependable data on the mineral content in plants is available for a site under consideration, it would be advisable to use such data. The reported results of the mineral content on plants in several studies are described in chapter 4.

Quantity of Minerals Removed by Crops and Plants

If the total annual utilisable dry biomass produce from agriculture in an irrigated are be A (in tonnes) and that from afforestation over an area P (in ha) at the rate of b tonnes/ha/year be $P \times b$, then the total quantity of minerals removed from the soil by crops and trees would be $(m_c \times A) + (m_p \times b \times P)$, where m_c and m_p are percentage mineral contents in crops and plantations respectively, that are grown in the area.

Balance

In case the annual import of minerals $(m_w V \times 10^{-6})$ exceeds the total annual extraction and removal which is $(m_c \times A) + (m_p \times b \times P)$ the salinity in the soil and / or groundwater would progressively increase. On the other hand, if the annual extraction and removal of minerals exceeds the quantity of annual import, the salinity in soil and/or groundwater should decrease progressively and there would be no threat to the sustainability of irrigated agriculture on account of the increase in soil salinity, if the rise of the water table is kept in control.

It is possible that even though the salt balance may be achieved by accounting for all minerals taken together, it may not be so in respect of all individual elements. The important common cations in water and plants are calcium, sodium, magnesium and potassium. An exercise for determining the salt balance in respect of each individual element can be conducted in the same manner as that for total minerals, by using the respective 'element content values' in place of the 'mineral content values' in the above equations.

In case of an imbalance, the rate at which the mineral (or element) content in groundwater would be expected to rise can be estimated as described in Chapter 2 and the period it is likely to take to reach threshold limits can be forecast. This in turn can help to assess the feasibility and expected lifespan of the proposed measures and to throw light on the additional supplementary steps that need be taken.

Area Under Plantation

It has been described earlier that for a stable water balance, the total annual withdrawal of water W_b $(= P \times A_{pan})$, where P is the area under plantation and A_{pan} is the surface evaporation from standard pan, should equal the total annual recharge R $(= R_c + R_p)$, where R_c is the net annual recharge from the water conveyance system and R_p that in the field during water application. It is assumed that tree water use equals A_{pan}. If not, a more appropriate higher or lower value may be used.

A_{pan} is largely dependent on the climatic conditions and can be determined by simple experiments for any region. The efficiency of the water conveyance system depends upon the method of water conveyance and physical conditions. In long unlined canals in permeable strata, conveyance losses can be very large. In lined canals and in impermeable strata, they are much less. In piped supply systems, the losses may be negligible.

The efficiency of water application in the field can vary widely. In well levelled or graded fields with small border strips or basins, high efficiencies of field application can be achieved. In case of sprinkler or drip methods of irrigation, the field losses can be minimised.

Some water is inevitably lost by evaporation during conveyance and field application. The net recharge to groundwater from a reasonably managed surface irrigation system may range between 20–40 per cent of the total irrigation supply. It would be significantly less with sprinkler and drip irrigation systems and can be higher in poorly managed surface irrigation systems. Let the ratio of net recharge of groundwater to the total irrigation water supply be called Recharge Factor, R_F.

Irrigation water is supplied to crops to meet evapotranspiration requirement (*ET crop*). Following the procedure described in FAO paper No. 24, the Reference Crop Evapotranspiration (ET_0) is generally determined by the modified Penman method. The appropriate value of crop coefficient (*Kc*), is determined on the basis of the crop characteristics, the time of sowing, stage of crop development and climatic conditions. *ETcrop* is then taken equal to $Kc \times ET_0$. The irrigation requirement is decided after allowing for effective precipitation (rainfall) during the crop period. The gross irrigation requirement (I_R) is determined taking into account the water conveyance and field application efficiencies.

All culturable land is not irrigated and put to agricultural use throughout the year. In dry arid regions, where the quantity of available water is limited and the area of available culturable land is relatively very large, the area under irrigation and agriculture, at any point of time, is less than one-half of the whole culturable area. Very often, two crops are raised in a year. The total cropped area, counting area under both winter and summer crops, may generally range between 80 to 150 per cent of the culturable area. In desert areas, lower intensities of irrigation of up to about 60 per cent have been practised. Let the intensity of irrigated agriculture be represented by the factor A_F.

Therefore, if the total culturable area be '*C*' , the annual irrigation water supply wold be $C \times A_F \times I_R$ and the net recharge to the groundwater would be $R_F \times (C \times A_F \times I_R)$.

If the entire quantity of the recharge is to be withdrawn by biodrainage, the requirement of the area under afforestation (P) would be:

$$P = \frac{R_F \times (C \times A_F \times I_R)}{T_R}$$

or

$$\frac{P}{C} = \frac{R_F \times A_F \times I_R}{A_{pan}}$$

if

$$T_R = A_{pan}.$$

where P/C represents the fraction of the culturable area that must be under afforestation;

R_F is the recharge factor, i.e., ratio of the net recharge (of groundwater) to the total irrigation water supply;

A_F is the area intensity factor of irrigated agriculture;

I_R is the gross irrigation requirement; and

A_{pan} is the surface evaporation from a standard pan.

T_R is the transpiration rate of tree plantations.

The position is depicted in Figure 8.1.

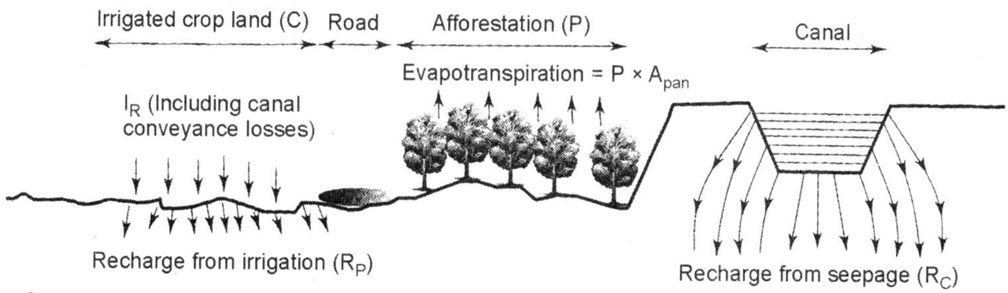

Groundwater table (GWT)

C: Total culturable command area — P: Plantation area

A_f: Annual intensity of irrigated agriculture — A_{pan}: The rate of evaporation from standard pan

I_R: Gross irrigation supply

Annual irrigation water supply = C × AF × IR

Recharge Factor, $R_F = \frac{R_C \times R_P}{I_R}$

Annual recharge of groundwater = $R_F \times C \times A_F \times I_R$

Annual transpiration by biodrainage = $P \times A_{pan}$

For water balance $R_F \times C \times A_F \times I_R = P \times A_{pan}$

or $P/C = \frac{R_F \times A_F \times I_R}{A_{pan}}$

Figure 8.1 *Water balance with biodrainage, schematic presentation*

As an illustration, if $R_F = 0.3$, $A_F = 1.0$, $I_R = 600$ mm and $A_{pan} = 1500$ mm, P/C would be 0.12, that is 12 per cent of the culturable area under afforestation can provide the needed biodrainage.

The plantation area can be along canals, roads, farm boundaries and in blocks of lands suitably scattered and distributed, within the boundary of the irrigated area.

The width and spacing between land strips for afforestation can be decided on the basis of the percentage area that is to be brought under plantations. For example, if 12 per cent area is to be covered with plantations, then the spacing between 50 m wide parallel plantation strips would be about 400 m.

The plantation areas can also be planned in the form of compact blocks in the middle of the irrigation area to provide the required drainage.

According to Morris (1991) within the irrigation areas in Victoria, recharge has been estimated to average in the order of 100 mm/year; inflow from the deep leads is additional to this. Even with our imperfect knowledge of tree water use rates, this rate of recharge seems manageable. For example at a density of 100 trees/ha, average water use of 30 litres/tree/day would remove 100 mm/yr. Based on presented water use and recharge estimates, this might be achievable by the establishment of trees on 5 per cent of farm land in the Shepparton area.

Turner Neil C. (2002) describe that under the conditions prevailing in south west Australia, when native perennial vegetation is cleared for development of agriculture systems, if 16 to 22 per cent area is planted with trees, then trees would use 100–150 mm water more than the annual rainfall and reduce annual deep drainage to 5 mm that was estimated to occur in the pristime state, and this would prevent waterlogging and salinisation of soils.

Water for Plantations

Biodrainage would be quite easy to practise, if agro-forestry yields good economic returns comparable to those from agriculture and horticulture. In such a case, there should be no problem in extending the area under tree plantations to remove the excess groundwater. However, if agro-forestry is not adequately remunerative and has to be practised as a biodrainage or environmental improvement measure, some water would need to be reserved for irrigating the plantations, particularly in the initial stages. To minimise the irrigation water requirement of tree plantations, it is suggested that in the area planned and reserved for afforestation, if the groundwater table is quite deep, such species of trees should be planted initially as would need very little water to grow. Desert trees like *Prosopis cineraria* are known to grow with little irrigation. With the development of irrigated agriculture, the groundwater table would begin to rise and would continue to do so till a water balance is reached. As and when the groundwater table comes up, to say 10 m below the ground level, the low water consuming tree species can be replaced by water hungry species like the *Eucalyptus*, which will take a period of 3–4 years for full growth. They may need some irrigation in the initial years, but as soon as their roots develop and go deep, and are able to draw water from the surface of the rising water table, no further irrigation water need to be supplied. Thereafter, the trees would transpire water almost wholly from the groundwater.

The time of plantation of biodrainage trees can be decided, taking into consideration factors such as the rate of the rise of water table, the species of trees to be planted, their growth period and other local conditions. The objective should be to ensure that the plantations are well-established, before the groundwater table comes up to the critical depth in any part of the area, to cause waterlogging.

Eucalyptus tree plantations in Victoria Australia, *Eucalyptus camaldulensis* and the *Acacia nilotica* species in Rajasthan, India have flourished for a number of years wholly on the basis of groundwater, without any surface irrigation.

There could be an apprehension, that underneath the biodrainage plantations, the salinity in the soil may progressively increase and after some years, reach such high levels as to jeopardise the survival of plantations. This may not happen if the conditions of salt balance, described earlier, are satisfied or it may take a very long period before that stage is reached. Assuming that the soil salinity underneath the plantations would indeed increase progressively, it should be possible to take care of the situation, by leaching down the salts by irrigation water before the salinity levels reach the threshold limit. A few waterings for leaching should be able to reduce the salinity levels significantly from the threshold limits. The leaching operations can be repeated periodically as may be needed. In such cases, some water would have to be allocated for leaching.

The requirement of water for irrigating plantations and for leaching is likely to be a small part of the total requirement of water for agriculture.

Groundwater Quality

In irrigation projects, the quality of irrigation water brought from storages or river diversions is generally quite good. The quality of groundwater is also generally good in humid regions, but in dry arid regions, the groundwater is quite deep and very often highly saline. With the introduction of surface irrigation, the water table rises and the quality of groundwater improves gradually and progressively due to the dilution effect. The percolating irrigation water, being less saline and of lighter density, accumulates over the more saline and dense groundwater underneath.

So long as the water table is more than about 2 m deep, its salinity level does not have any significant effect on agriculture, horticulture etc., receiving surface irrigation. But since the biodrainage plantations meet their water requirements from the groundwater, the quality of groundwater, when it comes up to about 10 m below the ground level, becomes important. The salinity of groundwater at that stage should be less than what the biodrainage plantations can tolerate.

In a desert area in Rajasthan, India, a study over an irrigated area spread river 4,790 sq. km shows that the percentages of area with groundwater salinity level above 8 ds/m, which was 28 per cent in 1983 reduced to 14 per cent in 1996, during a period of 13 years. Similarly the percentage area with salinity levels higher than 4 ds/m and 12 ds/m, came down from 47 to 29 and from 22 to 9 per cent respectively during the same period, as shown in Figure 2.7.

At the stage, when the quantity of drainage matches with the quantity of net recharge to the groundwater, and if the total salts removed equal the total salt brought in by the irrigation water, both the groundwater table levels as well as the salinity levels in groundwater would stabilise.

There are several species of trees that can tolerate high levels of salinity in water of up to about 12 ds/m. With salinity levels in groundwater being higher than the threshold limit, biodrainage by plants would be significantly reduced. Supplemental surface irrigation would be needed till such time that the salinity level of the groundwater does not come down below the threshold limit.

An illustrative example would clarify the position. In IGNP, Rajasthan, India the canal water contains mineral elements (average) as shown in Table 8.1.

Table 8.1 *Mineral Content in mg/l in Irrigation Water*

Sodium Na^{+1}	*Potassium* K^{+1}	*Calcium* Ca^{+2}	*Magnesium* Mg^{+2}	*Sulphur* *S*	*Chlorine* *Cl*	*Total*
12	3	19	9	3	22	68

An analysis of crop plants and afforestation trees on IGNP indicated average mineral element contents as shown in Table 8.2.

Table 8.2 *Mineral Element Content as Percentage of Dry Biomass, by Weight*

Name of crop	*Sodium* Na^{+1}	*Potassium* K^{+1}	*Calcium* Ca^{+2}	*Magnesium* Mg^{+2}	*Sulphur S*	*Chlorine Cl*	*Total*
Crop	0.1	0.8	1.9	0.6	0.1	NA	3.5
Plantation	0.1	0.7	1.7	0.4	0.1	NA	3.0

On the basis of annual water use of 9.3×10^9 cubic metres of water for irrigation, agricultural crops over an area of 1.52 million ha with an average dry biomass (grain and fodder) yield of 6.5 tonnes per hectare and afforestation dry biomass produce of 15 tonnes per ha per year from an area of 291 thousand hectares, the relative position of the annual import of mineral elements with irrigation water and the annual absorption of mineral elements by crops and plants emerges as shown in Table 8.3.

Table 8.3 *Estimated annual mineral elements import and extraction on IGNP in the tonnes*

Import/ Extraction	*Sodium Na^{+1}*	*Potassium K^{+1}*	*Calcium Ca^{+2}*	*Magnesium Mg^{+2}*	*Sulphur S*	*Total*
Import with irrigation water	111.6	27.9	176.7	83.7	27.9	427.8
Extraction by crops and plants	14.2	109.6	261.9	76.7	14.2	476.6

Since the total annual extraction of mineral elements by crops and plants exceeds the total annual import of mineral elements alongwith irrigation water, there should be no overall increase in groundwater salinity due to irrigation water. In case analysis is made on the basis of cation content in biomass, as explained in Chapter 4, the requirement of plantation area would be even less.

The concentrations of Na^+ , $Ca^{++}$and Mg^{++} in irrigation water in milli equivalents per litre are 0.53, 0.97 and 0.77 respectively which give the Sodium Adsorption Ratio (SAR) and Equilibrium Exchange-Sodium-Percentage (ESP) values of less than 1.0. Therefore, the possibility of salinity and sodium (alkali) hazards is also remote.

Kijne et al. (SWIM paper II, IIMI 1998) report that "pasture and cereal crops can remove 0.1–0.5 metric ton of Na^+ per hectare per year from moderately saline soils compared to up to 1 metric ton of Na^+ per hectare per year for halophytes such as salt bush from highly saline soils (Rengassamy and Olsson, 1993)".

Irrigation water containing 12 mg/ litre sodium, applied with a depth of 500 mm in a year, would cause an influx of less than 0.1 metric ton of Na^+ per hectare per year. It should therefore be possible for properly selected pasture and cereal crops to remove all sodium (Na^+) from the soils that is brought by irrigation water containing 12 mg/ litre sodium.

It follows that lands that are not already saline, should not pose the problem of sodium hazard due to irrigation with good quality water, so long as they are not water-logged.

Effect on Lowering Groundwater Table

Biodrainage would lower down the groundwater table underneath the plantation area. The objective, however, is not to lower down the groundwater table only underneath the plantation area, but to do so under the entire irrigated area. The lowering down of the groundwater table underneath the plantation area should be to such a depth, that the draw-down effect extends up to the farthest point from the plantation area, and the groundwater table, does not rise above the critical depth (say 2 m) to cause waterlogging anywhere under the irrigated area. Besides achieving water balance and salt balance, the plantations should be able to generate effective groundwater movement from all around under the irrigated area towards the plantation area. The quantity of subsurface water drainage towards the plantations should be adequate to prevent the rise of the water table above the critical depth anywhere and everywhere under the irrigated area. To achieve this, the plantation areas would have to be suitably planned and dispersed over the irrigated area.

If plantation areas are separated by distance L, the depression of the water table underneath them would result in groundwater flow behaviour similar to that seen in the case of flow towards parallel ditches penetrating an unconfined aquifer. On equilibrium, the position would be as shown in Figure 5.8 and the relationship between the depression of the groundwater table, the rate of recharge, hydraulic conductivity, depth to barrier layer and distance between plantations can be expressed by the Donnan equation:

$$L^2 = \frac{8kY_o h}{R} + \frac{4kh^2}{R}$$

Where, L = is the distance between plantations;

R = is the rate of recharge;

Y_o = is the height of water level above barrier layer underneath plantations;

k = is hydraulic conductivity; and

h = is the head difference.

It is seen that plantations are able to provide effective biodrainage to farther distances in permeable soils (in comparison to impermeable soils) and when the barrier layers are sufficiently deep. With a rate of recharge (R) equal to 0.5 mm/day, the head difference (h) of 10 m, depth to barrier layer under plantations (Y_0) as 10.0 m and hydraulic conductivity value of 100 mm/day, plantations can be spaced 500 m apart. With hydraulic conductivity values 1000 mm/day and 10 mm/day, the distance between plantations would work out to 1500 m and 150 m respectively.

If transpiration from the plantation area, surrounded by irrigated agriculture fields all around be considered like pumping from wells, the theoretical drawdown (in metres) at varying distances from the plantations (in m) can be estimated on the analogy of pumping from wells by the following Theiss non-equilibrium relationship:

$$S_C = \frac{4UTt}{r^2}$$

and

$$z = \frac{Q}{4\pi T} W(u)$$

where z is the draw-down in m;

T is transmissivity in m^2/day;

t is the time since pumping started, in days;

S_c is the dimension-less, storage coefficient;

r is the radial distance from the transpiring plantation area to the point of observation in m;

Q is the constant rate of transpiration from the plantation area in m^3/day;

U is $r^2S_c/4Tt$; and

$W(u)$ is the well function of U.

The values of W(u) for various values of μ are available in any standard book on groundwater. WAPCOS estimated theoretical draw-down as shown in Figure 5.6.

SALT BUILDUP IN CAPILLARY FRINGE AND TREE ROOT ZONE

An apprehension is often expressed that when trees use water from the water table, salts dissolved in the groundwater will concentrate in the root zone of the trees because roots exclude almost all the salt at the root surface during transpiration and this may jeopardise long term sustainability of trees. Stirzaker et al. (1999) state that unless there is some process to remove salt, the concentration of salt in the tree root zone will reduce and finally halt uptake from the groundwater. Heuperman. A (1999) expressed the view that long term sustainability of deep rooted tree plantings growing on shallow water tables is questionable and groundwater pumps or tile drains may be needed to sustain long term growth.

The phenomenon of concentration of salts is mostly confined to the capillary fringe which may be a layer of no more than 1.0 m thick. Lower down, below the capillary fringe the groundwater salinity is not affected much due to uptake of water by the tree roots. In the Kybram, Australia study where water table fluctuated between 3.5 m to 5.5 m depth Heupuman reports salt accumulation under tree plantations between 2.5 m to 5.5 m depth from the surface (Figure 6.5). This corresponds to the capillary fringe in the range of the fluctuating water table during the period 1982–1992. There was very small increase in groundwater salinity in layers at depths 0–2.5 m and 5.5 to 9.0 m

Since the thickness of capillary fringe in which the salt concentration occurs at any point of time, is no more than 1.0 m, while the tree root zone can extend up to 10 m depth or more, the trees should be able to draw their requirement of water from zones of unconcentrated salts. The apprehension on long term sustainability of tree plantations under such conditions, therefore, is misplaced. The survival and growth of Kyabram plantations that were established as long back as in 1976, is positive indicator. Moreover, the problem of salt accumulation in capillary fringe can be mitigated by rainwater leaching or occasional irrigation water application. In Asia, Australia and America, occasional heavy rain storms, once is 5 years are not uncommon.

Secondary salinisation in India and Pakistan is generally caused when previously dry lands are brought under irrigation and under such conditions non saline irrigation water can be used, once in a while to leach down the concentrated salts under the tree plantations.

According to Morris (1991) a continuous build up of salt around the roots to a concentration too great for continued water uptake is unlikely to occur under field conditions. This process is readily demonstrated in pot trials, with confined root systems and restricted drainage or leaching; but it is opposed in the real world by other processes operating in time scale from minutes to months. They include diffusion of salt down the concentration gradient away from roots; root growth in less saline soils; redistribution of water and salts when transpiration ceases during the night; daily and seasonal fluctuations in water table levels; and leaching by rain or applied water

However in locations where the shallow groundwater salinity is very high and exceeds the thresh-hold limit of tree tolerance, tree plantations cannot survive without outside supply of non-saline water. In such situations lowering of groundwater table with the help of tree plantation would not be feasible. Fortunately, most locations facing the threat of secondary salination have mild to moderate groundwater salinity.

Trees, while drawing groundwater, also uptake and absorb some salts. About 5 per cent of plant dry body weight is made up of minerals. Overall, trees extract salts from the groundwater, reduce its salinity and improve the soils.

Jeet Ram et al. (2004) report performance of 18 year old *Eucalyptus hybrid* at Dhob-Bali, Haryana, India. The soil is alluvial sandy loam and EC of groundwater varied from 0.66 to 22.10 dS/m. The average annual rainfall is 492 mm. The trees planted in year 1986, were not irrigated after the first year. In a 2.56 ha plot, 301 growing trees had an average height of 24.13 m and average girth of 100 cm. The surrounding lands were irrigated fields. The water table underneath the trees was deepened by about 1.0 m and the impact of trees in depressing the water table extended up to a distance of 730 m from the edge of the plantation area.

ROOT DEPTH AND DENSITY

The maximum draw-down by biodrainage can occur underneath plantations up to the depth to which the roots of the trees can extend effectively. The density of roots at different depths also plays a role. Under normal irrigated or rain-fed conditions, water and nutrients are available more in the upper layers of the soil strata than at the lower depths. Therefore the root density of plants is higher in the upper layers than at the lower depths. Nearly 40 per cent of the roots occur in the uppermost 25 per cent of the Vadose zone, 30 per cent in the zone range 25–50 per cent, 20 per cent in 50–75 per cent zone and 10 per cent in 75–100 per cent zone.

The root system has a remarkable ability to increase in length and to branch, the latter ability being over one-thousand-fold more developed in the roots than in the top system. The root system moves in the direction of food and water though the plant itself stays rooted in its place. To provide the plant with sufficient food and moisture, the roots have to spread wide and deep and grow numerous feeding roots and root hairs.

Kolesnikov reports that excavations at the Chaklov State farm in Crimea showed that the total length of all tree scaffold and fibrous roots of a 45 year old *Sary Sinap apple* equalled 2.7 km, of which the vertical roots accounted for 1.6 km and the horizontal roots for 1.1 km. In the same orchard, but on a site with a high water table, the excavations of 25 year old *Reinette de Champagne apple trees* revealed a root depth of only 1.3 m and a total root length of 523.6 m, the vertical roots accounting for only 77 m and the horizontal ones for 456.6 m.

In a desert area of Rajasthan in India, the perched water table had developed on both sides of the irrigation canal, and water pools appeared on the surface of the ground. Plantation work was taken up in the area and the water levels started receding very soon. Within a period of 6–7 years, the water level went down by 15 m. The plantations were not given any irrigation and drew their water requirement wholly from groundwater. Pits excavated up to a depth of 10 m, showed that tree roots were extending up to the excavated depth

Knight (1999) reports that on global average desert vegetation reaches a maximum rooting depth of 13.4 m. The root system of *Jerrah* trees growing in south-west Australia, was found to extend from 15 m to 20 m below soil surface to reach the groundwater table.

Study and research can provide useful data on the maximum depth to which plantations can draw down the groundwater table and their degree of effectiveness with the declining groundwater table, though available evidence shows that plantations can draw their requirement of water from a 15 m deep groundwater table.

▼ PRODUCE FROM BIODRAINAGE PLANTATIONS

Upadhyaya and Soni have reported results of plantations of over 31,000 ha area in the Indira Gandhi Nahar Project command (IGNP) Rajasthan, India. *Eucalyptus camaldulensis* plantations were raised in 1986 and the study was carried out in 1991. The reported total oven dry biomass was 185.12 tonnes per ha from five year old irrigated plantations. The production per ha per year was 37.02 tonnes out of which the utilisable biomass was 29.44 tonnes/ha/year. In a field trial in Israel, biomass yields of up to 57 tonnes/ha/year were achieved.

Landsberg (1999) estimates dry bromass produce of *E globulus* per unit of water transpired at five sites in western Australia in the range of 3.3 to 6.6 grams per kilogram of water with an average value of 4.76 g/kg. Using this value, total biomass produce from a *E plantation* with tree density of 1100/ha and assuming average tree water use of 40 litres/day would workout to 76.4 tonnes/ha/year.

Gill and Abrol (1991) report annual increments in biomass by *prosopis juliflora, A nilotica, C equiseti folia* and *E tereticornis* on average ranging between 15–20, 14–16, 10–15 and 5–7 t/ha/annum respectively. Field investigations show a great potential of *Sesbania* species in yielding huge amounts (120–150 t/ha) of biomass in relatively short time spans (30–36 months).

The quantity of utilisable biomass produce from plantations per unit area is important as the quantity of salt removed from the soil by the plantations would be in direct proportion to the yield of utilisable biomass. The biomass produce would depend upon the species of trees, their density, and the care and upkeep of the plantations. The aim should be to maximise the yields for better economic returns as well as for increasing the salt extraction capacity of the plantations.

Research and study is needed to determine the mineral content in dry biomass produce from plantations. Some species of trees have more mineral content than others, it is desirable to obtain the elementwise data of the total mineral content in the biomass for suitable plantation species.

Besides the plantation trees, the biomass produce from agricultural crops also contains some minerals which are obtained from the soil. Appropriate data on biomass produce from agriculture crops and its mineral content can enable a reasonable estimate of the quantity of minerals removed annually by the crops from the soil mass. Besides the total mineral content in the dry biomass of crops, analysis and data on the content of individual constituents (elements) in the total mineral component, pertinent to all growing crops in the area would be very useful.

SOME COMMONLY EXPRESSED CONCERNS ABOUT BIODRAINAGE

Some concerns and doubts are expressed regarding the feasibility and success of biodrainage as an effective anti-waterlogging measure. These pertain to:

- the availability of land for raising plantations;
- the irrigation water requirement for plantations;
- the progressive accumulation of salts in soils underneath the plantations;
- the progressive increase in the salinity of groundwater;
- the limited distance up to which the groundwater table can be lowered beyond the boundary of plantations;
- the cost of raising and maintaining plantations.

Availability of Land for Plantations

Land under command in irrigation projects is generally used for agriculture by farmers who own the lands or hold tenancy rights. Sometimes these holdings are quite small in the densely populated countries and regions like in India. It is apprehended that it may not be possible to persuade the farmers to part with their lands or to change its use from agriculture to forestry.

Land is no doubt scarce in the densely populated areas. But fresh water is even more scarce. Leaving aside a few exceptional cases, water is relatively much more scarce than land. In the dry arid regions, the disparity is so marked that land without water has little value. With limited quantity of available water, it is rarely possible to make the full productive use of all land. It is very common to leave part of the sand fallow every season because there is not enough water to irrigate the whole land. In large tracts of irrigated commands in India and Pakistan, irrigation water is hardly available to meet the crop water requirement over half the area during each of the winter and summer seasons. The rest of the land either remains fallow or produces much less than its potential. Therefore, it would not be right to say that land cannot be made available for forestry.

Generally, farmers receive their share of irrigation water on some rationing basis. The returns to a farmer depend more on the volume of his share of water rather than on the size of his land holding. In such a situation, it should not be very difficult to persuade the farmers to divert a part of their lands for tree plantation by assuring them that there would be no reduction in their share of water.

Tree plantations or agro-forestry can be as remunerative as or more than agriculture. This aspect has been discussed in Chapter 10. The apprehension that tree plantations on a small part of the farming area would have an adverse effect on farmer's incomes and would therefore be unacceptable to them is therefore misplaced.

It must be considered that without effective drainage measures, the farmers may eventually lose a much larger area due to waterlogging and salinisation. In the case of the horizontal subsurface drainage system too, land is needed for open ditch, sub-collector, collector and main drains, and if the drainage water is to be disposed of in evaporation ponds, then the total land area required in the horizontal subsurface drainage system would be no less than that under biodrainage scheme.

The constraint of land for practising biodrainage is therefore unlikely to be a problem that cannot be resolved by appropriate management and administrative measures without economic loss to the farmers.

Irrigation Water Requirement for Plantations

Water is harnessed and conveyed for agricultural use. The diversion of any quantity of water for raising tree plantation would have the effect of a corresponding reduction in the quantity of water for agricultural use. It is apprehended that it may not make economic sense to do so.

In the first place, tree plantations or agro-forestry can yield as good economic returns as agriculture. More importantly, the principle underlying biodrainage is that the tree plantations draw most of their water supply from the groundwater and survive on sub-irrigation. As and when the groundwater table comes up sufficiently high and plant roots can reach the surface of the groundwater table, the trees would draw most of their water requirement from the groundwater and little surface irrigation would be needed.

It is for this reason that the plantation of such species of trees is recommended in the first instance that can grow under the prevailing climatic condition, with very little or no need for water for surface irrigation. This is to minimise the pressure on the limited irrigation water supply. In due course of time, as and when the groundwater table rises sufficiently, the plantations can be replaced by water-hungry plantations that would draw their water requirement form the groundwater surface.

In this manner, the biodrainage plantation would make very little demand on the irrigation water supply that can be reserved for agricultural use.

Progressive Accumulation of Salts in Soils

Tree plantations would depress the groundwater table underneath. The water would then rise above the groundwater table in the capillary zone, where the roots of the plantations would absorb water, leaving salts behind. A view has been expressed that this can eventually cause accumulation of salts in the root zone immediately above the ground-water table to such an extent that survival of tree plantations may not be possible after some time. In the Kyabram plantation in Victoria, Australia, it was apprehended that the life of the plantations, on this account may not be more than 20 years.

It is true that there would be some concentration of salts in the capillary zone above the groundwater table. But if the groundwater is sufficiently high and the plantation roots can extend up to the groundwater table surface, the trees should be able to draw their water requirement from the groundwater table surface by passing the saline soil zone, and the accumulation of salts in the soil in the capillary zone should not prove a hazard. It has been explained in Chapters 2 and 8, that under water and salt balance conditions, the salinity of groundwater is not likely to increase with the passage of time and even if it does so, the rate of increase would be too slow to be a matter of concern.

There is another way of taking care of the accumulated salts in the capillary zone. As and when the salt concentration reaches unacceptable levels, the salts can be leached down by good quality irrigation water to acceptable levels that would pose no threat to the growing tree plantations. Such leaching operation may be needed, if at all, only once after an interval of many years.

For these reasons, the accumulation of salt underneath the plantations should not pose a threat to the tree plantations.

Progressive Increase in Salinity of Groundwater

Irrigation water always has some salts. It is often stated that crops and plants do not remove any minerals from the soil or do so in negligible quantities so that almost all the salts contained in irrigation water ultimately go to the groundwater reservoir alongwith the deep percolating water. It is further inferred that this must cause a progressive increase in groundwater salinity.

It is explained in Chapters 4 and 8 that all plants absorb minerals from soil water up to an extent of about 5 per cent of their dry body weight. Minerals to this extent are removed form the ground alongwith the utilisable biomass. It is shown that if the irrigation water is of good quality, as is generally found in natural rivers, it can be possible to strike a salt balance such that total annual removal of minerals from the soil water mass matches the total input of minerals in the ground with irrigation water. There should therefore be no abnormal increase of salinity in groundwater due to irrigation. It is explained in Chapter 2 that in case minerals removed annually through biomass are less than the minerals imported through irrigation water, the rate of rise of groundwater salinity is likely to be too slow to be a matter of concern.

Limited Distance up to Which Groundwater Table can be Lowered Beyond the Boundary of Plantations

It has been stated on the basis of the Kyabram, Victoria, (Australia) study and a few other experiences that plantations depress the groundwater table directly underneath but have little effect on the groundwater table in the surrounding areas. The effect on the goundwater table would depend upon the quantity of recharge and the quantity of transpiration. If the rate of transpiration from an area exceeds the rate of recharge, groundwater table under the area, would go down. This lowering of the groundwater would continue till a water balance is struck. The depressed groundwater table underneath plantations would cause the flow of groundwater from the surrounding area towards depressed groundwater table underneath the plantations, resulting in the lowering of the water table even in the surrounding area. The hydraulic conductivity of ground strata, depth of impermeable layer and the rates of recharge and transpiration would determine the extent of groundwater table depression and the distance up to which it would occur.

In case a small plantation area is surrounded by large irrigated area from which recharge is taking place, the depression of the groundwater table and the distance up to which it would occur would be small. On the other hand, if the plantation area is reasonably large, the groundwater table would be depressed sufficiently and the lowering effect would extend up to a longer distance. This position is explained in detail in Chapter 8.

Cost of Raising and Maintaining Plantations

It is said that raising tree plantations to serve the purpose of biodrainage would cost money and this cannot be ignored. Tree plantations can provide as much return as agriculture. The subject is dealt in detail in Chapter 10. Even if the studies and projections are over-optimistic, there should not be any doubt that plantations can provide returns that are at least equal to the costs. Government agencies are least suited to undertake plantation work. This can best be done by involving beneficiary farmers with their active participation or through private agencies.

In addition, the plantations help to improve the environment which yields an invaluable and immeasurable benefit.

AGROFORESTRY SYSTEMS

In the agroforestry systems like the one tried in California USA (Experimental Project 1987–95) and in Victoria, Australia (SBC, Chapter 6) where tree plantations are raised outside and away from the irrigated area of the farm and are irrigated by the saline drainage effluent from the irrigated farm land, the problem of increasing salinity is likely to be encountered inspite of the installed subsurface drainage system.

The total salts evacuated by drainage water as a fraction of salt input with irrigation water can be estimated by the following equation:

$$\frac{\text{Salt output}}{\text{Salt input}} = \frac{V_d}{V_i} \times \frac{EC_d}{EC_i} \ (approx.)$$

where,

V_d = Volume of drainage water

V_i = Volume of irrigation water

EC_d = EC value of drainage water

EC_i = EC value of irrigation water

If EC of irrigation water be 10 ds/m and that of drainage water 17 ds/m (as in SBC, Australia and San Joaquin valley USA studies) and the volume of drainage water be 20 per cent of the volume of irrigation water, the salt output-input ratio would be 0.34 (approx). A major part of the incoming salts being left behind in the plantation area, long term sustainability of such irrigation systems in doubtful.

There is a qualitative distinction between the agroforestry systems of the type described above and the biodrainage system in which trees are planted within the farm area and non-saline water is used for irrigating the crops on the farm. In the biodrainage system, the agriculture crops and the trees together can achieve 'salt balance' which is not feasible in the agroforestry systems.

Soil Quality Under Tree Plantations

In an agroforestry trial near Mendota in California USA, the trees received the drainage water of EC 10 dSm^{-1} from irrigated farm land. It was found that the reuse of this drainage water for irrigating trees resulted in steady increase in soil salinity under the trees. The soil conditions were monitored during the period 1987–1995 (Cervinka et al.) and are shown in Table 8.4.

Table 8.4 *Soil quality under tree plantations, Mendota experiment*

		1987	*1990*	*1991*	*1994*	*1995*
EC	dS/m	10	29	20	14	11
Cl	mg/l	1003	2159	2126	1114	897
SO_4	mg/l	3904	11361	7175	6969	5716
Mg	mg/l	96	172	223	132	147
Na	mg/l	465	1441	1719	3477	2466
Ca	mg/l	2161	5584	4242	471	465
B	mg/l	5	20	27	11	11
Se	mg/l	0.118	1.888	0.551	0.134	0.117
SAR		9.8	19.0	25.7	141.6	99.7

The soil salinity increased from 1987 to 1991. After installing the drains in 1991–92, the soil salinity started to return to initial conditions. This experience motivated to develop the IFDM system.

It is interesting to note that while the EC of soil declined from 29 dS/m (year 1990) to 11 dS/m (year 1995), as a result of the provided subsurface drainage, the SAR of the soil increased from 19.0 to 99.7 during the same period. The results indicate that irrigation of trees by saline water (10 dS/m) caused increase in total cations (Na, Ca, Mg) in the soil underneath. The bulk of Calcium cations was removed by precipitation or by drainage water but removal of Na cations was relatively much less, resulting in very high SAR. A question that remains unanswered is how long can trees be sustained under conditions of such high SAR in the soil. Research and experiments are needed to ascertain the causes of increase in SAR and find remedial measures.

If, instead of irrigating the tree plantations by the saline drainage water (EC 10 dS/m), the trees are made to draw water requirement directly by biopumping water from the shallow groundwater table, the accumulation of salts would be limited within the capillary fringe above the groundwater table. This should not affect the sustainability of trees so long as a ground-water of EC upto 10 dS/m is available at shallow depth which can be accessed by the tree roots. The rate of transpiration would correspond to the degree of salinity of the groundwater, which may be about 0.5 Apan when EC of groundwater is 10 dS/m, but the survival of the tree plantations would not be threatened. The salinity of groundwater at shallow depth under irrigated crops in known to decrease with passage of time due to dilution effect when non-saline or low-saline surface water is used for irrigation. Transpiration by trees, planted over a relatively small area within a large irrigated cropped area, would have a marginal impact on increasing groundwater salinity, which is more than neutralised by the less saline infiltrating water from surface irrigation of crops in surrounding areas. At the Mendota trial, the quality of drainage water from crops into trees, and therefore also that of the groundwater table under the cropped area, has remained almost unchanged during the period 1988–1995. The EC ranged between 8–12 dS/m and Na(mg/l) 1200–1600. This suggests that trees planted over an area adjacent to the cropped area and irrigated by sub-irrigation, (not surface irrigation), are unlikely to face the threat of increasing salinity and sodicity.

IRRIGATED FORESTRY IN DESERT AREA

In IGNP, Rajasthan, India, some experience has been gained on irrigated forestry, since 1973. More than 150 thousand hectare area has been planted with the primary objectives of: (a) protection of canals, roads and colonies from drifting sand, and (b) provision of fodder, fuel wood and timber to meet the needs of the local population.

The main activities of afforestation consist of canal side plantations, block plantation, sand-dune stabilisation, pasture development, road-side and environmental plantation. It has been decided that an area of about 10 per cent is to be reserved in each irrigation block for afforestation to provide biodrainage.

Different species of trees were planted including *eucalyptus camalduiensis, Delbergia sissoo, Acacia nilotica, Ziziphus mauritiana, Tecomnella undulata* and *Prosopis cineraria*. After four years, the maximum growth in height as well as girth was attained by *Eucalyptus camaldulensis* and minimum by *Prosopis cineraria*. For this reason Eucalyptus provided a more effective shelter belt in checking sand deposition in the canals. However the biomass produce of *Delbergia sissoo* and *Acacia nilotica* was comparable to that of Eucalyptus. *Ziziphus mauritiana* started producing fruit at a rate of 20 kg/plant from the third year. *Tecomnella undulata* and *Prosopis cineraria* attained heights of 4.56 and 3.87 m respectively in four years. *Prosopis cineraria* started producing pods in three years at a rate of 5–6 kg per plant and *Tecomnella undulata* started flowering and growing fruits from the third year. The latter two plants can grow under extremely water-scarce conditions, without any irrigation but with a much slower rate of growth.

It was found that the growth of forage on the ground under *Prosopis cineraria* was the maximum followed by *Acacia nilotica, Delbergia sissoo, Tecomnella undulata* and *ziziphus mauritiana* and almost negligible under Eucalyptus camaldulensis.

The results of the study are shown in Table 8.5

Table 8.5 *Growth of plantations in IGNP, Rajasthan, India*

S.No.	Species	Height (m)	Collar girth (m)	Girth at breast (cm)	Per ha produce of forage on ground (tonnes)
1.	Eucalyptus camaldulensis	12.85	55	45	0.4
2.	Delbergia sissoo	9.90	40	33	2.1
3.	Acacia nilotica	7.65	37	29	2.4
4.	Ziziphus mauritiana	6.10	34	28	1.6
5.	Tecomnella undulata	4.56	30	24	1.8
6.	Prosopis cineraria	3.87	22	16	4.4

The selection of species of trees for biodrainage plantations can be made from several varieties depending upon the local conditions of soil, climate, economic returns and social preferences.

GUIDELINES FOR PLANNING LOCATIONS OF PLANTATION PLOTS FOR BIODRAINAGE

In IGNP, Rajasthan, India, land strips on both sides of all canals have been reserved for plantations. The width of the strips on either side of the canals varies between 200–300 m on the main canal, 100–150 m on branch canals and 40–70 m on distributary canals.

In addition, for selecting the location of plots for afforestation to provide biodrainage within irrigation blocks, the following guidelines have been prepared:

- The minimum size of the plantation plot is to be 12.6 ha.
- The area to be reserved for tree plantation plots is to be kept about 10 per cent of the culturable area as follows:

Irrigation block size (ha)	Area to be reserved for biodrainage plantations (ha)
100–150	12.6 (2 units of 6.32 ha)
150–200	19.0 (3 units of 6.32 ha)
200–250	25.0 (4 units of 6.32 ha)
250–300	31.5 (5 units of 6.32 ha)

- As far as possible, the tree plantation plots should be selected at the lowest surface ground levels in the irrigation blocks.

- The distance between plantation plots may generally vary between 1 km – 2 km.
- The plantations may be effective in lowering down the water table up to a maximum distance of 1 km.
- While planning the location of plantation plots in any block, the position of plantation plots in the adjoining blocks and along canals is to be taken into consideration.
- In case of irrigation blocks, where the distance of the farthest irrigation field from the nearest plantation exceeds 1 km, field investigations are to be carried out to determine the hydraulic conductivity of the sub-strata up to 10 m depth and the depth to barrier layer. The distance between plantations can be determined using the Donnan equation.

Illustrations of the selection of biodrainage plots with the above guidelines pertaining to a few irrigation blocks are shown in Figure 8.2 to 8.8.

It is not necessary that plantations are planned in compact blocks in the middle of irrigated area. Parallel afforestation strips 50 m or more wide, suitably spaced, if feasible, can provide better and more effective drainage arrangement.

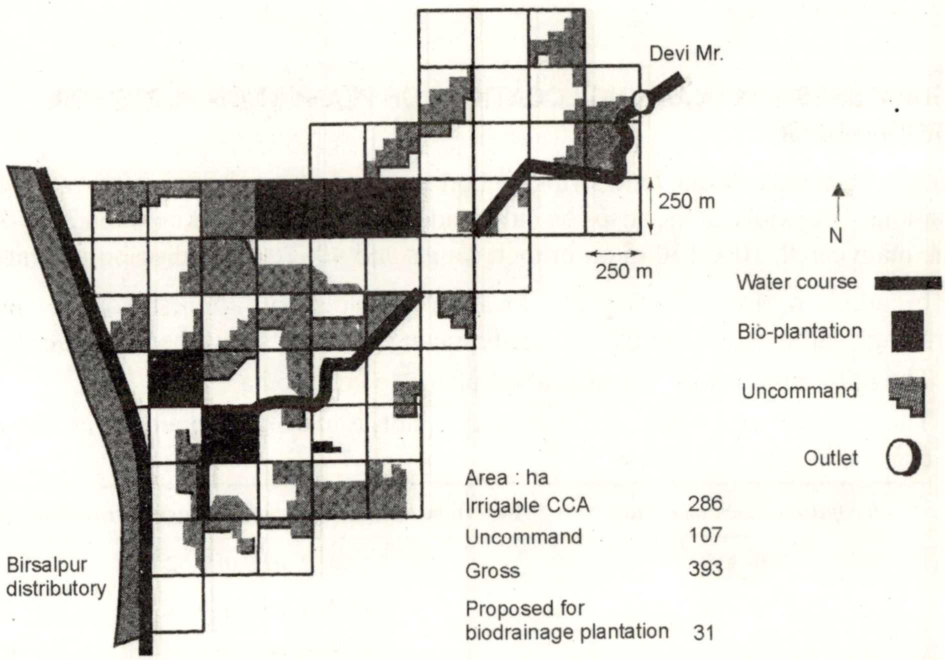

➲ **Figure 8.2** *Planning of biodrainage plots for afforestation, Chak No. 4 DM IGNP*

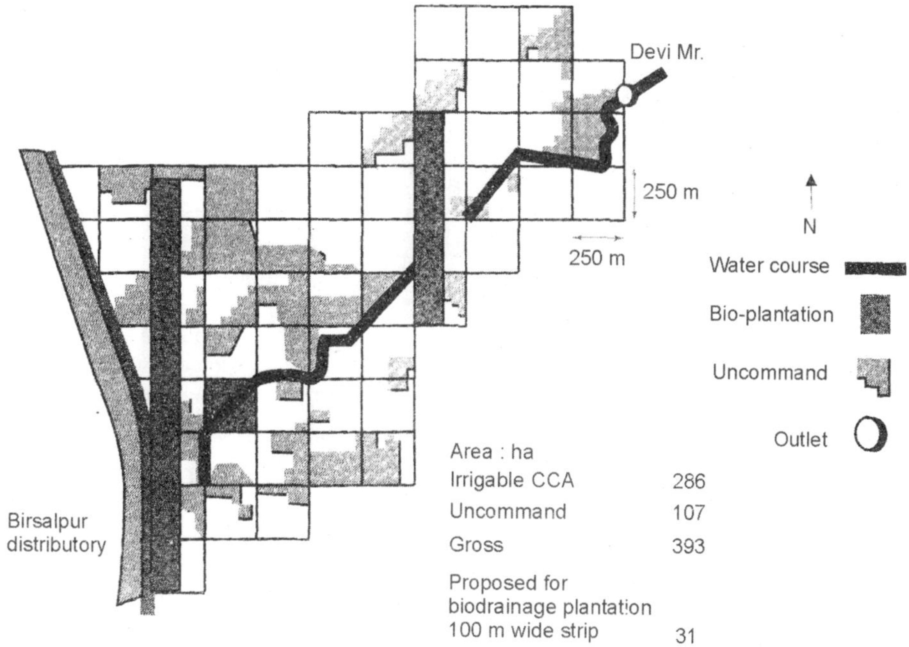

Figure 8.3 *Planning of biodrainage plots for afforestation, Chak No. 4 DM IGNP*

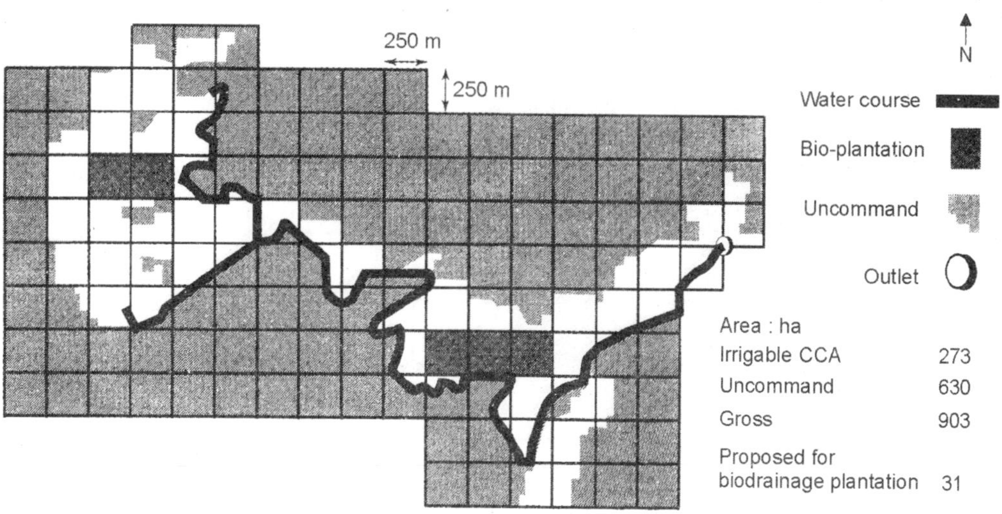

Figure 8.4 *Planning of biodrainage plots for afforestation, Chak No. 3 RM - II IGNP*

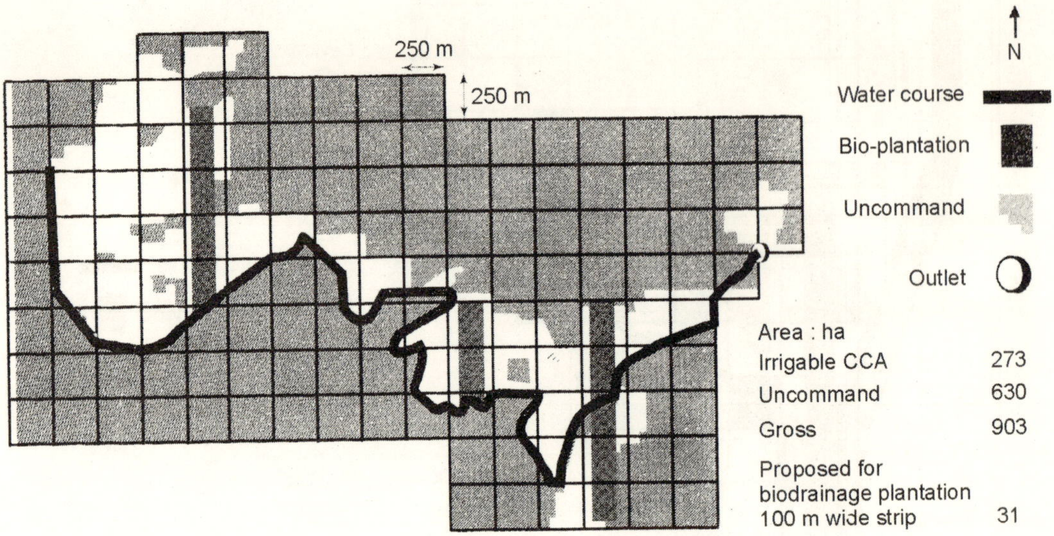

Figure 8.5 *Planning of biodrainage plots for afforestation, Chak No. 3 RM - II, IGNP*

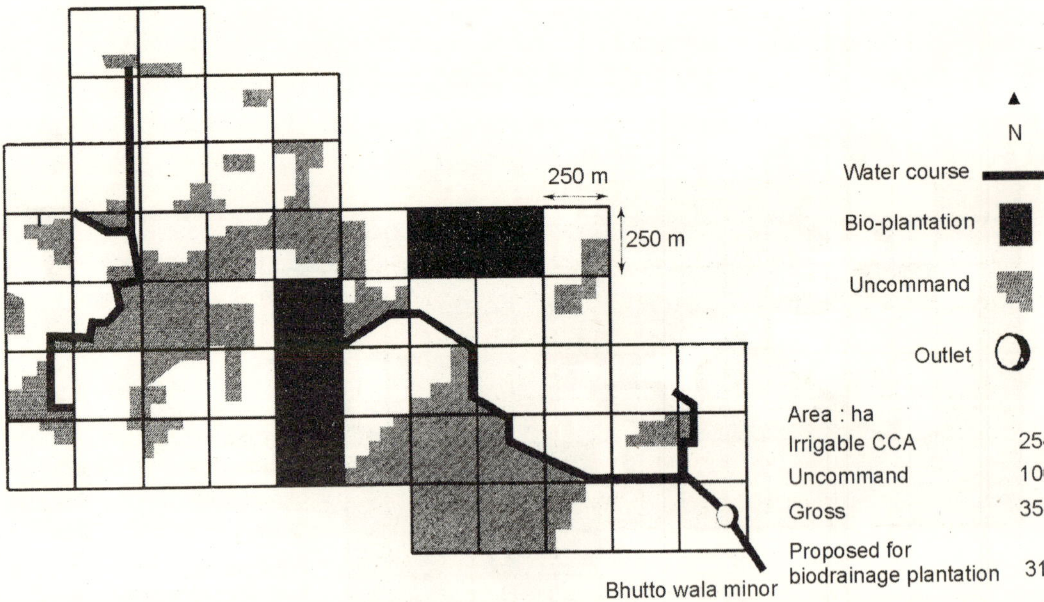

Figure 8.6 *Planning of biodrainage plots for afforestation, Chak No. 4 BWM IGNP*

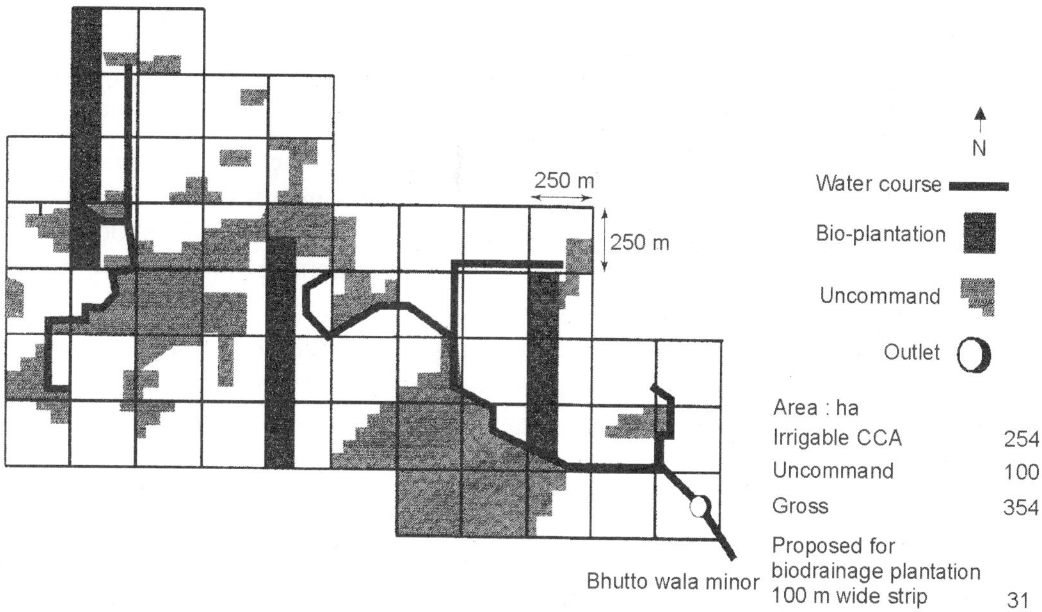

Figure 8.7 *Planning of biodrainage plots for afforestation, Chak No. 4 BWM IGNP*

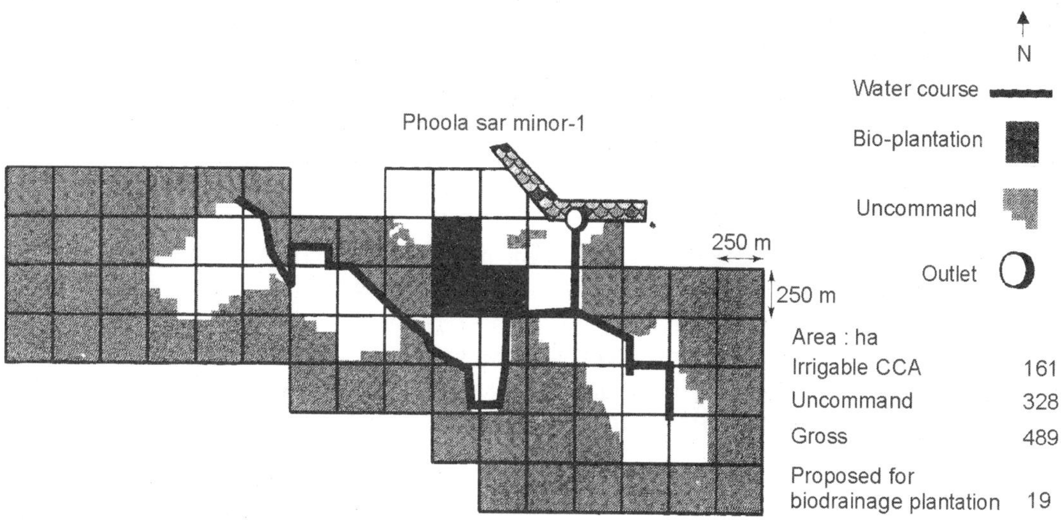

Figure 8.8 *Planning of biodrainage plots for afforestation, Chak No. 14 PSM - I IGNP*

CONCLUSIONS

The following conclusions can be drawn after studying the various principles of biodrainage planning and design:

1. To be successful biodrainage should achieve suitable water and salt balance. The area to be brought under plantations should be limited and the demand of irrigation water for the plantations should be small. The groundwater quality should not be allowed to deteriorate and the water table must be maintained at a safe depth at the farthest point from the plantations.
2. The knowledge of tree water use rates and irrigation recharge can help to determine the area needed for plantations to provide effective biodrainage.
3. If the rate of recharge, hydraulic conductivity of the ground sub-strata, depth of the barrier layer etc. are known, it would be easy to estimate the required spacing between plantation plots to provide effective drainage.
4. The composition of mineral content in crops and trees growing in the area and the quantity of minerals brought in annually with irrigation water, can help in determining the extent of imbalance, if any. The use of good quality irrigation water can preclude the threat of deterioration of groundwater quality to an unacceptable level.
5. Tree roots can go quite deep in search of water (15 m or more) and draw water from the surface of the groundwater table.

REFERENCES

Central Ground Water Board, India (1995). *Studies on the Conjunctive Use of Surface Water and Ground Water Resources in IGNP Stage–I*, Rajasthan.

Cervinka, V., C. Finch, M. Martin, F. Menezes, D. Peters and K. Buchnoff (2001). *Drainwater, Salt and Selenium Management Utilizing IFDM/Agroforestry Systems*, Report Prepared by USBR, p. 44.

Department of Science and Technology, Government of Rajasthan (1984). 'Application of Science and Technology for Afforestation', Proceedings of the National Seminar.

Donnan, W.W. (1946). *From Hydrogeology of Different Types of Plains*, Vol. I, International Institute for Land Reclamation and Improvement, The Hague.

Gill, H.S. and I.P. Abrol (1991). 'The Utilization of Salt-affected Soils and Water Resources for Tree Based Land Management System in India', Proceedings of the Symposium Held at the Institute of Sustainable Agriculture at Tatura, Australia, July 23.

Heuperman, A.F. (1999). 'Hydraulic Gradient Reversal by Trees in Shallow Water Table Areas and Repercussions for the Sustainability of Tree Growing Systems', *Agricultural Water Management Journal*, p. 162 and pp. 153–54.

Kijne, J.W., S.A. Prathapar, M.C.S. Wopereis and K.L. Sahrawat (1998). 'How to Manage Salinity in Irrigated Lands: A Selective Review with Particular Reference to Irrigation in Developing Countries', SWIM Paper 2, International Irrigation Management Institute, Colombo, Sri Lanka.

Kapoor, A.S. (1997). 'Biodrainage of Lands Under Irrigation in Dry Arid Regions with Particular Reference to the IGNP, Rajasthan' (Unpublished).

— (1998). 'Biodrainage: Principles of Planning and Design' (Unpublished).

— (1998). 'Rate of Increase in Ground Water Salinity Due to Irrigation with and without Biodrainage' (Unpublished), Agriculture University, Bikaner.

Knight, J.H. (1999). *Root Distributaries and Water Uptake Patterns in Eucalyptus and Other Species*, Published in RIRDC (Australia) Publication No. 99/37, titled 'The Ways Trees Use Water', pp. 57–59.

Kolesnikov, V. (1966). *Fruit Biology*, Mir Publishers, Moscow.

Lal, P. (1997). 'Information on Biomass of Wheat Crop' (Unpublished), Chief Scientist, Agriculture University, Bikaner.

Landsberg, J.J. (1999). *Tree Water Use and Its Implications in Relation to Agroforestry Systems*, Published in RIRDC (Australia) Publication No. 99/37, titled 'The Ways Trees Use Water', pp. 47–48.

Morris, J.D. (1991). 'The Water Use Processes and Salinity Management in Irrigation Areas', Proceedings of the Symposium Held at the Institute of Sustainable Agriculture at Tatura, Australia, July 23, pp. 18–20.

Palladin, V.I. (1992). *Plant Physiology*, Edited by B.I. Livingston, 2nd American Edition, Arihant Publishers, Jaipur.

Pandey, S.N. and B.K. Sinha (1995). *Plant Physiology*, 3rd Edition, Har-Anand Publications, New Delhi.

Ram, Jeet and Vineet Garg (2004). 'Biodrainage for the Reclamation of Waterlogged Areas: A Case Study', Paper Presented to the Ministry of Water Resources, India.

Stirzaker, R.J., F.J. Cook and J.H. Knight (1999). 'Where to Plant Trees on Cropping Land for Control of Dryland Salinity: Some Approximate Solutions', *Agricultural Water Management Journal*, 39: 125.

Todd, David Keith (1980). *Ground Water Hydrology*, University of California, Berkeley.

Turner, Neil C. and Philip R. Ward (2002). 'The Role of Agroforestry and Perennial Pasture in Mitigating Waterlogging and Secondary Salinity', *Agricultural Water Management Journal*, 53: 271–75.

Upadhyaya, A.K. (1994). *Irrigated Forestry in IGNP Stage–II: Planning and Sustainability in Irrigation.*

Upadhyaya, A.K. and R.G. Soni (1997). 'Growth, Biomass Production and Dry Matter Distribution Pattern of Eucalyptus Camaldulensis Grown Under Irrigated Conditions in the GNP Area of Rajasthan', *Indian Forester*.

WAPCOS (1995). *Feasibility Report, IGNP Stage–II.*

9

Status and Comparison of Drainage Methods

▼ INTRODUCTION

Salinity is one of the most serious problems facing the world's irrigated agriculture. Recounting the damages inflicted by salinity, Hall (1974) writes, 'History is on their side, as are the laws of physics, chemistry and biology. Today, every arid land region of the world is in some immediate or final stage of this process and nowhere, it would seem, has there been established a genuine detente with these deceptively simple destroyers of man's vaunted accomplishments'. But there can be no two opinions that our long-term economic survival depends upon how we come to terms with the problems raised by soil and water degradation. The three major ways to fight the ever-growing problem of waterlogging and salinisation of lands are:

- Improving irrigation management to reduce the amount of groundwater recharge;
- Ameliorating the effect of salinity through shallow groundwater management, land use changes, and the development of more salt-tolerant crops; and
- Taking action to lower the water table and leaching and evacuation of saline water through drainage.

▼ SUBSURFACE DRAINAGE SYSTEMS

Any drain or well which is installed to control or lower the high water table in an area is considered to be an element of the subsurface drainage system. The high water table may be caused due to percolation, seepage from canals and surface water bodies located at a higher elevation, irrigation water, leaching water and leakage from an artesian aquifer. In the arid and semi-arid areas a small portion of the excess water comes from precipitation. Elsewhere the major source of excess water is precipitation. The major sources of excess water in irrigated areas are percolation losses from irrigation, the leaching water applied and seepage losses from irrigation canals. In humid, arid and semi-arid areas, the sources of excess water may be groundwater moving through shallow aquifers and emerging as springs or groundwater under artesian pressure. When the total quantity of water introduced into the surface in an area from the various sources exceeds the total quantity disposed of through natural drainage processes, the water table will rise. It is then necessary to install artificial drains to remove the surplus water in order to maintain the water table at some per-determined level which is not harmful to crops.

CLASSIFICATION OF SUBSURFACE DRAINAGE

From a functional point of view, subsurface drainage falls into two classes: *relief*, and *interception drainage*. *Relief drainage* is used to lower a high water table which is generally flat or of a very low gradient. *Interception drainage* is used to intercept, then reduce the flow and lower the flow line of the water in the problem area. While planning a subsurface drainage system, the site conditions should be evaluated to decide whether to use relief of interception drainage.

Relief Drainage

In a relief drainage system, one can identify three categories of drains: field laterals, collectors and main drains, which can be either buried pipes or open ditches or a combination of both. Ditches used for subsurface drainage may carry both surface and subsurface water. Ditches are best adapted to large flat fields where the lack of grade, soil characteristics or economic conditions do not favour buried drains. Following are the advantages of using ditches:

- They usually have lower initial cost than drains.
- The inspection of ditches is easier than that of drains.
- It is possible to build them in some organic soils where drains are not suitable due to subsidence.
- Ditches may be used on a very flat gradient where the permissible depth of the outlet is not adequate to permit the installation of drains having the minimum required grade. The gradient required for water transport is approximately 0.01 per cent in ditches and 0.1 per cent in pipe drains.

The disadvantages in using ditches are as follows:

- Ditches require considerable right of way which reduces the area of land available for cropping. This is particularly true in the case of unstable soils where flat side slopes are required.
- Ditches usually require more frequent and costly maintenance than drains because of weed growth and erosion.

Buried Drains

Buried drains refer to any type of buried conduit with open joints or perforations which collect and convey drainage water. Buried drains may be fabricated form clay, concrete, bituminised fibre, metal, plastic or any other material of suitable quality. These drains, if properly installed, require little maintenance. They are usually preferred by land owners as they are buried, do not require any removal of land from cultivation and maintenance is considerably less than for ditches.

Heupuman A. (1991) states that subsurface drainage reclaims salt affected soils quickly but is expensive, both from the investment and operational point of view (pumping costs). Disposal of saline drainage water without incurring a downstream 'penalty' is only possible by pumping it to the sea. This technique has a high economic cost both from the investment and operational point of view.

Interception Drainage

The first step towards solving an agricultural drainage problem is to identify its cause. Sometimes what appears to be a major drainage problem can be solved by a simple remedy once the cause is recognised. An example of this is waterlogging which is caused by water coming in from outside the wet area. This problem can be resolved through a single drain which intercepts the flow. On sloping land it is worth trying the effect of an interception drain before applying a relief drainage system.

Vertical Drainage

The rise of the groundwater table can be prevented and it can be lowered by pumping groundwater through tube wells. Tube well with a pumping equipment comprises the vertical drainage system. Power should be available at a reasonable cost for operating a vertical drainage system. Tube well drainage systems are suitable under the following conditions:

- in flat land with a high water table position in a large area;
- where the groundwater is under an artesian condition;
- in the absence of an adequate gravity drainage outlet;
- if the groundwater needs to be lowered beyond 2.5 m and an impervious layer does not exist at a shallow depth; and
- if the groundwater is not very saline.

Before the installation of a tube well drainage system, however, it is necessary to ascertain the hydraulic inter-connection of the pumped aquifer and the shallow water table aquifer. If the saline water table exists at a shallow depth, a vertical drainage system cannot skim the thin layer of good quality water overlying the saline water. For skimming the good quality water, a horizontal drainage system is more appropriate. A vertical drainage system can arrest the rise of the saline water table which cannot be effected by a horizontal drainage system.

The use of vertical drainage is much more restricted by the geohydrological conditions of the area and it cannot be regarded merely as a substitute for gravity drainage. Unlike gravity which has been practised in various forms for long, the technique of well drainage is a comparatively recent development and the number of projects where well drainage is being applied is still small. The design of a vertical drainage system should suggest the location of the wells, their number and pumping schedule, and should also suggest the utilisation of the pumped water if it is of good quality or its disposal if it is of poor quality.

The conjunctive use of canal and tube well water is practised in many areas.

PRESENT STATUS

About one-third of some 260 mha of irrigated area worldwide is threatened by waterlogging and salinity. Thatte et al. using 1996 FAO data, estimate that an area of 60 m ha is waterlogged and about 20 mha is salt-affected. ICID (Schultz, 1990) estimated that about 150 m ha of the world's irrigated area has been provided with drainage facility, out of which about 30 m ha has been equipped with subsurface pipe drainage system (horizontal drainage). It has been installed on a large scale in more than 35 countries including Canada, Egypt, Pakistan, Iraq, Mexico, Turkey, Malaysia and Uzbekistan (Chedieng and Visvanathan, 1997).

In the fifteen countries of the European Union (Austria, Belgium, Finland, France, Germany, Greece, Ireland, Italy, Luxemberg, the Netherlands, Denmark, Portugal, Spain, Sweden and the United Kingdom), the climate is mostly humid and oceanic temperate, with precipitation evenly distributed throughout the year. The winter storage of precipitation in shallow soil cover over an impervious barrier results in perched water tables, more often in Belgium, France, Germany and the United kingdom. In some regions, the groundwater tables are already high due to the influence of rivers or the sea and precipitation causes further rise resulting in the permanent problem of waterlogging along the coastal areas and alluvial valleys in the Netherlands. It is only in the southern part of Europe that irrigated areas may encounter a salinisation hazard.

Les and Zimmer (1995) report that the agriculture intensification drive in Western Europe resulted in land upgradation and more than 50 per cent of the waterlogged areas were reclaimed by subsurface drainage. This resulted in over-production of cereals and a policy change was made in 1992 to set aside roughly 15per cent of the arable land.

The detrimental effects of drainage towards environment were recognised and drainage was restricted, even prohibited in wet biotopes, such as uncultivated marsh lands, small ponds and alluvial plains. The annual rate of installation of subsurface drainage that was about 3,00,000 ha during the 1980s came down by about one-half in the 1990s. In some places (Germany and the UK), drained areas were converted back to marshlands. In several countries (Netherlands and Switzerland and Germany for instance) investments on drainage are restricted to the rehabilitation of ancient systems.

In the wet humid regions of North America, drainage is required to remove excess soil water, especially in poorly drained soils. Of the 53 million ha crop land drained in North America (45 million in the US and 8 million in Canada), about one-third is subsurface drained. The total annual precipitation (snow-melt and rain) exceeds evapotranspiration.

Egypt has succeeded in implementing subsurface drainage on a large scale. A special authority called the Egyptian Public Authority for Drainage Projects (EPADP) was established in 1973. The EPADP is responsible for surveying, planning, designing, constructing and overseeing the financial aspects of the drainage schemes. Since the 1970s, around 1.9 m ha have been equipped with subsurface drainage out of a total of 2.7 m ha of irrigated areas. At present, drainage installation on 60,000 ha/annum is undertaken. The EPADP was given the target to equip 2.5 m ha of cultivated land with drainage by 2005. Drainage is financed by the State, but the farmers have to pay back the investment over 20 year period at no interest which amounts to more then 50 per cent subsidy. The drainage installation machinery is not developed locally but purchased from outside by the local contractors who have developed enough experience in installation. Drainage material is locally manufactured in the public sector industries. Approximately 10 per cent of the investment is available for crop compensation. It is likely that rehabilitation of the early drainage system will have to be undertaken in the foreseeable future. A Drainage Research Institute (DRI) and a Drainage Training Center (DTC) have been established by the Government of Egypt as national level organisations for carrying out relevant research and capacity-building programmes.

In Pakistan, waterlogging and salinity became such a threat to food production that the Government had to put the need for drainage on the political agenda. In 1958,the Water and Power Development Authority (WAPDA) initiated 51 salinity control and reclamation projects (SCARP). Drainage has been implemented on some 1 m ha out of a total irrigated area of 15.4 m ha. An additional 4.6 m ha still requires drainage facilities. In Pakistan, drainage provision consists of vertical well drainage and horizontal pipe subsurface drainage. Since the physical conditions are not favourable for pipe drainage, only 0.23 m ha are provided with pipe drainage, while 0.5 m ha are equipped with vertical well drainage system. Vertical drainage is used when the groundwater is of good quality and used for irrigation. In saline groundwater areas, horizontal pipe drainage is preferred to evacuate and dispose of the poor quality water. Drainage investment is fully financed by the government and farmers are not required to pay any investment cost. There is no private consultancy sector as yet. Drainage machinery is imported. However, now with the initiative of the government, the private industry has taken up the production of drain pipes. the design, installation and supervision are undertaken by international contractors. Pakistan has established the International Waterlogging and Salinity Research Institute (IWASRI) for assisting WAPDA in providing valuable research information.

On the basis of his observation of drainage development in Egypt and Pakistan, Croon (1997) arrived at the following conclusions:

- Long-term, large volume, well-financed programmes are a pre-requisite for the effective development of a drainage tradition and industry. Large scale implementation of these programmes is required to make the entire exercise cost-effective.
- The part of investment cost to be paid by the farmers must be acceptable and affordable.
- The drainage programme must be a part of the political agenda. Considerable direct indirect financing by the government is needed over a long period of time.
- The development of a private sector drainage industry is desirable for flexibility, construction and effectiveness.

In Western USA 25–30 per cent irrigated area is reported to be provided with subsurface drainage (Rao, 1998).

In India, which had 50 m ha of irrigated area in 1993 (CWC, 1996), subsurface drainage has been installed in less than 0.02 per cent area (Rao, 1998).

Pumping of groundwater for use in agriculture is a very common practice the world over. In India nearly 50 per cent of the irrigation is from groundwater followed by 43 per cent in USA, 27 per cent in China and 25 per cent in Pakistan. The mainstay of irrigation in Austria and Germany is 80 per cent and 90 per cent respectively. In surface irrigated areas, where the source of water is outside the region, irrigation by canals causes additional recharge to groundwater. When water supply from canals is restricted, and groundwater quality is good, the farmers resort to installation of wells to pump and use groundwater for supplementing irrigation. In the Gangetic plains and the Indus basin in India (also in Pakistan), thousands of wells have been installed in hundreds of thousand hectares area, in this manner. The pumping from such well besides supplementing irrigation, helps to control the rate of rise of the groundwater table and is a very effective measure to control waterlogging. Data on the total area, provided with vertical drainage by the conjunctive use of water is not readily available, but if compiled, it would show that quite a large area in the world is served in this manner.

Chandra (1995) claims that the quality of drainage water and its safe disposal will be one of the biggest challenges in future. In addition to agricultural chemicals, drainage water from irrigated area contains common salt. The latter does not constitute a problem, if there is an adequate outfall drain to the sea. The failure to construct an adequate outfall drain to the sea, the use of a water fowl habitat for drainage water disposal and the lack of water quality monitoring can lead to disastrous results as in case of the Kesterson reservoir of California's San Joaquin Valley.

Where the disposal of drainage water to the river or sea is not possible, evaporation ponds are being tried but these may eventually lead to other environmental problems. The Agro-forestry concept, in which saline drainage water is reused for irrigating salt-tolerant crops and trees, has been tried in California. Drainage water is continually reused for increasing salt-tolerant tree species. The trees/crops are of economic value. Ultimately, the drainage water, with its volume significantly reduced, is disposed of in a solar evaporator, where the salt is converted into a crystalline form.

Dinar et al. (1993), dwelling on the need to restrict irrigation drainage flows, recommend the levy of taxes or quotas on the drainage produced at the farm or regional level. Indirect policies can restrict drainage by discouraging inefficient water use, either through taxes or a quota on purchased water, or cost sharing for water conserving technologies.

Biodrainage is yet to be accepted and practised as a planned drainage measure. There are examples and experiences in USA, Australia, India, and Israel where swampy lands have been reclaimed for agriculture with the help of plantations. In California, USA, agro-forestry has been used to consume some of the saline drainage discharge from the subsurface pipe system and to reduce the net quantity to be disposed of. In India, biodrainage has shown promising results on the IGNP, Rajasthan.

COMPARISON OF DRAINAGE METHODS

Each drainage method, viz. horizontal, vertical and biodrainage has some limitations. Each has its own merits and demerits. The choice of the most appropriate drainage measure would depend upon specific site and climatic conditions. Sometime, it may be most appropriate to adopt two or all the there methods to complement one another.

The subsurface piped drainage is a proved and tested technology. Its demerit is its high cost and the pollution by drainage effluent. Vertical drainage by pumping form wells is feasible only where the groundwater is non-saline. Moreover, energy is needed for pumping water which, particularly in the developing countries, is not readily available. Biodrainage holds lot of promise but has not yet been adequately tested in the field under different physiographic and climatic conditions. Appropriate planning and design standards are yet to be evolved for it.

An important consideration in the choice of the drainage measure is its cost. This would vary from region to region depending upon the prevailing conditions and circumstances. On the basis of some experiences in India, Gupta (1998), gives estimates of cost as shown in Table 9.1.

Table 9.1 *Cost of different drainage systems*

Drainage method	*Cost Rs[3] ha*
Subsurface piped (horizontal) drainage	20,000–30,000
Well-pumped (vertical)drainage[1]	5,000–7,500 + Rs 1,00,000 per annum
Biodrainage[2]	11,000

Source: S.K. Gupta, "Drainage Alternatives for Management of Waterlogged Areas" (1998).
Notes: 1.The cost/ha is estimated under the assumption that a tubewell serves an area of 5 ha. The value shown separately with the plus sign is the operational and maintenance cost of a government tubewell.
2.It includes the maintenance cost for two years and the surface drainage cost.
3.The Indian Rupee Rs – US Dollar exchange rate was about Rs 32 to 1 US Dollar when these figures were estimated.

In needs to be mentioned here that the reported cost of subsurface piped drainage in Table 9.1 probably does not include the cost of open collector and main drains. The reported cost of biodrainage is perhaps the cost of plantations. In Rajasthan, India, an expenditure of Rs 6.822 million was incurred on plantations over an area of 600 ha. Since an area under plantation can provide drainage cover to an area about 10 times the plantation area, the net effective cost on biodrainage on the basis of served area would be Rs 1,100 per ha (not Rs 11,000 per ha).

Diwan (1998) in a study of area that was already waterlogged or was threatened with waterlogging in North West India, estimated area needing different type of drainage measures as:

Type of drainage measure	*Area in thousand hectares*
Horizontal – Subsurface	10
Vertical – through wells	162
Biodrainage	598

An attempt is made to compare the three drainage methods and comparative position is depicted in Table 9.2.

Table 9.2 *Comparison of drainage methods*

Parameter	Type of drainage method		
	Horizontal drainage	Vertical drainage	Biodrainage
1	2	3	4
(A) Efficiency and Dependability	Tried and tested method, where an outfall is available. Evaporation ponds have shown mixed results. No standards available for evaporation ponds	Tried and tested method	Tried and tested at many locations with success. But so far not adopted as planned and designed whole drainage method on irrigation projects
(B) Cost	Very high cost, almost as high as construction costs of canal system	Medium cost	Low cost
(C) Benefit	Reclamation of waterlogged area. Would keep the water table rise under control	Reclamation of waterlogged area. Would keep the water table rise under control. Can make additional water available for conjunctive use.	reclamation of waterlogged area. Can keep the water table rise under control. Would provide shelter belts against wind hazards. Would provide additional wood and forest produce.
(D) Expertise requirement for installation	Special expertise and equipment needed for design, planning and installation.	Normal expertise and equipment for installation of wells/tubewells would be needed.	Normal expertise for planning and development of afforestation would be needed.
(E) Operation and maintenance	Subsurface drains would need periodical cleaning. open drains require regular maintenance. Clearance of weeds in open drains is sometimes problematic. Evaporation ponds would require periodical excavation or scrapping or disposal of salts.	All wells would need periodical maintenance . Pumping sets would need repair maintenance and replacements. Power transmission lines would need maintenance	The plantation would need up-keep and care, normally needed on afforestation works.

Parameter	Type of drainage method		
	Horizontal drainage	Vertical drainage	Biodrainage
1	2	3	4
		Interruption in power supply would affect operation. Pipelines and channels carrying pumped water need normal repairs and maintenance	
(F) Requirement of			
(i) Land	Nil for subsurface drains. Open drains and evaporation ponds would require land.	Very little	Land would be needed for afforestation. In IGNP, land requirement is estimated as about 10 per cant of the CCA.
(ii) Water	Nil	Nil	Water would be needed for establishing plantation in the initial years.
(G) Energy requirement	Nil, if no pumping is involved.	Energy needed for pumping water.	Nil
(H) Flexibility	If results are not satisfactory, can be augmented or remodelled with extra costs.	If results are not satisfactory, can be augmented or remodelled with extra costs.	If results are not satisfactory, can be supplemented by horizontal and/or vertical drainage in selected areas at an extra cost.
(I) Environmental impact	Adverse. The drainage discharge is highly saline requiring suitable disposal. If discharged into natural streams, would pollute them. Evaporation ponds, too, can pose environmental problems	Adverse. Pumped water is of inferior quality. Causes deterioration in the quality of stream or canal water on mixing	Positive. There is all round positive impact on the environment particularly in dry and arid regions.
(J) Social acceptability	Negative-Positive	Negative-Positive	Positive

CONCLUSION

Biodrainage has several advantages over other conventional methods of subsurface drainage. It is cost-effective and can serve a very useful social purpose. Most importantly, it is environment-friendly and does not cause pollution or the degradation of natural resources, which other subsurface drainage measures do.

REFERENCES

Cervinka, V. (1990). 'A Farming System for the Management of Salt and Selenium on Irrigated Land (Agroforestry)', California Department of Food and Agriculture, Sacramento, USA.

Chandra, A. Madramootoo (1995). *Sub-surface Drainage in North America: 150 Years of Progress and Challenges*, McGill University, Montreal, Canada.

Chedieng and Vishwanathan (1997). CWC (1996). Quoted from C.D. Tahte et al. (1998). 'Degradation of Irrigated Area Due to Waterlogging and Salinity: Status and Prospects'.

Dinar, Ariel, Marcel P. Aillery and Michael R. Moore (1993). *Water Resources Research*, Vol. 29, No. 6, p. 1535.

Diwan, P.L. (1998). 'Prevention and Remedial Measures in the Management of Waterlogged Areas', WAPCOS National Workshop, India.

Gupta, S.K. (1998). 'Drainage Alternatives for Management of Waterlogged Areas', National Workshop, India.

Hall, W.A. (1974). 'Salty Solutions to Salty Problems: Salinity in Water Resources', Proceedings of 15th Annual Western Resources Conference, University of Colorado, Merriman Publishing Co., Boulder, USA, July 1973.

Heuperman, A.F. (1991). 'Overview of Potential Benefits and Problems Associated with the Growing of Trees in Irrigated Areas', Proceeding of the Symposium Held at the Institute of Sustainable Agriculture at Tatura, Australia, July 23, p. 4.

Kapoor, A.S. (1997). 'Biodrainage of Lands Under Irrigation in Dry Arid Regions with Particular Reference to IGNP, Rajasthan' (Unpublished).

Les, B. Affer and D. Zimmer (1995). *Environmental Engineering: Review of Western European Experience in Sub-Surface Drainage*, French National Institute for Agricultural Research, Paris, France.

National Institute of Hydrology (NIH) (1995). *Drainage Manual*, Chapter 2, pp. 23–30.

Rao, K.V.G.N. (1998), 'Waterlogging in Irrigated Agriculture and Its Management', National Workshop, India.

Schultz (1990). Quoted from C.D. Tahte et al. (1998). 'Degradation of Irrigated Area Due to Waterlogging and Salinity: Status and Prospects'.

Thatte, C.D. and S.A. Kulkarni (1998). 'Degradation of Irrigated Area Due to Waterlogging and Salinity: Status and Prospects', National Workshop, India.

Tyagi, N.K. (1998). *Improvement in Irrigation System for Salinity Control in Agricultural Salinity Management in India,* Chapter 20, CSSRI, Karnal, India, p. 309.

10

Agro-Forestry and Silviculture

A major part of the description in the preceding chapters relates to biodrainage as an option to control waterlogging and salinity in irrigated sands. Various factors such as feasibility, planning, design and comparison with other drainage methods have been considered in this context. Forests and plantations are important for ensuring environmental stability and maintenance of the ecological balance is vital for the sustenance of all life. Plantations are also needed to meet needs of firewood, timber, fodder, etc. There is an urgent need to increase forest cover, particularly in the regions where forests have suffered serious depletion during recent times and the available cover is inadequate. India is an outstanding example in this regard where forest cover constitutes only about 19 per cent of the land area against the world average figure of 33 per cent.

Forests are the world's air conditioners and earth's blankets. Without forests, the world would be an inhospitable place to live in. Dwivedi (1992) categories the functions performed by forest as:

(a) Productive
(b) Protective and ameliorative
(c) Recreational
(d) Developmental

PRODUCTIVE FUNCTIONS OF THE FORESTS

Forests performs the following productive functions:

1. Forests are valuable natural resources. The goods provided by the forests are of immense importance. Wood is a major forest produce and is used extensively for various purposes. In India, most the wood produced is used for the construction of houses, and the building of agricultural implements, bridges, sleepers, etc. About

12.5 million cubic metres of timber is produced form forests in India. Many species such as *teak, sal, deodar, sissoo, babul, chir, haldu, axlewood, rosewood* and *dipterocarps* yield valuable timber.

2. Wood is a universal fuel. For thousands of years until the advent of coal, oil, gas and electricity, wood constituted man's chief source of fuel. Even today, more than half of the total world consumption of wood is for fuel. Wood remains a major source of domestic fuel in India. Approximately 175 million cubic metres of wood is used as fuel in the country, most of which comes from forests (Anon, 1976).
3. Forests provide raw material to a large number of industries, such as paper and pulp, plywood and other boards, saw mills, furniture making, packing cases, matches and toys.
4. A large number of the non-wood products are also available from forests. These are commonly called minor forest products (MFP) not because these are of minor significance but because they are harvested in smaller quantities. Some of the important minor forest products are:
   (i) **Fibres and Flosses:** Fibres are obtained from the base tissues of certain woody plants which are used for making ropes. Flosses are obtained from *semal (Bombax ceiba)* and *kapok (Ceiba pentandra)*.
   (ii) **Grasses and Bamboos:** A large variety of grasses are found in the forests. About 20 per cent of the 419 million livestock in India graze in the forests. Among valuable grasses, *sabai (Eulaliopsis binata)* is harvested annually to the tune of about 80,000 tonnes. About 5.5 million tonnes of bamboo is harvested from forests every year. Approximately 10 per cent of this bamboo is used for housing, 25 per cent for rural/agricultural works, 25 per cent for paper-pulp and the remaining quantity for packaging and other uses.
   (iii) **Essential Oils:** India produced about 1500 tonnes of essential oils during 1980. Many species, e.g., *Eucalyptus spp, Bursera spp, Cymbopogon spp, santalum album,* etc. produce these oils, which are utilised in making soaps, detergents and chemicals.
   (iv) **Oilseeds:** Many tree species, e.g., *Madhuca indica, Pongamia pinnata, Shorea robusta, Azadirachat indica, Schleichera oleosa,* and *Vateria indica* produce oil-bearing seeds which are commercially important. Some of these oils can *be* made fit for consumption. Presently these seeds are used in the soap industry. Tribals use these oils for various purposes. There is a potential for the production of about one million tonnes of oil every year form forest tree seeds.
   (v) **Tans and dyes:** A variety of vegetable tanning materials are produced in the forests. Important vegetable tanning materials are the myrobalan nuts and bark of wattles (*Acacia mearnsii, A. decurrens, A. nilotica, and Cassia auriculata,* etc.). Other tanning materials include leaves of *Emblica officinalis* and *Anogeissus Latifolia*, bark of *Cleistanthus collinus, Cassia fistula, Terminalia*

alata, T. arjuna, and fruits of *Zizyphus xylocarpa, Kattha* and *Kutch* are obtained form *Acacia catechu* trees.

(vi) **Gums and resins:** Gums and resins are exuded by trees as a result of injury to the bark or wood. Gums are collected from several tree species, viz. *Sterculia urens, Anogeissus Latifolia, Lannea coromandelica, Acacia nilotica, Cochlospermun religiosum, Pterocarpus marsupium,* and *Butea monosperma*. Resin is obtained from *Pinus roxburghii.*

(vii) **Drugs, spices and insecticides:** Out of a total of about 2000 items of drugs mentioned in the Indian Medica, over 1800 are of vegetable origin. A large number of these are obtained from forests. All parts of the plants, e.g., roots, shoots, leaves, fruits, seeds and barks are used for making drugs. The important species which yield drugs include *Rauvolfia serpentina, Ephedra spp., Hemidesmus cus, Swertia chirata, Dioscorea spp, Podophyllus spp., Atropa spp. and Datura innoxia.* A variety of spices like *kalajira* (seeds of *Carum carvi*), *dalchini* (bark of *Cinnamomum zeylanicum*), cardamom (*dried capsules* of *Ellatteria cardomomum*) and *tejpat* (leaves of *Cinnamoum tamala*) are obtained from forests. *Pyrethrum* and *neem* are used as insecticides.

(viii) ***Tendu* leaves and other leaves:** *Tendu* leaves are used to produce *'bidis'* and therefore, these are also called *bidi* leaves. The annual collection of *tendu* (*Diospyros melanoxylon*) leaves is about 90,000 tonnes in the country. In Madhya Pradesh alone, about 45per cent of this quantity is produced. Leaves of trees such as *Bauchinia spp*. and *Butea spp*. are used for making plates, *dona*, etc.

(ix) **Edible products:** The fruits, flowers, seeds and tubers of several forest species are eaten. The fruits and seeds of *Anacardium occidentale, Tamarindus indica, Syzygium cumini, Emblica officinalis* and *Buchanania Lanzan*, flowers of *Madhuca indica*, green pods of *Maringa Oleifera* and new shoots of bamboo are in great demand. There are more than 213 species of trees, 14 species of palms, 128 species of shrubs, 166 species of herbs, 4 species of ferns and 15 species of fungi in the Indian forests which yield edible products (Solanki, 1981).

(x) ***Lac* and other products:** *Lac* is a resinous secretion of the *lac* insects which feed on forest trees, particularly on *Butea monosperma*. Similarly, silkworm is raised on the *Terminalia alata* and *Morus alba* Plantations. Honey is another product which is obtained from forests.

(xi) **Fodder and grazing:** Forests provide fodder leaves and grazing facilities to rural animals. About 20 per cent of the livestock population depends upon forest grazing and leaf fodder supply. Leaf fodder of several tree species is almost as nutritious as that of agricultural fodder crops. Good fodder yielding tree species include *Ailanthus excelsa, Moringa oleifera, Sesbania spp, Morus alba, Albizia lebbeck, Leucaena leucocephala, Pongamia pinnta* and *Hardwickia binata*.

PROTECTIVE AND AMELIORATIVE FUNCTIONS OF THE FORESTS

The protective and ameliorative functions of the forests are of great significance to mankind. Some of these functions are as follows:

(i) Forests play a significant role in maintaining the CO_2 balance in the atmosphere. Without sufficient forest cover all the CO_2 which is released in the atmosphere will not be utilised resulting in higher per cent of CO_2 in the atmosphere. This, according to scientists, will result in environmental disasters such as the warming of world temperature, disturbance in the climate, melting of polar ice caps and increasing sea levels. The CO_2 per cent in the atmosphere has already reached 0.042 as against the normal figure of 0.030 per cent. If this increases continuously, higher temperature and other disturbances on the earth may bring about unimaginable miseries to man-kind.

(ii) Forests increase local precipitation by about 5 to 10 per cent due to their orographic and micro-climatic effects and create conditions favourable for the condensation of the clouds.

(iii) Forests reduce temperature and increase humidity. The temperature in the forests is 3°C to 8°C less than in the adjoining open areas. The reduced temperature adds comfort to life. It also reduces evaporation losses. The effect of forests is not limited to the forested areas but is also experienced in the nearby areas.

(iv) Forests maintain the productivity of the soil by adding a large quantity of organic matter and by recycling of nutrients. The leaves of trees are used as manure. The supply of firewood from forests releases *dung* for use as manure.

(v) Tree crowns reduce the violence of rain and check splash erosion. Forests increase infiltration and the water-holding capacity of the soil resulting in much lower surface run-off. This, in turn, results in the checking of soil erosion.

(vi) Forests check floods. They intercept 15 to 30 per cent the total rainfall. They increase infiltration and the water-holding capacity of the soil. This results in reduced surface run-off and checking of erosion. Most floods are caused due to siltation of river channels resulting form erosion and higher peak discharges caused due to greater surface run-off.

(vii) Forests conserve both soil and water. They prolong the water cycle form its inception to the final disposal into streams and oceans. The longer the water is retained in the land, the greater is its usefulness in nurturing crops and trees and in maintaining the regular supply of water in stream throughout the year. Forests increase subsurface run-off which is much slower than surface run-off and does not cause erosion.

(viii) Forests and trees reduce wind velocity considerably. The reduction of wind velocity causes considerable reduction in wind erosion, checks shifting of sand dunes and halts the process of desertification.

(ix) Forests by reducing erosion check the siltation of irrigation and hydel reservoirs. The rapid siltation of various reservoirs in the country is the result of deforestation in the catchment areas of these reservoirs.

(x) Forests are the storehouse of genetic diversity. Several unknown plants may be of potential use for making medicines and food. Deforestation eliminates several species of plants and animals. About 5000 species of plants, 180 of species mammals, 180 species of birds and several thousand species of insects are already on the verge of extinction due to the loss of forests.

(xi) Forests protect us form physical, chemical and noise pollution. Dust and other particulate and gaseous pollutants cause serious problems.. Forests protect us from these pollutants.

(xii) Forests and trees provide shelter belt and windbreak effect which is beneficial for agricultural crops, particularly in the arid and semi-arid areas. Shelterbelt and windbreak increase agricultural production.

RECREATIONAL AND EDUCATIONAL FUNCTIONS OF THE FORESTS

(i) Forests provide recreational facilities to the people. A large variety of trees and shrubs, animals and birds attract a large number of people. National parks and sanctuaries, which are rich in flora and fauna, are visited by a large number of people.

(ii) Forests function as experimental fields and laboratories for learning for college and university students. They also provide sites for ecological studies.

(iii) Forests offer a natural healing effect for a number of diseases. There are thus a number of sanatoria in the wooded areas.

DEVELOPMENTAL FUNCTION

(i) Forests provide employment to a large number of people. In India, during 1985,forests provided employment to about 10 million people in the primary sector. Even greater employment was generated in the secondary and tertiary sectors.

(ii) Forests and various forest activities help tribals to improve their socioeconomic condition through the collection, processing and marketing of various forest products and by offering them gainful employment.

(iii) Forests provide a good sum as revenue to the Government which is used for various developmental works. During 1985, forests in India provided revenue worth R s 14,500 million.

OTHER FUNCTIONS

Trees are capable of utilising solar energy throughout the year. Forestry systems maintain soil fertility. They prevent soil erosion. Trees fix nitrogen from the atmosphere and return much more in leaf-fall than they take form the soil.

Verinumbe (1987) reported higher yields of dry matter, particularly maize and sorghum, in forest-influenced soils in Sahel (Africa) than in ordinary field soils. Higher crop yields were obtained in the soil under '*neem*' followed by *Prosopis* and *Eucalyptus* than in ordinary fields. In the semi-arid areas of Africa, 30–40 per cent higher yields of maize have been found with *Leucaena leucocephala*, *Gliricidia sepium* and *Cassia sepium* inter-cropped as hedgerow and prunings used as mulch (ICRAF, 1988). The International Institute of Tropical Agriculture (IITA) has developed a system of alley cropping in which *Leucaena* is inter-cropped with corn, yams and rice.

In Haryana and Uttar Pradesh (India), farmers have planted *Eucalyptus hybrid* and *Populus deltoides* in rows in field boundaries, with better total returns from agricultural crop and wood than from simple agriculture without trees.

Supplement food and fodder: Several trees like *Madhuca indica (mahua tree)* yield substantial quantity of food materials in the form of flowers, fruit and seeds. A list of important tree and shrubs yielding food is given in Table 10.1

Table 10.1 *Important forest species yielding food*

Species	*Part used as food*
(A) Trees	
Aegle mannelos	Pulp of ripened fruits
Anacardium occidentale	Fruit; seeds very nutritive
Annono squanosa	Fruit
Artocarpus communis	Fruit
A.heterophyllus	Ripened fruit eaten as fruit; unripened used as vegetable
Bauhinia malabarica	Tender shoots used as vegetable
B.purpurea	Flower buds used as vegetable
B.variegata	Flower buds used as vegetable
Buchanania lanzan	Seeds edible, very nutritive, yield edible oil
Cardia dichotoma	Fruit used as vegetable or pickled
Dillenia indica	Ripened fruit eaten as fruit; unripened used as vegetable
D.pentagyna	Flower bud and young fruit
Dendrocalamus stricaus	Young shoots cooked for vegetable or picked
Emblica officinalis	Fruit eaten or pickled
Ficus hispida	Green fruit used as vegetable
F.palmata	Fruit

Species	Part used as food
F. recemosa	Ripened fruit eaten as fruit; unripened cooked as vegetables
Madhuca indica	Flowers and ripened fruit eaten; seeds yield edible oil
Mangifera indica	Fruit
Moringa oleifera	Flower and fruit used as vegetable
Morus alba	Fruit
Phoenix sylvestris	Fruit
Syzygium cumini	Fruit
Tamarindus indica	Fruit
(B) Shrubs	
Asparagus filicinus	Tuberous foots and tender shoots eaten as vegetable
Capparis decidua	Fruit and flower bud
Carissa carandis	Fruit
Salvadora oleoides	Fruit
Ziziphus spp.	Fruit

(Solanki 1981; Biswas and Bhuyan 1983; Singh 1988).

Trees of several species yield palatable and nutritious fodder. Fodder yields form some trees are indicated in Table 10.2.

Table 10.2 *Fodder yields form some trees*

	Species	Fodder yield per tree per annum (kg)
1.	*Acacia nilotica*	50–120
2.	*Ailanthus excelsa*	100–200
3.	*Bauhinia variegata*	30–50
4.	*Grewia optiva*	25–50
5.	*Prosopis cineraria*	25–50

AGRO-FORESTRY IN SALT AFFECTED SOILS

Singh et al. (1993), in a study of the potential of agro-forestry in the salt-affected soils in India divide the salt-affected soil in two groups: alkali (sodic) and saline. About one-third of the reported salt-affected soils in India are alkali and are characterised by high pH of more than 8.2, large percentage of carbonates and bicarbonates of sodium, exchangeable sodium percentage above 15, variable E C, low fertility and very poor physical properties, which support very little vegetation. The main causes that render saline soils unfavourable for plant growth, include excess amounts of neutral soluble salts chiefly chlorides and sulphates of sodium, magnesium and calcium causing osmotic stress, specific ion toxicity and nutritional disorders. Often the high water table conditions in such soils cause impeded drainage. Coupled with this, poor quality of groundwaters are responsible for the desertification of saline areas.

Large tracts of salt-affected soils in India are confined to villages as community lands, as public properties near cities, along the railway tracks and roads, and government lands reserved for specific purposes. Today, these marginal lands contribute least to the national output and present a picture of desolation. There is a need to revegetate these wasted lands and to prevent their further degradation. Agro-forestry is an option of great promise for the use of such lands, in view of the growing demand for fuelwood and fodder and also of environmental considerations. The establishment of a permanent vegetative cover will stop further degradation of these land resources.

The National Commission on Agriculture (1976), reported a wide gap between the requirement and availability of fuelwood and industrial wood in India. The present availability of firewood is about 25 million m^3 against the requirement of 184 million m^3. This gap is expected to widen further in the near future. According to the available data, based on satellite imageries, the forest cover in India is about 19 per cent as against the stipulated figure of 33 per cent of the geographical area, accepted as ideal for maintaining an ecological balance. Alarmed at the gravity of the situation the National Agriculture Commission recommended afforestation on all types of lands otherwise considered unsuitable for arable crops. Salt-affected soils represent one such category of lands. Although several agencies in different parts of the country have initiated rigorous tree plantation drives, the lack of well-tested technologies severely restricts the success of such afforestation programmes.

Researchers at the Central Soil Salinity Research Institute, Karnal, have developed agro-forestry systems for the optimal use of salt-affected soils. The results of green house and field studies describe salt-affected soils and their scope for agro-forestry, suitable tree species and intercrops, practices for raising trees, the role of agro-forestry in water conservation, the ameliorative effects of trees and the economics of trees plantations.

Considering the vast scope that salt-affected soils offer for agro-forestry, a number of tree species have been evaluated for their tolerance to sodic and saline conditions. Among these, the most tolerant species for alkali soils identified are: *Prosopis Juliflora, Acacia nilotica, Casuarina equisetifolia, Terminalia arjuna, Pongamia pinnata* and *Tamarix articulata*. The moderately tolerant species are: *Albizzia lebbek, Eucalyptus tereticornis, Sesbania sesban, Sesbania aegyptica* and *Dalbergia sisoo*. While in alkali soils, the adaptation is aided by the mode of site preparation; under saline conditions, tolerance is governed by the ability of trees to withstand salt stress and/or waterlogging. In areas where the soil salinity is not accompanied by a shallow water table, species like *Acacia auriculiformis, Terminalia arjuna* and *Leucaena Leucocephala* have showed promise. When high salinity and shallow water table co-existed, *casuarina equisetifolia, prosopis juliflora* and *Tamarix articulata* performed better than the former group.

The adverse adaphic environment of salt-affected soils usually restricts the choice of crops to be grown alongwith trees. For raising good inter-crops some primary colonising plants could be selected, or alternatively the soil between tree rows could be reclaimed by physical and chemical means before planting the inter-crops. Both these approaches have been found to be practical. *Leptochloa fusca (Karnal grass/Kallar grass)*, has proved to be an excellent primary coloniser of salt-affected system. *Leptochloa* produced 46.5 t/ha of green fodder in 15 cuttings over a 50 month period without fertiliser or soil amendments. The soil was improved to such an extent that it was possible to plough under the grass after four years to grow less tolerant but more palatable fodder species like *Trifoloum resupintum, T. alexandrinum* and *Medicago sativa*. Other suitable primary colonising grasses are *Brachiaria mutica* (*para-grass*) and *Panimum antiotale* (*Rhoades grass*).

The second method requires greater monetary investment as inter-crops between tree rows are planted after the reclamation of alkali soils through amendments, or through drainage in case of saline soils. Inter-crops like rice, wheat, mustard, lentil, berseem and gram can be successfully grown alongwith the established plantations of *Eucalyptus, Acacia, Populus, Tectona, Ailanthus* and *Tamarindus* in alkali soils. Inter-crops favoured faster growth of trees like *Eucalyptus* and *Populus* but slightly retarded the growth of Acacia.

Normal silvicultural practices may not favour the establishment of trees in salt-affected soils. Suitable methods for the establishment of trees and related management practices have thus been identified. Special planting techniques have been developed for both saline and alkali soils. An ideal method for site preparation, termed the 'augerhole method' of planting trees in highly alkali soils with hard-pan subsurface layers, has been devised. It involves preparing auger holes 100–120 cm deep and 20–25 cm in diameter using a tractor operated auger. Each augerhole is refilled with a mixture of original alkali soil +3 kg gypsum + 8 kg FYM before planting tree seedlings. In saline soils, where the maximum salt accumulation occurs on the soil-surface, a subsurface planting technique has been found to be ideal wherein the saplings are planted in 30 cm deep furrows or trenches which could also be used for irrigation. This technique has been successful in the afforestation of saline areas.

Studies on the irrigation requirement of woody species have shown a three-fold increase in biomass when the saplings received some irrigation during the first two years of their growth. Fertiliser application at the rate of 25 g nitrogen per plant at planting and an annual repetition of the same up to seven years of plant growth increased eucalyptus growth significantly as compared to plants without N fertilisation. Nitrogen did not favour the growth of tree species capable of fixing N through symbiosis with rhizobium or frankia. Phosphorus and potassium are not normally needed for tree plantations as these are available in sufficient quantities in the soils of the Indo-Gangetic Plains. The application of farm yard manure supplemented the N requirements and production for other essential plant nutrients.

A significant increase in biomass production in *Prosopis, Acaia* and *Casuarina* species was observed with proper plant spacing and tree training practices. For firewood production, a spacing of 2 m between rows and in between plants was found ideal for tree plantations alone, which for plantation in conjunction with intercrops, a spacing of 4 m ×4 m was found to be optimum. The pruning of side branches, which become inactive photo-synthetically, produced better tree growth and significantly improved the yied of the inter-crops. In 2 × 2 m Prosopis plantings a firewood yield of 13t/ha was obtained by cutting side branches during a 40 month growth period. The amount of biomass thus generated is sufficient to supplement the fuelwood needs of a rural family. In the ridge trench system of planting in alkali soils where trees are planted in ridges, run-off is prevented and the rainwater so conserved, promotes good growth of the alkali-tolerant inter-crop of Leptochloa grass.

Studies on the amelioration of alkali soils through agro-forestry have provided comprehensive information on litter fall, nutrient cycling, soil fertility build-up, nitrogen fixation, water conservation and improvement in general environment. The short- and long-term effects of *Prosopis Juliflora, Acacia nilotica, Casuarina equisetifolia, Terminalia arjuna, Albizzia lebbek, Sesamia spp,* and *Eucalyptus tereticornis* on soil properties show the superiority of leguminous tree on the non-leguminous ones. The growing of salt-tolerant crops in association with trees further hastens the amelioration process. The ameliorative effect of certain trees in alkali soils is in the following order. *Prosopis Juliflora>Acacia nilotica>Sesbania spp>Terminalia arjuna>Albizzia lebbek>Eucalyptus tereticornis.* Nitrogen fixation by *Sesbania Sesban* is about 320 to 360 kg/ha/annum in partially reclaimed alkali soils.

The cost-benefit ratio of afforestation of alkali soils by applying the recommended planting technique was worked out for the biomass production obtained form the 6–8 year old plantation of *prosopis juliflora, Acacia nilotica* and *Casuarina equesitifolia*. The approximate wood production (firewood+small timber)/ha/annum was 15 tonnes for *Casuarina,* 20 tonnes for *Acacia* 25–30 tonnes for *Prosopis*. The likely net benefits/ha/annum were worth around Rs 8165. This technology for raising tree plantations has been adopted by agencies like forest departments and other organisations in their afforestation programmes. So far, more than 20,000 ha of barren alkali soils have been revegetated in the States of Haryana, Uttar Pradesh and Punjab.

Experimental evidences accumulated so far have identified agro-forestry as an economically viable land use system for the utilisation and improvement of salt-affected soils, especially the alkali soils of the Indo-Gangetic plains. In many salt affected soils where good quality irrigation water is not available , afforestation with hardy species is the only practical solution for their proper utilisation. Moreover, agro-forestry is an alternative to check wide spread waterlogging and salinity in the canal command areas.

Agro-forestry should not only be viewed as a source of wood/fuel, fodder and food production system, but also as an alternative form of the biological reclamation of the salt-affected soils.

Snowdown (1984) is convinced that trees reduce salinity problems on farms. As an example, one farmer with the guidance of Matt Mathews of the National Resources Conservation League, Australia, almost twenty years ago, planted a three-row belt of salt tolerant *Eucalyptus* and *melaleucas* down the middle of an unproductive back paddock that was virtually 'white ground'. That plantation is now well established and when the author inspected the site he found quality pasture extending for more than forty metres on each side of the belt before it deteriorated into bead bush. This farmer wished that he had put tree belts at 100 m interval across that section of his farm. Many farmers were including tree belts and clumps in their re-layout plans.

TREE SPECIES SUITABLE FOR SALINE SOILS

Saline soils suffer due to the excessive concentration of soluble salts and high water table often leading to waterlogged conditions and are met in areas with poor quality groundwaters. The poor aeration of the root zone and the adverse effects of excess salts, which may operate simultaneously, prohibit the establishment of good plantations on saline soils. Therefore only those plant species which can withstand high salt content and thrive under high water table conditions should be selected for planting. The high salt contents in soil solution cause physiological drought condition. To cope with salinity stress, plants such as *Salvadora persica*, *Tamarix aphylla*, and *Melaleuca leucodendron* accumulate salts in their tissues.

Firmin (1968) reviewed the works on the salt-tolerance of trees and shrubs in the areas with high saline water table in the root zone and reported the woody species of the genus *Atriplex, Prosopis, Tamarix, Ziziphus, Casuarina* and *Acaia* to be most tolerant to underground saline water situations. Similarly, the salt-tolerance of about 200 plant species, mainly trees and shrubs, irrigated with saline water with salt contents of 2000 to 6000 mg/1 or more, was studied over a 15 year growth period by Boyko and Boyko (1986). Le Houeron (1986) has compared many native and exotic tree species in the Mediterranean region for salt-tolerance. While *Atriplex* proved most tolerant up to a salinity level of 60 ds/m, *Casuarina, Eucalyptus, Acacia, Tamarix, Populus, Salvadora* and *Halosylon* species were moderately tolerant.

The greenhouse studies of Tomar and Gupta (1986) have shown that tree species differed in their ability to withstand salinity and aeration stresses individually and simultaneously due to their generic differences (Table 10.3). Based on 50 per cent yield reduction and The salinity level at which the satisfactory growth of many species occurred, Tomar and Gupta (1986) grouped various tree species for their salinity-tolerance. *Tamarix* and *Prospis Juliflora* were identified as tolerant; *Acacia nilotica, Acaia tortilis, Callistemon Lenceolatus, Casuarina equisetifolia* and *Eucalyptus camaldulensis* as moderately tolerant; *Pongamia pinnta Eucalyptus hybrid*, and *Terminalia arjuna* as moderately sensitive and *Salix spp, Syzygium cumini* and *Tamarindus indica* as sensitive tree species. Tomar and Gupta (unpublished) have recently compiled data based on 6 to 8 years of experiment on a saline soil and reported on the salinity levels ideal for the satisfactory growth of trees (Table 10.4). Species such as *Acacia farnesiana, Tamarix (articulata), Prosopis Juliflora, Pithecellobium dulce* and *Parkinsonia aculeata* were rated as the most tolerant among the species studied.

Table 10.3 *Effect of salinity and moisture regimes on the survival of saplings 40 days after transplanting in pots*

Tree	Salinity levels at			
	Field moisture (EC_e)		Saturation moisture (EC_e)	
	4.6	26.0	4.6	26.0
Acacia auriculiformis	3	2	1	1
Acacia catechu	2	0	0	0
Acacia nilotica	3	2	3	1
Eucalyptus hybrid	3	0	2	0
Cassia siamea	2	0	2	0
Casuarina equisetifolia	3	3	3	3
Prosopis Juliflora	3	3	3	3
Dalbergia sissoo	3	1	2	0
Leucaena Leucocephala	3	1	0	0
Pongamia pinnata	3	1	3	1
Terminalia arjuna	3	2	3	0

Ece in dS/m
0,1,2, and 3 denote complete mortality, poor, moderate and good tolerance respectively.

Table 10.4 *Salinity levels for some fuelwood tree species up to which satisfactory growth takes place*

Tolerant (Ec 25–35 dS/m)	*Tamarix troupii, T. articulata, Prosopis Juliflora, Pithecellobium dulce, Parkinsonia aculeata, Acaia farnesiana*
Moderately tolerant (Ec-15–25 ds/m)	*Callistemon lanceolatus, Acacia nilotica, A.pennatula, A. tortilis, Casuarina glauca 13144, C. glauca 13987, C.obesa 27, C.glauca (FRI), C.equisetifolis (FRI), Eucalyptus camaldulensis, Leucaena leucocephala, Eresecntia alata*
Moderately sensitive (Ec 10–15 dS/m)	*Casiarina cunninghamiana (FRI), C. cunninghaniana) (Aust), Eucalyptus tereticornis, Acacia auriculifoumis, Terminalia arjuna, Pongamia pinnata, Guazuma ulmifolia, Leucaena shannoni, Samanea saman, Albizzia caribea, Senna atomaria*
Sensitive (Ec 7–10 dS/m)	*Syzgium cumini, S.fruticosum, Tamarindus indica, Salix spp., Acacia deami, Albizzia quachepele, Aleleia herbertsmithi, Caesalpinia eriostchys, C.velutina, Haematoxylon Brasiletto*

PRACTICES FOR RAISING TREES IN SALINE SOILS

Saline soils suffer form disabilities caused by the excessive concentration of salts, high water table often leading to waterlogging and then occurrence of poor quality ground-water in many areas. Poor root zone aeration caused by the high water table and excess presence of salts, which operate simultaneously, impair the success of plantations on such soils. Saline soils need to be leached of excess salts at least in the root zone for crop production. This can be best achieved through suitable drainage systems which will remove salts from the root zone as well as lower the water table. Large scale subsurface drainage is, however, expensive and at times not feasible. Afforestation affords the alternate land use system for the speedy rehabilitation of soils degraded through salinity and waterlogging. Suitable planting techniques and the choice of appropriate tree species are prerequisites for reducing mortality and for effective initial establishment of the saplings. The planting technique should be such that salt concentration in the root zone remains at a low level and the plants are able to escape the adverse affects of high salinity. The following afforestation practices in case of saline and waterlogged soils need special emphasis.

RAISING SEEDLINGS

With germination and the early stages of seedling being more sensitive, the plantations in saline soils are better established by transplanting seedlings grown in a mixture of normal

Table 10.5 *Effect of saline irrigation and fertiliser use on shoot growth in seedling of three tree species*

Fertilisers	Saline water (E_c dS/m)			
	0.4	3	5	7
	cm			
	Eucalyptus hybrid			
Control	29.5	3.7	3.0	—
N	26.7	4.8	3.3	—
P	25.0	7.2	—	—
	Acaacia nilotica			
Control	33.9	23.2	8.5	—
N	30.0	26.0	10.6	4.3
P	29.6	23.6	8.0	–
	Prosopis Juliflora			
Control	23.0	21.9	15.9	5.3
N	23.2	20.8	16.3	5.7
P	26.9	20.8	16.1	6.4

soils, sand and well-rotten farmyard manure. The tolerance of forest tree species to saline water irrigation greatly vary at the germination and seedling stages. *Eucalpytus* hybrid is more sensitive and may fail to survive when irrigated with water of Ec exceeding 2 ds/m while *Acacia nilotca* and *Prosopis Juliflora* are observed to be more tolerant (Tomar and Yadav, 1980). In saline soils, the seedlings become chlorotic, the foliage appears yellowish and despite good manuring, the growth will be very poor.

Less seedling failure occurs if initial irrigations are done with good quality water and saline water is used at a later stage. Tomar and Yadav (1985) irrigated the young seedlings of *Prosopis Juliflora, Acacia nilotica* and *Eucalyptus hybrid* with water of Ec 0.4,3,5 and 7 dS/m. Seedlings of these species could tolerate only moderate soil salinity (Table 10.5).

▼ PLANTING METHODS

A suitable planting technique is necessary for raising successful plantations in saline soil. This technique should ensure the maximum use of rainwater to help lower the salt concentration in the active root zone. In waterlogged saline soil, it should prevent excess salt accumulation in the surface soil and allow adequate aeration of the root zone. Sowing or planting *Prosopis Juliflora* on the high ridges instead of pits has produced good results in little Rann of Kutch because the salt content is reduced through leaching of the ridges (Shah, 1957).

An experiment with ten tree species and two planting methods was initiated in March, 1984 at CSSRI Farm, Sampla, to evaluate their relative salt-tolerance and to compare two planting methods (Tomar and Gupta, 1986). In the subsurface planting method, saplings were transplanted in an‘ auger hole’ of 30 cm diameter and in 30 cm deep trenches, and in ridge plantation method, saplings were planted on 40 cm high ridges. The height, diameter and survival of all tree species were observed to be higher in the subsurface method than in ridge planting (Table 10.6). It was observed earlier that salt accumulation in the root zone (30–90 cm) was substantially higher in ridge planting than on the surface level (Tomar and Gupta, 1986). Ridge planting exposed the sapling to higher salt concentration in the root zone. Survival and height data of the different tree species (Table 10.6) show that most species performed better when planted by the subsurface method. Besides high salinity, ridges also increased the depth to groundwater table in a high water table area like Sampla Farm of CSSRI Karnal. Hence, the water requirements of ridge planted saplings are mostly met from the less salinised trans-mission and phereatic zone. Tomar and Gupta (1986) conducted another experiment at the same site at Sampla in which nine tree species were planted under the subsurface and furrow planting methods. In the former, saplings were transplanted in 15 cm diameter and 30–45 cm deep holes and in the latter, they were transplanted at the sole of 60 cm wide and 90 cm deep furrows made by a tractor-driven furrow plough. The survival of saplings as monitored four months after planting is shown in Table 10.7. The soil salinity was monitored along each sapling before and after applying irrigation water. The data indicates that furrow planting had an edge over the subsurface planting method.

In another experiment, Tomer and Gupta (1987) planted five tree species under the subsurface, ridge-trench and furrow planting methods. After 27 months, furrow planting showed significant improvement in survival and height as compared with the other two methods. The distribution of the root zone salinity in the three planting methods showed that furrow planting provided the least hostile saline environment to the roots and therefore ensured better survival and plant growth.

Table 10.6 *Plant height, diameter and survival percentage of tree species at different growth periods*

Tree	Planting methods	Months after planting					
		6		27		69	
		Height	Diameter	Height	Diameter	Height	Diameter
		cm					
Acacia nilotica	S	95 (100)	2.7	326(69)	5.2	586(50)	12.5
	R	153(100)	3.0	286 (100)	4.4	380 (13)	6.5
Prosopis Juliflora	S	185 (100)	3.1	372 (100)	4.5	725 (100)	11.7
	R	196 (100)	2.9	330 (100)	3.7	549 (100)	8.4
Acacia tortilis	S	118 (81)	2.9	290 (75)	4.5	500 (56)	10.8
	R	82 (100)	1.4	180 (75)	2.5	291 (25)	4.2
Eucalyptus camaldulensis	S	110 (81)	2.2	224 (50)	3.3	593 (19)	8.9
	R	116 (69)	1.8	157 (38)	2.3	177 (0)	—
Terminalia arjuna	S	99 (100)	2.2	183 (81)	4.0	414 (50)	12.7
	R	71(100)	1.1	140 (25)	3.2	165 (0)	—
Leucaena leucocephala	S	152 (100)	2.4	277 (69)	4.7	628 (63)	9.2
	R	103 (100)	0.7	168 (19)	2.0	370 (13)	5.4
Eucalyptus hybrid	S	79 (63)	2.2	287 (31)	4.1	513 (0)	—
	R	87 (69)	1.9	181 (31)	2.6	181 (0)	—

S= Subsurface planting method R= Ridge-trench planting method
Figures in parenthesis are percentage of survival of the tree species

Table 10.7 *Effect of planting methods on the survival of some salt tolerant tree species*

Tree species	Planting methods		Survival	
	Subsurface	Furrow	Subsurface	Furrow
	Ec (dS/m)		(%)	
Acacia nilotica	10–18*	4–6*	30	100
Acacia auriculiformis	8–15	7–9	40	85
Casuarina cunninghamiana	7–25	8–12	60	95
Casuarina equisetifolia	9–20	5–9	90	95
Casuarina glauca	8–27	11–17	80	90
Eucalyptus camaldulensis	6–18	8–11	50	100
Parkinsonia aculeata	7–21	6–10	75	90
Tamarindus indica	8–29	10–14	10	50
Terminalia arjuna	7–18	5–7	15	100

*Range of soil profile salinity (30–90 cm depth) in the saturation extract, before and after irrigating the saplings. Limit values are based on 20 observations. The higher survival rate of saplings in furrow planting method was due to the greater desalinisation of the soil profile with the uniform application of irrigation water.

Jeet Ram et al. (2002) report successful planting and growth of *Eucalyptus Bhadrachallan clones* in fully waterlogged land with groundwater EC values up to 10 dS/m by making earthen ridges 0.70 m high and 2 m wide at top and planting two rows of saplings at a spacing of 1 m × 1 m. The earthen ridges were spaced 30 m apart. Two rows of pits 45 cm × 45 cm × 45 cm size at a spacing of 1 m × 1 m were dug. 2 kg gypsum and 5 kg farm yard manure per plant was added in the soil of all the ridges as well as in pits. The ridges were covered with 5 cm layer of mixture of sand, gypsum and farm yard manure on the top. Peripheral bunding was done on top of ridges so that rainwater collects and percolates into the soil of ridges to leach down the salts. The trees attained a height of about 10 m and a girth of 20 cm in 2 years time.

WATERING

Watering ensures better survival and growth and therefore, wherever possible, plants should be watered during the dry and hot weather. Salinity effects are also reduced by irrigation as it prevents the excessive accumulation of salts in the root zone. Afforestation of salt-affected community lands requires the provision of adequate fresh water supplies. The establishment of tree saplings in saline soil often suffers for want of sufficient canal water resulting in the use of highly saline groundwater for irrigation. Tomar and Gupta (1989) evaluated the furrow planted *Acacia nilotica* and *Prosopis Juliflora* irrigated with canal water of EC 0.5 ds/m, and saline water of EC 18–29 ds/m received as effluent from the drainage experiments. Irrigation with canal water drastically reduced the root zone salinity below the sill up to a depth of 90 cm. The soil salinity increased with the use of saline water but in the root zone below the sill of furrow, it was still lower than the values observed near the furrow edges and in the centre of the bed between two furrows. Salts accumulated in the 90–150 cm soil layer below the sill could be removed by providing a higher leaching fraction.

The initial survival and establishment of the saplings of tolerant tree species can be possible with saline water as well but the detrimental effects of saline irrigation will eventually show up (Table 10.8).

Table 10.8 *Survival percentage of two forest species after six months of planting*

Tree species	*No. of irrigations*	*Total depth of water applied (cm)*	*Quality of irrigation water*	*Survival (%)*
Acacia nilotica	7	12.6	Canal	98
	7	12.6	Saline	73
Prosopis Juliflora	7	12.6	Canal	100
	7	12.6	Saline	98

SEEDLING/TRANSPLANTING

The stand of plants for fuelwood can be established either by sowing or drilling seeds or by transplanting individual small plants. Each method has its own advantages and disadvantages. Transplanting is more successful where the annual precipitation is less than 25 mm. Kulkarni (1959) reported that under saline situations, transplanting gave better results than the direct sowing of *prosopis Juliflora*. The choice of the method will thus depend upon the stress conditions of the planting site.

AMELIORATION OF SALT-AFFECTED SOILS THROUGH TREE PLANTATIONS

Trees are known for their beneficial interactions with the surroundings in which they grow. Soil, which is an important natural resource, is influenced by trees in many ways. The establishment of tree plantation on salt-affected soils sets in an amelioration process by favourably affecting the physical, chemical and biological properties of the soil. The ameliorative effects are mediated through:

- Litter production and nutrient cycling, addition of matter and rise in the humus status of the soil,
- Initiation of biological activity in the otherwise almost lifeless mass of salt affected soil. The biological activity produces CO_2 in the soil which, in combination with water, mobilises the insoluble calcium carbonate.

(i) $CaCO_3+CO_2+H_2O \longrightarrow Ca\ (HCO_3)_2$

(ii) $(2Na)$ Clay+ $Ca\ (HCO_3)_2 \longrightarrow$ Ca Clay+2Na HCO_3

- Reduction of salt deposition in the upper layers of the soil profile by cutting down the evaporation losses of soil water.
- Lowering of the water table in waterlogged and salt-affected areas as trees act as biopumps;
- Opening up the soil and improvement in water permeability by deep and sturdy root systems of trees which facilitates the leaching of salts; and
- Exclusion of large amount of salts through absorption by trees.

The degree to which the salt-affected soils get ameliorated by trees depends upon the tree species, their growth habits, the quantity and quality of litter, planting density, the age of the plantation, ability to fix nitrogen and management practices. Trees having the ability to fix dinitrogen ameliorate salt-affected soils at a much faster rate in comparison with the non-N fixing trees.

LITTER PRODUCTION AND NUTRIENT CYCLING

Litter is formed of tree parts like dead leaves, flowers and fruits which fall on the ground, Upon their decomposition, organic acids, essential nutrients and a few other chemical substances are released which help in soil reclamation. The accumulation of litter on the surface also checks weed growth, protects the soil from the erosive impact of rain and this reduces the surface fun-off. The poor humus status of salt-affected soils makes the litter production potential of forest species quite important. Not much is, however, known about the quantity of litter produced and the amount of nutrients recycled through the litter of forest species raised on salt-affected soils.

In a four-year study on *Acacia nilotica* and *eucalyptus tereticornis* plantations raised on a highly alkali soil (pH_2 10.5, ECe 3.98 and ESP 96 in the 0–15 cm surface layer), Gill et al. (1987) found that the litter production in *Acacia* plantation was significantly higher than in the *Eucalyptus* plantation of the same age and stocking. The higher litter yield in the *Acacia* plantation was attributed to the nature and extent of canopy growth in *Acacia*. The litter yield in *Eucalyptus* was low mainly because of its poor growth in the highly alkali soil. The variation in the total annual litter yield during the observed period ranged from 1.03t/ha to 1.13t/ha *Eucalyptus* and from 2.54 to 5.75t/ha in *Acacia*.

There are seasonal changes in litter fall, and winter accounts for a major fraction of the annual litter in both species. This is unlike deciduous plantations which yield maximum litter during the summer months (Kikuzava et al., 1984). The seasonal litter yield in *Acacia* increases with a peak occurring in the winter season.

Concentrations of N, P, S and K were significantly higher in *Acacia* litter than in *Eucalyptus* (Table 10.9). The opposite was true for Ca, Mg, Na, Fe and Mn contents.

The annually recycled quantities of all nutrients were significantly more in *Acacia* than in *Eucalyptus*. The concentration of most of the nutrients in the litter of both species tended to decrease with the successive growth years notwithstanding seasonal fluctuations. This may be ascribed to the dilution effect.

Table 10.9 *Mean concentration of nutrients in litter of acacia and eucalyptus from 1982 to 1986*

Nutrients	*Acacia*	*Eucalyptus*
	(m mol/kg)	
N	243	160
P	59	20
K	213	100
Ca	134	171
Mg	147	159
S	72	49
Na	37	125
Fe	4.00	5.79
Mn	0.80	28.59
Zn	0.27	0.24
Cu	0.11	0.12

Gill and Samra (1989) studied the nutrient recycling through litter production in *Leucaena leucocephala var, K 8* and *Sesbania sesban* raised by direct seeding, in soils of PH_2 8.4–8.5. The litter (dry) yields increased progressively with the advancement of the growth period in *L. leucocephala*. But in *S. sesban*, the increase was observed up to 42 months only. The amounts of important nutrients recycled through the litter are given in Table 10.10 .The litter composition of about 20 tree species raised on reclaimed alkali soils showed marked variation amongst the species (Table 10.11).

Table 10.10 *Litter (dry) yield and amounts of important nutrients recycled in leucaena leucocephala and sesbania sesban planting in a given period of their growth span*

Item	*Leucaena leucocephala*	*Sesbania sesban*
Average litter (kg/ha) Nutrients recycled (kg/ha)*	5406	5940
N	379.2	508.8
P	27.7	45.4
K	352.7	428.5
Ca	385.8	—
Mg	182.2	221.1

Item	Leucaena leucocephala	Sesbania sesban
S	60.9	85.6
Na	5.26	12.6
Fe	11.47	11.44
Mn	1.92	3.12
Zn	0.43	1.08
Cu	0.26	0.33

*Nutrients recycled in a growth period of 78 months in Leucaena and 66 months in Sesbania plantation.

Table 10.11 *Leaf litter quality in selected tree species in a reclaimed alkali soil*

Tree specie	Nutrient composition								
	P	K	Ca	Mg	Na	Fe	Mn	Zn	Cu
	(%)						(ppm)		
Acacia nilotica	0.24	1.01	1.30	0.17	0.03	510	37	35	34
Aciacia auriculiformis	0.05	0.61	2.30	0.71	0.13	600	40	25	20
Acacia catechu	0.14	0.75	1.30	0.40	0.03	970	35	21	18
Ailanthus excelsa	0.15	1.39	2.40	0.56	0.04	480	90	36	16
Albizzia lebbedk	0.11	0.78	1.70	1.25	0.04	390	58	34	24
Azadirachta indica	0.18	1.57	3.50	0.64	0.04	370	39	36	22
Butea monosperma	0.07	0.45	3.95	0.45	0.05	1080	55	24	20
Casuarina equisetifolia	0.14	0.39	1.60	0.26	0.22	750	99	23	14
Caassia siamea	0.15	0.48	3.15	0.26	0.04	580	74	27	24
Delbergia sissoo	0.01	0.48	3.20	0.70	0.12	500	83	27	22
Eucalyptus terticornis	0.11	0.55	1.55	0.35	0.12	760	75	24	24
Grevillea robusta	0.39	0.45	1.70	0.35	0.04	480	167	19	12
Melia ozadiarch	0.25	0.78	2.65	0.42	0.05	1850	114	79	24
Populus deltoides	0.20	2.16	1.30	0.42	0.08	340	60	74	26
Sesbania spp	0.22	1.51	1.75	0.73	0.04	560	94	43	18
Tractona grandis	0.14	0.58	2.90	0.74	0.04	380	107	51	26
Termarindus indica	0.17	0.78	3.15	0.36	0.03	600	47	25	32
Terminalia arjuna	0.11	0.30	3.45	0.68	0.04	1420	67	39	34

SOIL IMPROVEMENT

Salt-affected soils are very low in organic matter and biological activity. Tree growth provides a green cover and improves the soil environment for enhanced biological activity, better water relations and increased organic matter and fertility status. Tree roots increase the carbon dioxide levels in the soil which help mobilise soil calcium carbonate and the released Ca exchange with Na on the soil exchange complex. Further, the litter produced also hastens the process of reclamation.

The Central Soil Salinity Research Institute, Karnal ,studied the ameliorative role of trees in salt-affected soils. Yadav and Singh (1970) observed that a *Prosopis Juliflora* plantation on an alkali soil near Aligarh led to a decrease in pH and soluble salt content and an increase in organic matter and nitrogen content of the surface 15 cm soil. However, soluble salts increased in the lower depths due to translocation through leaching. Shah (1957) also reported decreased salinity and alkalinity in soils under the prolonged occupation of *Prosopis juliflora* and *Tamarix articulata*. In *soda solonetzes* of *wooded steppes*, Sumbur (1963) observed an increase in the exchangeable Ca equal to twice the amount of Ca applied through gypsum. Gill et al. (1987) observed that in a span of five years, *Acacia nilotica* and *eucalyptus tereticornis* decreased soil pH_2 and EC and improved the organic carbon of a highly alkali soil. The decrease in soil pH was more in the surface layer and the rate of decrease declined with depth. The increase in organic carbon was about three times under *Acacia* and two time under *Eucalyptus*. There was a significant improvement in the soil water permeability under both plantations.

The high density plantation of *Sesbania sesban* resulted in a marked decrease in pH_2 and EC and an improvement in the organic carbon, total and available N during a growth period of about 30 months (Rao and Gill, 1988). The decrease in soil alkalinity and improvement in soil fertility was mainly confined to the surface 30 cm layer (Table 10.12).

Table 10.12 *Changes in soil properties in 30 months in sesbania sesban planted at selected spacings*

Spacing between		*Soil depth*	*pH_2*		*Organic carbon*		*Total Nitrogen*		*Available N*	
Rows	*Plants*	*cm*	*1985*	*1988*	*1985*	*1988*	*1985*	*1988*	*1985*	*1988*
cm					*(%)*				*kg/ha*	
50	100	0–15	8.80	8.75	0.71	0.82	0.08	0.09	74	109
		15–30	9.50	9.20	0.50	0.54	0.07	0.07	73	27
100	100	0–15	8.65	8.30	0.76	0.84	0.08	0.10	72	105
		15–30	9.35	9.15	0.49	0.58	0.06	0.07	66	67
200	100	0–15	8.61	8.50	0.82	0.90	0.08	0.10	76	105
		15–30	9.28	9.20	0.47	0.60	0.07	0.06	74	84
300	100	0–15	8.64	8.20	0.82	0.82	0.09	0.10	75	116
		15–30	9.25	9.00	0.50	0.64	0.07	0.07	71	77

Sesbania sesban produced a 4 t/ha/year litter with a high mean N content of 2.3 per cant showing great promise for planting in semi-reclaimed alkali soils for firewood production and soil improvement. Chhabra et al. (1987) reported an appreciable decrease in pHe, ECe and SARe and an improvement in organic matter and nitrogen content in the top 15 cm soil layer when Sesbania aegyptica was grown for more than four years (Table 10.13)

Table 10.13 *Properties of surface 15 cm soil as affected by plantation of Sesbania Aegyptica*

Soil properties	*Original soil*	*After 54 months*
pHe	10.10	8.95
ECe (ds/m)	18.54	1.44
Organic matter (%)	0.20	0.65
$CaCO_3$ (%)	0.85	0.76
Available N (kg/ha)	21.00	85.50

However, no change in the physical properties of the soil was observed. They opined that a period of 54 months is not sufficient to expect any improvement in the physical properties of the soil under a situation where the soil is not amended and root activity is not encouraged in the surface layers. The ameliorating effects of 20 year old plantation of *prosopis juliflora, Acacia nilotica, Eucalyptus terticornis, Albizzia lebbek* and *Terminalia arjuna,* on an alkali soil (pH_2 10.5 to 10.2 and ECe 1.75 to 0.45 dS/m) were examined by Singh et al. (1990). The average pH_2 of the 0–12 cm soil profile decreased form 10.2 to 9.1 in *Eucalyptus,* to 9.03 in *Acacia,* to 8.67 in *Albizzia,* to 8.15 in *Terminalia* and to 8.01 in *Prosopis* during their 20 years of growth (Table 10.14). the reduction in pH throughout the profile was low in *Eucalyptus* and *Acacia* plantations probably due to less litter fall. Though the EC in profile came down drastically from their original levels, differences among species were small. The organic carbon content in the soil increased considerably, with the maximum being under *Prosopis* and the least under *Eucalyptus*. In the latter, the increase was only in 0–45 cm depth , while in all the other species, the increase was observed throughout the 120 cm depth. P and K contents were maximum under *Prosopis juliflora*. The ameliorating effect on the soil appear to be in the order: *Prosopis*> *Acacia*> *Terminalia*> *Albizzia*> *Eucalyptus*. Soil improvement took place to the extent that agricultural crops could be grown without any soil treatment after removing the trees.

Table 10.14 *Ameliorating effects of 20 year old tree plantation in an alkali soil*

Tree species	Soil depth (cm)	pH_2	ECe (dS/m)	OC (%)	Available P (kg/ha)	Available K (kg/ha)
Original soil (1972)	0–15	10.2–10.5	0.75–175	0.12–0.29	32	310
Acacia nilotica	0–15	8.39	0.25	0.85	59.0	498.6
	15–30	8.77	0.31	0.61	33.9	280.5
	30–45	8.96	0.42	0.51	27.7	266.7
	45–60	9.11	0.56	0.46	27.1	255.8
	60–90	9.37	0.72	0.42	26.2	257.1
	90–120	9.56	0.90	0.42	29.6	231.3
Eucalyptus tereticornis	0–15	8.47	0.44	0.66	33.3	358.9
	15–30	8.80	0.38	0.46	31.3	208.2
	30–45	9.10	0.45	0.28	34.7	262.7
	45–60	9.39	0.62	0.22	38.5	233.7
	60–90	9.55	0.77	0.18	36.3	224.5
	90–120	9.75	0.93	0.18	33.8	172.3
Prosopis juliflora	0–15	7.27	0.51	0.93	110.5	701.5
	15–30	7.63	0.31	0.64	62.8	428.3
	30–45	7.81	0.25	0.58	31.7	296.6
	45–60	8.04	0.27	0.48	25.8	199.5
	60–90	8.37	0.36	0.46	40.7	178.7
	90–120	8.95	0.40	0.41	29.2	175.7
Terminalia arjuna	0–15	7.85	0.32	0.86	67.8	409.8
	15–30	8.00	0.39	0.76	52.7	288.9
	30–45	7.93	0.41	0.50	40.2	212.5
	45–60	7.99	0.44	0.49	27.9	182.2
	60–90	8.24	0.50	0.43	28.7	181.2
	90–120	8.93	0.67	0.44	28.9	161.6
Albizzia lebbek	0–15	7.86	0.32	0.62	42.6	387.0
	15–30	8.30	0.35	0.54	45.1	296.8
	30–45	8.73	0.50	0.44	40.3	183.1
	45–60	8.88	0.56	0.43	47.4	165.8
	60–90	9.04	0.68	0.42	48.5	169.8
	90–120	9.20	0.65	0.36	41.3	151.1

DINITROGEN FIXATION

About one thousand species of tropical woody legumes are known to develop dinitrogen fixing root nodules. Some non-leguminous trees like *Casuarina* can also fix nitrogen. Although some estimates are available on the amounts of nitrogen fixed by trees in normal soil, information is lacking for salt-affected soils. The adverse edaphic environment prevailing in the salt-affected soils may not be hostile to the N-fixing micro-organisms, but may have adverse affects on the host tree-legumes.

Rao et al. (1989) quantified the N fixation by *Sesbania sesban* grown on a moderately alkali soil (pH 8.6 in 0–15 cm and 9.4 in 15–30 cm soil depth). The nitrogen fixation calculated by the 'nitrogen balance method' ranged from 358 to 466 averaging 398 kg/ha year. In another experiment in inter-row spacing, the N fixation by *sesbania sesban* ranged between 312 to 398 kg N/ha/yr. Such a high rate of N fixation could possibly be due to its excellent growth, the presence of highly effective rhizobia strains and adequate phosphorus and molybdenum nutrition to promote N fixation.

In a greenhouse experiment, Batra (1988) observed that at pH_2 10.0 modulation was absent in *Casuarina obesa, C. glauca, C. cunninghamiana* and *C. equisetifolia*. The nodule weight was significantly less at ph_2 9.3 than at pH_2 8.2. Among the species, *C. obesa* and *C. glauca* had more nodule weight. The inoculation of *Casuarina* with *Frankia* culture did not affect nodulation but was beneficial for its growth and N uptake. The nodulation process of *Casuarina-Frankia* symbiosis appeared more sensitive to high pH levels. Further studies on the isolation of *Frankia* strains capable of producing effective nodules in Casuarina in the face of adverse conditions in salt-affected soils, may yield more positive results on the symbiotic functions.

MICRO-CLIMATE MODIFICATION

Tree plantation improve micro-climate, reduce the detrimental effect of extreme weather and check unproductive evaporation. Gill et al. (1990) observed that in the established plantation of *Acacia nilotica* and *Eucalyptus tereticornis* raised on an alkali soil, the mean air temperature under the *Acacia* canopy was lower by 2–5°C during summer and higher by 2–4°C during winter as compared to that of the adjacent open fields. Similar changes were also found in the *Eucalyptus* canopy but these were of smaller magnitude than in *Acacia*. Both canopies also modified thermal regime of the soil at 5 cm and 15 cm depths. This indicates that tree plantations provide a blanketing effect during summer and a sheltering effect during winter.

ECONOMICS OF AGRO-FORESTRY SYSTEMS

Agro-forestry systems can yield significant economic gains. Dwivedi lists the following factors that influence the economic returns from agro-forestry:

(a) Land-man ratio;
(b) Availability of wood products;
(c) Environmental awareness; and
(d) Wood consumption pattern.

Where the land holdings are small and productivity from the land is low, the main aim of small and marginal farmers is to produce enough foodgrains for their sustenance. They are reluctant to divert any piece of agricultural land for any other purpose. The remedy would lie in increasing the productivity of land with assured irrigation and other inputs so that the net return from agriculture-cum-plantations exceeds that under the existing conditions. Irrigation water is an important input and can be used as an incentive tool for increasing agricultural income as well as to promote plantation of trees.

It was estimated (1985) that the prices of agricultural commodities in India during the past 25 years increased by 3 to 5 times while the prices of firewood, timber and charcoal increased by 15 to 20 times during the same period. The future long-term scenario also indicates a greater likely shortage of wood products in the future vis-à-vis that of the agricultural produce.

There is growing awareness about the need of environmental conservation. The felling of trees and cutting of forests has been prohibited in India. Afforestation on waste-lands and planting of tree of trees on farm lands and on the sides of rails, roads and canals is receiving greater attention.

Wood consumption studies carried out in different parts of the world indicate that with an increase in income, there is almost always an increase in commercial and industrial wood consumption. India is facing a shortage of timber and firewood and has taken to importing timber.

The lack of adequate marketing facilities is sometimes the reason for the inadequate returns in terms of the produce of tree plantations. Chambers et al. (1989) pointed out that the under-development of wood markets is the cause of the low price for the producers while consumers pay a substantially higher price.

Agro-forestry systems take a long time span to yield returns. This is a handicap, where long-term investments are not possible.

Woody annuals and perennials can be a valuable source of biomass and nitrogen. Rao (1998) reports that *Sesbania cannabina* yielded woody stem yield (dry) of 20 tonnes/ha at 200 days after sowing with a calorific value of 4730 K cal kg^{-1} . An additional green matter production of 17 annual accessions amounting to 26 tonnes/ha (5.9 tonnes dry matter) at 54 days growth is also reported.

In high density plantations managed for 6 years, dry biomass production was 25–35 tonnes/ha/yr.

Tararix artocilata has been recorded yielding 7–10 tonnes dry matter ha^{-1} yr^{-1} with a final density of 600 tree ha^{-1} after 5 years.

Prosopis juliflora is reported to produce up to 52.3 tonnes ha^{-1} biomass in 6 years in alkali soils.

Tomar and Minhas (1998) report biomass production from afforestation in highly alkaline soils with appropriate technologies, of 197, 162 141 tonnes ha^{-1} from 8–10 years growth of *P.Juliflora, A. Nilotica, C equisetifolia* and *E. tereticornis* respectively. They also report biomass yields from *Sesbania* species of 120–150 tonnes ha^{-1} in a period of 2.5 to 3 years.

Singh and Dogar (1998) report *Leptochloa fusca* (*Kallar grass*) in association with *Prosopis juliflora* (*mesquite*) produced 46.5 tonnes green fodder in 15 cuttings over a 50 month period without any fertiliser or other amendment. Similarly *mesquite* produced 160 tonnes ha^{-1} air dried fire wood in 6 years when planted at 2 m × 2 m spacing.

Studies conducted by NCAER 1998; USAID 1988 noted the phenomenal growth rate of farm forestry in several parts of the country (India), with economic returns being much higher than those from agriculture. But soon thereafter there was a collapse in the market prices, particularly that of *Eucalyptus* and this made farmers averse to planting *Eucalyptus* trees. The lesson to be learnt is that tree plantation should be diversified, to match with market demand and conditions. Some researchers – Chaturvedi (1981), Mathur and Sharma (1983) and Patel (1988), have reported much greater profitability form plantations-cum-agriculture crops as compared to that from agriculture alone.

The Central Soil and Water Conservation Research and Training Institute, Dehradun, carried out a study on the economic evaluation of watershed management projects in India. Some of the project had 'forestry ' as an important component of the watershed projects.

The Kandi Watershed Management Project was taken up by the Punjab State Government with financial assistance from the World Bank. The benefit-cost analysis of the two sub-watersheds, namely Maili watershed with an irrigation cum attenuation dam and Chohal watershed without a dam, in the Kandi tract of Punjab, was done by Singh et al. (1991), on the basis of three rounds of surveys conducted from 1980–81 to 1987–88. The result is reported in Table 10.15.

The benefit-cost ratio for the forestry component of the project at 12 and 15 per cent discount rates were respectively 2.74 and 1.55 for the Maili Watershed Project and 2.27 and 1.32 for the Chohal Watershed. After horticulture, the forestry gave better returns than all other watershed programmes.

The evaluation of soil conservation works in the catchment of Matatila reservoir was done by the Agriculture Finance Corporation (1988). The reported result is shown in Table 10.16

Table 10.15 *Net Present Worth (NPW), Benefit-Cost Ratio (BCR) and Internal Rate of Return (IRR) for different components, Maili and Chohal Watershed, Punjab*

Particulars	NPW (Rs lakhs)		BCR at		IRR (%)
	12%	15%	12%	15%	
A. Maili Watershed					
Forestry	252.95	74.03	2.74	1.55	18.2
Animal husbandry	3.37	0.35	1.14	1.02	15.6
Soil Conservation	–3.6	—	0.09	—	10.8
Horticulture	11.19	6.02	7.05	4.65	25.3
Irrigation	Negative	—	—	—	1.22
Overall (including irrigation)	–5.54	—	0.99	—	11.9
Overall (excluding irrigation)	213.60	24.35	1.81	1.10	15.6
B. Chohal Watershed					
Forestry	68.91	16.18	2.27	1.32	17.3
Animal husbandry	5.38	2.17	1.31	1.14	18.5
Soil Conservation	5.45	-1.28	1.16	0.96	14.4
Horticulture	68.98	39.98	7.44	5.01	26.4
Overall	99.16	16.66	1.66	1.12	16.8

Note: One 'lakh' is one hundred thousand

Table 10.16 *Evaluation of soil conservation measures matatila*

Sector	Project life (yrs)	Discount rate (%)	B:C ratio	IRR (%)
Soil conservation treatment of agricultural land	12	12	1.68:1	41
Water potential created through erosion control/mini storage structures (Bundhies)	12	12	1.38:1	32
	20	12	1.84:1	
Forest/tree plantation cover (mixed plantation of *Eucalyptus, Acacia catechu* and *Dalbergia sissoo*)	—	—	4.48:1	

The forest/tree plantation cover of mixed plantation of Eucalyptus, Acacia Catechu and Dalbergia sissoo gave a benefit cost ratio of 4.48:1, which was better than that obtained from all other activities.

Economic analysis of research conducted in several research farms in India for the effective utilisation of degraded land through fuel fodder plantations was made in several studies and results are presented in Table 10.17

Table 10.17 *Economics of utilisation of degraded lands for non-agricultural uses*

Land use	*Project life (yrs)*	*Data period*	*discount rate*	*B:C ratio*	*IRR*	*Remarks/ references*
			Dehradun			
Dalbergia sissoo+Chrysopsgon fulvus (19.15 m × 9.15 m)	17	17	10	2.75	—	Mathur et al. (1979)
Acacia catechu+Eucaliopsis binata (4.5 m× 4.5 m)	17	17	10	3.45	—	–do–
Acacia catechu+Chrysopagon fulvus (18.30 m × 18.30 m)	17	17	10	2.02	—	–do–
Chrysopogon fulvus	17	17	10	3.02	—	–do–
Eucalyptus (2 m× 2 m)15	15	11	10	1.68	—	Ram Babu et al. (1982)
Agave sisalana+Eulaliopsis binata (50% harvesting intensity of leaves)	15	12	10	1.67	32.8	Gupta et al. (1989)
			Ootacamund			
Eucalyptus plantation for fuel	25	10	10	2.20	—	Secker et al. (1989)
Eucalyptus plantation for fuel and timber	25	25	10	2.38	—	–do–
			Chandigarh			
Eucalyptus+Eulaliosis binata	—	9	10	2.25	—	Sud et al. (1989)
Acacia catechu+Eulalipsis binata	50	19	8	1.79	16.0	Vimal Kishore et al. (1983)
			Kota			
Acacia nilotica+Dichanthium+annultum Prosopisjuliflora and (on ravine top and Cenchrus ciliaris slopes) and grasses Dendrocalamus strictes (On ravine bottom)	20	20	6	1.55	—	Puri and Khybri (1975)
			Agra			
Acacia nilotica (60%)+Prosopis juliflora (20%)+Miscellaneous species (20%)	23	15	10	1.81	16.2	Ram Babu et al. (1984)
Second rotation of above plantation	23	8	10	1.61	27.0	–do–
			Vasad			
Acacia nilotica + Cenchrus ciliaris+Dichanthium annulatum	15	15	12	1.38	15.0	Vimal Kishore et al. (1982)

		U.P Hills				
Bhimal	20	Survey	10	1.64	17.27	Tiwari and Singh (1984)
Chir	100	Survey	10	1.29	12.16	–do–
Bhimal + Chir	100	Survey	10	1.34	12.50	–do–
Lemon (Eureda round)	14	14	10	1.4	20.1	Ram Babu and Arora (1985)
Mandrin (Kinow)	13	13	10	2.13	27.5	–do–

In all the studies, the plantation provided attractive returns.

It can be inferred that the benefits in the form of timber, fuel and fodder from biodrainage plantations, can exceed all the costs of raising and maintaining such plantations. The contribution by such plantation in providing effective drainage would, therefore, be a bonus, free of cost.

CONCLUSIONS

The following conclusions can be drawn from the above studies:

1. Biodrainage plantations, besides providing effective drainage, confer all the benefits that come from agro-forestry and silviculture.
2. Well planned agro-forestry systems can provide equal or better returns than those obtained from agriculture, in which case biodrainage would be a bonus, free of cost.

REFERENCES

Anon (1976). Quotes from A.P. Dwivedi (1992). *Principles and Practices of Indian Silviculture*, Surya Publications, Dehradun.

Babu, Ram, B.L. Dhyani, M.C. Agarwal and J.S. Sharma (1997). 'Economic Evaluation of Watershed Management Projects', Bulletin No. T–33/D–23 of the Central Soil and Water Conservation Research and Training Institute, Dehradun, India.

Batra, L. (1988). 'Performance of Casuarina Species, Their Nodulation Pattern and Nitrogen Fixation', *Ann. Rep.*, CSSRI, Karnal, pp. 122–24.

Buyko, H. and E. Boyko (1968). *Salinity and Aridity*, Dr I.W. Junk Publishers, The Hague, pp. 214–82.

Cervinka V., Finch C., Martin M., Menzo F., Peters D., Buchruff (2001). Drain water salt and selenium management utilising IFDM/Agroforestry systems, Five points near mendata California USA.

Dwivedi, A.P. (1992). *Agroforestry: Principles and Practices*, Oxford and IBH Publishing, Delhi.

— (1992). *Principles and Practices of Indian Silviculture*, Surya Publications, Dehradun.

Firmin, R. (1968). 'Forestry Trials with Highly Saline or Sea Water in Kuwait', in H. Boyko (ed.), *Saline Irrigation for Agriculture and Forestry*, N.V. Junk Publishers, The Hague, pp. 107–32.

Gill, H.S., I.P. Abrol and J.S. Samra (1990). 'Effect of Selected Tree Plantations and Their Under Store Vegetation on Important Soil Water Characteristics in a Sodic Soil', in N.K. Tyagi, P.K.

Joshi, R.K. Gupta and N.T. Singh (eds), *Management of Irrigation System*, CSSRI, Karnal, pp. 255–64.

— (1987). 'Nutrient Recycling Through Litter Production in Young Plantations of Acacia Nilotioca and Eucalyptus Tereticornis in a Highly Alkaline Soil', *Forest Ecol. and Mgmt*, 22: 57–69.

Gill, H.S. and J.S. Samra (1989–90). 'Evaluation of Afforestation Techniques and Tree Soil Interaction in Salt-affected Soils', *Ann. Rep.*, CSSRI, Karnal, pp. 150–57.

Houeron, Le (1986). 'Salt-tolerant Plants of Economic Value in the Mediterranean Basin', *Reclamation and Revegetation Res.*, 5: 319–41.

Kikuzawa, K., T. Asai and M. Fukuchi (1984). 'Leaf Litter Production of Alnus Inokumea', *J. Ecol.*, 72: 993–99.

Ram, Jeet, Vineet Garg and S.K. Dalal (2002). 'Biodrainage for Salt-affected Waterlogged Soils', Paper Presented at the Seminar at Rohtak, Haryana, India, November 9, pp. 11–12.

Rao, D.L.N. and H.S. Gill (1988). 'Nitrogen Fixation and Biomass Production by Sesbania Species', *Ann. Rep.*, CSSRI, Karnal, pp. 124–27.

— (1989). 'Regional Experience with Perennial Sesbania in India', Proceedings of the Workshop on 'Perennial Sesbania in Agroforestry Systems', ICRAF, Nairobi, Kenya, 27 March–2 April.

Rao, D.L.N. (1998). 'Microbiological Processes', in P.S. Minhas and N.K. Tyagi (eds), *Agricultural Salinity Management in India*, Chapter 9, CSSRI, Karnal, India, pp. 133–34.

Snowdown, V.G. (1984). 'Tree Planting in Northern Irrigation Areas (Australia)', *Water Talk*, 5 (December): 14–16.

Shah, S.A. (1957). 'Propagation of Mesquite' (Letters to the Editor), *Indian Forester*, 83: 472.

Singh, Gurbachan and J.C. Dogar (1998). 'Agroforestry in Salt-affected Soils', in P.S. Minhas and N.K. Tyagi (eds), *Agricultural Salinity Management in India*, Chapter 28, CSSRI, Karnal, India, p. 247.

Singh, Gurbachan, N.T. Singh and O.S. Tomar (1993). 'Agroforestry in Salt-affected Soils', Publication Bulletin No. 17, Central Soil Salinity Research Institute, Karnal, India, pp. 1–4 and 12–14.

Sumbar, G.N. (1963). 'Mobility of Exchangeable Sodium and Determination of Rates of Application of Gypsum During Reclamation of Solonetzes Pochvovedenie', Quoted by Kapoor A.S. in 'Biodrainage – A Biological Option for Controlling Waterlogging and Salinity', p. 2777.

Tomar, O.S. and R.K. Gupta (1980). 'Effect of Saline Irrigation Water of Varying EC, SAR and RSC Levels on Germination and Seedling Growth of Some Forest Species', *Indian J. Agric. Res.*, 3: 306–14.

— (1985). 'Effect of Saline Water Irrigation and Fertiliser Applications on Growth of Tree Seedlings', *Annals of Arid Zone*, 24: 94–100.

— (1985). Performance of some forest tree species in saline soils under shallow and saline water table conditions. Plant and Soil 87: 329–35.

— (1986). 'Performance of Tree Species in Saline Soils', *Ann. Rep.*, CSSRI, Karnal, pp. 119–23.

— (1987). 'Performance of Some Forest Tree Species in Saline Soils', *Ann. Rep.*, CSSRI, Karnal, pp. 101–3.

Tomar, O.S. and P.H. Minhas (1998). 'Afforestation in Salt-affected Soils', in P.S. Minhas and N.K. Tyagi (eds), *Agricultural Salinity Management in India*, Chapter 28, CSSRI, Karnal, India, p. 463.

Yadav, J.S.P. and K. Singh (1970). 'Tolerance of Certain Forest Species to Varying Degrees of Salinity and Alkali', *Indian Forester*, 96: 587–99.

11

Biodrainage: Management Aspects

INTRODUCTION

There is a wide variation in the topography, soils, climate and water availability in different regions of the world. These features have a profound impact and influence on the population. The local vegetation and food habits of the human beings also vary in different ways. But everywhere the need for food is met primarily by agriculture, which largely depends upon the land and water resources. These resources are not unlimited. Good quality water which is fit for human use and for agriculture is particularly limited. The population growth is continuously putting increasing pressure on this valuable but limited resource. The source of all water is precipitation, which does not occur uniformly nor as and when it is needed by vegetation. Irrigation is practised to make adequate water available for agriculture as and when needed, in order to obtain the maximum crop yields.

Land productivity improves manifold with the help of irrigation, improved varieties of seeds and fertilisers and the use of new agricultural technologies. But irrigation has one severe adverse effect. It has led to waterlogging and salinisation of lands to such an extent that it has become a cause for worry. The commonly practised method of horizontal subsurface drainage to control the rise of the groundwater table and waterlogging in irrigated lands has helped but it has limitations. In developing countries, the high costs are a constraint while the pollution of surface water resources by the saline drainage water is posing problems at many places. Biodrainage can be a biological option to overcome the problem in a cost-effective and eco-friendly manner. Its main objective is to prevent waterlogging and salinisation of otherwise fertile lands with the help of tree plantations and to make sustainable use of land for irrigated agriculture.

It is not adequate to merely establish the viability and feasibility of a new technology like biodrainage. To make it acceptable and adaptable, several other aspects have to be attended. People generally get accustomed to the prevailing conditions and attitudes and are slow to adapt to new techniques and measures. New technologies, however good, can stumble against pre-conceived notions and attitudes. Good management requires a proper study of all the aspects that have a bearing on the merits of a new technology as well as its adaptation and finding pragmatic answers and solutions that would be acceptable to people.

The importance of good management can never be overstated. It aims to achieve the desired objectives with the available resources within a reasonable period of time with general consensus. No programme or activity can be carried out successfully in isolation. There are a large number of external factors that would materially affect any particular activity or programme. Similarly the conduct of a new activity can influence other established systems in a favourable or disagreeable manner. To ensure its success, all the prevailing situations and circumstances and their influence and impact on any new programme or activity should be carefully considered and accounted for before launching the new activity. Similarly, the likely effect of the new programme to be launched and the manner in which disagreements or opposition to the programme are to be met should be well thought of in advance. Some of the important subjects to be considered while planning drainage measures in irrigated lands include preliminary assessment, surveys, investigations, planning, design, management, review, monitoring and research.

PRELIMINARY ASSESSMENT

A preliminary study of the land and water resources of the region would indicate whether the available water is in abundance or is adequate or scarce in comparison to the land resource. Water distribution and management strategies would depend upon the emerging scenario. In case of abundant or adequate water availability, after meeting the essential non-agricultural requirements, all land can be put under irrigated agriculture. Two or more crops can be raised in a year on almost all the agricultural land. On the other hand, in case water is in short supply or scarce, it may not be possible to use all land for agriculture and some land may have to be left fallow. In most regions where irrigation is practised, water is in short supply or scarce in comparison to the requirement of water for the available land. The land-water ratio is an important factor to be considered for deciding the most suitable management practices.

CROPS

Agriculture crops everywhere, depend upon the soils, climate, water availability, food habits of the people and prevalent market and economic conditions. The introduction of irrigation can alter the conditions altogether and an entirely different crop pattern, which is far more profitable for farmers, can be adopted. But this does not always happen in practice and where it does happen, the pace of change is often slow. Agro-forestry is an example, which in spite of its potential and proven advantages, is not being practised adequately.

The excessive use of irrigation water initially after the introduction of irrigation and the appropriation of extra water in excess of their authorised share by the upstream users at the cost of the downstream users are widespread maladies which contribute to waterlogging. Experience has shown that corrective measures are very difficult to implement in such cases.

Where water is in short supply, high water consuming crops should be discouraged and effort should be made to maximise the overall agricultural produce value from the limited quantity of available water. But this is in conflict with the interest of individual farmers who want to maximise agricultural produce from their land holdings and therefore strive to obtain more than their due share of water. Paddy and sugarcane cultivation in areas with short water supply are examples of such distortions. The excessive and injudicious use of irrigation water is very often the root cause of waterlogging and salinisation of lands.

LAND HOLDINGS

The size of land holdings has an important bearing on management practices. In densely populated countries and regions, the land holdings are generally quite small, many times just a fraction of one hectare. The farmers are economically quite weak. They have neither the means nor the knowledge to improve their lot. They are unable to obtain good yields from their lands and are also averse to making any changes in the land pattern. Comparatively, farmers with large land holdings have a better economic status and are in a better position to accept and adopt changes.

LAND AND WATER LAWS

The prevailing land laws shape the practices and attitudes of people. In some countries, private owners of very large holdings sub-let the land for cultivation. In other countries, land laws prohibit consolidation of very small unviable holdings. Neither situation can be commended.

In most countries, an individual has an absolute right on the underground as well as the surface water on his property. In countries like Israel, however, this is not so and the State exercises complete control on all water resources and their distribution.

WATER MANAGEMENT AND DISTRIBUTION

In many countries Government agencies construct and operate water resources projects. The maintenance of conveyance channels as well as the distribution of water is carried out by Government agencies. There is very little participation of farmers and beneficiaries in the management and distribution of water. There is no satisfactory mechanism for ensuring fair water distribution and of resolving water disputes and conflicts. Government agencies are incapable of providing good administration and farmers have no incentive or motivation to participate in the management. This result in much less output from the land and water resources in comparison to their production potential.

SURVEYS

Land and Topography

The extent and type of land and its topography are important factors. The agriculture and irrigation practices in undulating hilly tracks are quite different from those in large alluvial plains. Their problems are also quite different. The land contours determine the surface water flow paths as well as the layout and alignments of irrigation channels. Different criteria are used for deciding the suitability of lands for agriculture, irrigation and forestry.

SOILS

Soils of varying characteristics are found in nature. Somewhere the soil cover is very thin while at other locations it is quite thick. Different soils have different binding, swelling and infiltration properties. Some soils permit the easy flow of water through them while others are impervious. Some soils are compact while others are loose. Mineral contents in soils vary from place to place. Standard soil classification criteria have thus been evolved to classify soils in terms of their suitability for agriculture.

WATER

The quality of surface and groundwater is very important for agriculture. The quality of water imported from outside the region may be quite different from that of locally available water. The suitability of water for irrigation depends upon its degree of salinity as the less the salinity the better it is for the crops. Excess presence of most minerals, particularly sodium, makes water unsuitable.

In humid regions, both the surface and groundwater are generally of good quality. But in dry arid regions, the groundwater is generally saline.

CROPS

It is imperative to survey the extent and type of crops grown in particular area, in order to obtain information on farming practices, methods of irrigation, crop yields, biomass produce and their mineral content. Knowledge about changes in crops under pre-irrigation and post-irrigation stages is also relevant.

METEOROLOGY

Temperatures, their day-night and seasonal variations, annual precipitation and its distribution, wind velocities, sunshine and humidity – all have a profound impact on transpiration rates, irrigation water requirement and the selection of suitable crops.

HYDROLOGY

Groundwater management is important for planning all subsurface drainage measures. Transmissivity, hydraulic conductivity, the depth of the barrier layer, porosity etc., all have a bearing on the selection and suitability of drainage measures.

IRRIGATION

The method of irrigation and efficiency of water use are very important. Water saving devices and good water management can reduce water wastage as well as deep percolation losses. Drainage requirements can be significantly reduced by good water management. Wasteful practices in surface irrigation have aggravated the waterlogging problem in many locations. This is generally due to very low water fees and lack of control on regulation and distribution of irrigation water. It would be unwise to make investments on drainage without first preventing wastage and the excessive use of water.

DRAINAGE

The prevailing conditions of surface and subsurface drainage need to be carefully studied. Water standing on the ground surface always contributes to deep percolation and recharge of groundwater. Many a time the accumulation of water is due to the blockage of natural flow paths due to the construction of embankments for roads, canals and other development activities. Surveys should identify all such obstructions as well measures that can facilitate quick drainage of water that collects on the ground surface due to a heavy downpour or floods.

The sources of both the groundwater inflow into and outflow from the area under study should be identified. A quantitative assessment of inflow-outflow pattern and groundwater balance should be attempted.

INVESTIGATIONS

Preliminary assessment and surveys can enable to identify gaps in information and data that are needed for the appropriate planning and design of drainage measures. Field investigations may be needed in the following areas:

- groundwater balance and movement;
- hydraulic conductivity, porosity, soil rock characteristics of sub-strata formations, and depth of barrier layers;
- change in groundwater quality due to irrigation;
- mineral contents in irrigation and undergroundwater in surface soils and the underlying sub-strata;
- analysis of soils and water to determine the chemical composition of the minerals;
- yields from wells and tubewells in the area, and the draw-down effect of existing wells;
- electrical conductivity (*EC*) values of irrigation and groundwater and of soils under various degrees of saturation;
- irrigation water losses that contribute to groundwater recharge;
- meteorological data on temperatures, precipitation, wind velocity, solar radiation and humidity;
- rate of evaporation from standard pan (A_{pan}), E_{To}, E_T crops etc. on a monthly and annual basis;
- rate of evapotranspiration from trees in a plantation area and its comparison with A_{Pan}, particularly, under conditions of abundant water supply (no water constraint) with the groundwater table at varying depths;
- impact of salinity levels in irrigation or groundwater on tree water use;
- biomass produce from trees of different species at different ages and the determination of utilisable biomass (dry) from different species of trees;
- mineral content and its composition in different parts of the tree plants (including the stem, leaves, roots, etc.);
- mineral content and its composition in the crops grown in the area;
- natural vegetations growing in the area and their relative usefulness to the inhabitants;
- root depths of trees growing in the area;
- market prices of crops, timber, forest produce and fuel in the region as also the available marketing facilities;
- crop and agro-forestry yields and economic returns for the farmers;
- credit facilities and availability of seeds, fertilisers and extension services;
- economic condition of the farmers as well as the social status of different communities, their customs and habits;
- research facilities;
- status of beneficiary participation in the management of water resources;

- prevailing environmental conditions, the ratio of farming land to land under afforestation, and pollution levels in air and water bodies;
- livestock, their feed and drinking water;
- existence and nature of animals and birds in the region;
- healthcare and prevalence of diseases;
- the role of women in society and gender equations; and
- inadequacies in the existing laws and in their implementation and practice.

Some or all of the above listed items may be of relevance at any particular location and the relative importance of the items may vary from place to place. The planners should find out and decide on the importance and priority of different items to be considered and accounted for in the planning and design of appropriate systems.

PLANNING

BIODRAINAGE: PREVENTIVE OR CURATIVE

Irrigation projects are under different stages of construction and operation in almost all countries. Many schemes are under planning and design with no physical work having been undertaken on the ground. Others are under implementation but not yet operative. Some are partially operative while others have become fully operational in recent times. Then there are schemes which have been under operation for a long time and their deficiencies particularly relating to drainage problems, are well known. There are also cases where irrigation schemes have outlived their utility because of waterlogging and salinity problems and are plagued by serious drainage problems to which solutions are yet to be found. There are very few irrigation schemes that do not or would not need attention in terms of drainage.

Biodrainage planning should therefore be classified into two categories, viz. *preventive* and *curative*, the 'preventive' planning to be done for schemes that are not yet fully operative and 'curative' in respect of those operative schemes in which lands have been rendered saline and unproductive on account of waterlogging.

IRRIGATION: INTENSIVE OR EXTENSIVE

Irrigation is practised to supply water to plants to meet their evapotranspiration requirement as and when needed. There are places, though not too many of them, where the land area under agriculture is limited while water is available in abundance. Here the need is for making water available in the desired quantity at the required time and to maximise the crop yield per unit of area. All the available land is brought under agriculture and two or more crops may be raised in a year on the same field. It is known as *intensive* irrigation, and if two crops are taken in a year from all the land, the intensity of irrigation can be up to 200 per cent.

But more commonly the area of irrigable land is much larger than what can be irrigated from the limited water available. Here, the area to be put under irrigated agriculture has to be restricted. This can be done in two ways. The size of the service area can be restricted to what is considered appropriate in order to enable the desired intensity of irrigation of 'say' in the range of 100 to 150 per cent. This would give a reasonable, thought not the maximum crop yield per unit area of holding.

In dry arid regions, the land areas available are generally quite vast but the available quantity of water for irrigation is relatively scarce. Here the limited quantity of water is supplied to farmers who can practise irrigated agriculture in no more than one-half of their land holding at any given point of time and the annual intensity of irrigation is very often less than 100 per cent. This is known as *extensive* irrigation. In this case, the annual yield of crops per unit area of holding is significantly less than the potential but the benefit of irrigation is provided to a larger number of farmers.

'Intensive' and 'extensive' systems of irrigation have their own merits and demerits and the choice between the two has to be made after considering all the aspects. 'Intensive' irrigation systems provide better economic returns to a fewer beneficiaries. The water conveyance system is required to be built over a smaller area and therefore they are less expensive to construct and operate. High intensity of irrigation and water use results in larger deep percolation of water and recharge of the groundwater table and consequently a quicker rate of rise of the groundwater table and a higher threat of waterlogging.

'Extensive' systems of irrigation provide comparatively smaller returns from irrigated agriculture to a large number of beneficiaries. Many times this enables what may be called 'sustenance' agriculture and not 'productive' agriculture. With the command or service area being large, the construction and operation costs are higher. On the other hand, the rate of rise of the groundwater table is slower and the threat of waterlogging is also less in the case of extensive irrigation.

The planning of drainage measures, including biodrainage, depends upon whether intensive or extensive irrigation is to be practised.

WATER MANAGEMENT PRACTICES

Before undertaking drainage measures, it is necessary to critically examine the water management practices to ensure that there is no avoidable wastage of water or excessive use of irrigation water. Excessive use or waste of irrigation water aggravates the problems and unnecessarily increases the drainage requirement. Therefore, it is necessary to identify the causes of excess use or wastage of irrigation water and find remedial measures. Very often, it is quite difficult to plan and implement water-saving practices wherever farmers get accustomed to wasteful ways. Wasteful irrigation practices are quite common in the developing countries and regions, where farmers pay very low water fees which are highly subsidised, and that too not on the basis of the volume of water consumed. Another contributing factor is the inefficient operation and management of water distribution by Government agencies which do not provide an assured supply of water to the farmers as and when needed. In such a situation, farmers tend to use excessive water for irrigation whenever possible.

Suitable remedies and solutions must be found to improve water management practices, in order to save water for productive use as well as to reduce an avoidable burden on the drainage measures.

CROP PATTERN

Generally the crop pattern to be followed is decided by the farmers themselves. This is as it should be and the Government or managing agencies should not try to impose any particular crop pattern which they consider to be the most suitable. The choice of crops depends upon soil and climatic conditions, market prices, economic conditions, traditions and local habits. An individual farmer tries to obtain the maximum financial return from agriculture from his field. Very often, there is a conflict between the interest of an individual farmer or a group of farmers and the society or community taken as a whole, when available water is short of total demand. The overall interest of the community is served best by distributing the available water in a rational manner, with each farmer sharing in the shortage in an equitable manner. This can help in obtaining agricultural produce of maximum value from the limited quantity of water. On the other hand, an individual farmer's interest is best served by obtaining the full quantity of water that he can gainfully use on his field even if this is at the cost of other farmers. On a large number of irrigation projects in the world, particularly in the developing countries, vested interests prevail and some farmers are able to obtain more water than their due share at the cost of other farmers. This can be illustrated by an example. Paddy and sugarcane crops are relatively quite remunerative on the basis of the produce value from a unit area and there can be no objection in growing these crops under conditions of no water constraint. But paddy and sugarcane are high water-consuming crops and generally do not give best produce value per unit of consumed-water. It may not be in the overall interest of the community to promote paddy or sugarcane under water constraint conditions though an individual farmer's interest may be best served by growing these very crops. There are several examples where vested interests prevail and the overall good of the community is sacrificed or compromised in their favour.

High water-consuming crops inevitably aggravate the drainage problem. Therefore, good management must aim to ensure an equitable distribution of water among all farmers. This would discourage farmers from growing high water-consuming crops or using excess water. This in turn would provide relief from the drainage problem.

WATER PRICING

Water pricing is an economic tool that can be used for the betterment of the farming community. This is usually done by supplying irrigation water at subsidised prices. But it can be counter productive if done in excess. Unfortunately, this is prevailing situation in most countries. The water fee paid by the farmers bears no relationship to the actual cost incurred in bringing water to their fields. In some parts of India, the water fee paid by a water user for irrigation is less than 2 per cent of the total actual cost incurred by the Government. The lack of revenue leads to grossly inadequate budget allocations by the Government for the operation and maintenance of dams and canal networks, which deteriorate with the passage of time and provide below par service. This ultimately hurts the farmers, who are supposed to be the beneficiaries of subsidised system. It is a vicious cycle which serves no one's interest.

There is another manner in which very high subsidies in water prices have an adverse effect. There is hardly any deterrent or penalty on the wastage or excessive use of water. This has two adverse impacts. One, the scarce and valuable water resource is wasted and secondly the drainage problem is aggravated.

It is thus extremely desirable that reasonable water pricing policies are evolved and implemented.

WATER POLLUTION

The natural river waters are generally of a very good quality. But with the increasing population, industrialisation, the use of chemicals in industry and fertilisers in agriculture and the lack of pollution control, natural river and lake waters are getting more and more polluted. Already, the quantity of water in most rivers and lakes has considerably deteriorated. In the low flow seasons and in the down stream reaches particularly, the river waters are highly polluted.

The sources of all irrigation water are rivers, reservoirs and lakes. Generally reservoirs are located in the upper reaches of rivers and fill when the flows in the rivers are large and therefore the quality of stored water is reasonably good and water pollution occurs during transit through water ways.

Drainage waters from agriculture including those from subsurface drainage are quite saline and pollute the water bodies into which these are discharged.

The use of polluted or saline water for irrigation, accentuates the problems of irrigation as well as of drainage. It is thus imperative that effective steps are taken to prevent the pollution of natural water surface bodies by industry, agriculture and other human activities.

WATER USERS PARTICIPATION

Water users must play an active, participatory role in the management of water resources. The operation of reservoirs and canal networks and distribution of water and its control should be in their hands. Governmental control should be minimum. There is a general awareness about this aspect, without much dissent. But, not much headway has been made towards this objective in many regions and countries of the world. Attempts have been made, here and there, from time to time to remedy the situation but the overall results are far from satisfactory.

The care and efficiency of water use, equity in water distribution, maintenance of water conveyance systems, prevention of water wastage and control on pollution can all improve significantly with the active participation of the beneficiaries in management. In many regions, drainage ditches get silted up or blocked due to weed growth. Neither Government agencies nor water users attend to the problem. Management solutions must be found to such problems.

IMPORTANCE OF AFFORESTATION

The land area under cultivation has been consistently increasing, mostly by clearing the forests. It is difficult to stop or reverse this trend without controlling population growth. Taking an overall view, the total land area under cultivation in the world is about one-third of that under forests. This may be considered as reasonable and acceptable. But in densely populated regions, there is a gross imbalance. In India, the land area under cultivation is three times of that under forests. To prevent environmental degradation and to meet the requirement of fuel and timber, this imbalanced situation needs to be corrected. This cannot be achieved by merely making stringent laws against the clearing of forests and cutting down of trees. Promotional activities and programmes that encourage afforestation and planting of trees are needed. Afforestation for biodrainage can be one such programme which would serve the dual purpose of plantation alongwith drainage.

SMALL LAND HOLDINGS

In thickly populated areas, individually owned land holdings are generally small, many times smaller than one hectare. On such small land holdings, farmers are unable to produce enough to make a living. Any programme that would call upon such farmers to earmark a portion of their land for planting trees or afforestation, in place of agriculture is unlikely to be accepted by them. Doubts are expressed that biodrainage, on account of the need of land for afforestation, may encounter this difficulty.

In this context, it is to be kept in mind that biodrainage would generally be practised in irrigated areas where the source of water lies outside the concerned area. Biodrainage plantations may not be needed in areas irrigated by open or tube wells. No such plantations may be needed in areas where conjunctive use is being made of surface and groundwater. The pumping out of groundwater through wells can meet the drainage requirement. In locations where irrigation water is brought from outside the area, for agricultural use, the quantity of available water is generally limited and not enough to meet the full potential requirement of land under agriculture. The productivity of a farmer, even with a small holding, depends more on his share in irrigation water rather than on the size of his holding. To motivate farmers to take to agro-forestry the following actions can be taken:

- development and propagation of agro-forestry systems and technology that would provide equally good or better returns to farmers in comparison to agriculture alone;
- provision of an extra share in irrigation water to farmers who reserve a part of their land for biodrainage plantations to serve as an incentive; and
- enactment of land and water laws that would promote tree plantations and penalise those who do not meet the minimum prescribed standards.

The requirement of privately owned lands for biodrainage forestry can be minimised by first taking up plantations along roads, canals and public lands. On new irrigation projects, where settlements are to be made in future, the reservation of lands for biodrainage can be made in advance. In dry arid regions, which are most susceptible to the waterlogging and salinity hazards, the total requirement of land for biodrainage forestry may be no more than 10 per cent. The demand on private land would therefore be small and should be manageable.

Alternative drainage measures like the horizontal subsurface drainage would also require land for open sub-collector, collector and main drains and in case the drainage water is to be discharged into evaporation ponds, the total requirement of land with horizontal subsurface drainage system is likely to be more than that under biodrainage. And finally, in case no drainage measures are taken for want of land, the loss of land and productivity due to waterlogging and salinisation of lands is likely to be much more.

The problem of the small size of holdings and the reluctance of farmers to take to tree plantations on small portion of their land, as and when needed, has to be overcome and can be resolved by a suitable system of incentives and disincentives. The failure of the management to do so would have a severe adverse impact on the land and land owners.

CHOICE OF PLANTATION SPECIES

Great care is needed in choosing the species of trees to be planted for biodrainage. The chosen species should be able to meet the following requirements:

- They should be able to grow under the prevailing climate and soil conditions;
- The tree water use capability and transpiration rates should be high;
- The tree species should give a good utilisable biomass yield;
- The plant biomass should be rich in mineral content;
- The plantations should have a good tolerance to grow under saline conditions;
- The plant produce should give good financial returns;
- The plant roots should be able to go deep into the ground and be capable of draining water from the groundwater table;
- The plantations should have a favourable environmental impact on the surroundings and on the human inhabitants; and
- The plantations should be socially acceptable.

In case quick results are needed to reclaim the already waterlogged areas, fast growing trees like the *Eucalyptus* may be most suitable. Otherwise, it should be possible to make a choice of the plantation species on other considerations. In the dry arid regions, tree species that grow under natural un-irrigated conditions like the *Prosopis cineraria* or *Acacia* varieties can be grown. As and when the groundwater table comes up sufficiently, and if such plantations are unable to transpire enough water to stabilise the upward movement of the groundwater table, these can be supplemented or replaced by high water consuming plantations.

There are some species like *Salvadora oleoides* which have very high mineral contents and can be planted to serve a special purpose. In water scarce regions, bio-plantations should not compete for irrigation water with the agricultural crops. They should be able to survive and grow on natural precipitation and by drawing their water requirement from the surface of the groundwater table through roots which are deep enough to reach the groundwater table.

Good management requires that action for planning and undertaking planting operations is taken as much in advance as possible without waiting for the groundwater table to come up. In new projects, this can be done as soon as the implementation of the project is taken in hand. In operational projects, this may be done wherever the groundwater table has a rising trend. In any case, urgent attention should be given wherever the groundwater table comes up to, say, 10 m below the ground surface and shows a rising trend.

LOCATION OF BIODRAINAGE PLANTATIONS

The areas to be irrigated may be small or large in size. In case of small tanks that store a limited quantity of water which is all used up within a few months, the problem of waterlogging in the command areas is not anticipated. In large projects, the command areas may be undulating with high grounds and low valleys or they may be flat with easy gradients. The areas in low valleys or large flat areas with easy gradients are more susceptible to waterlogging in comparison to high grounds or lands with sufficient gradient. The groundwater profile, its gradients and the shape and position of underground barrier layers are other important features which decide whether and which portions of an irrigated command area would be susceptible to waterlogging. Drainage measures may therefore, not be necessary everywhere. Areas susceptible to waterlogging should be identified. An assessment should be made of areas that are likely to get waterlogged in the shortest period of time. Such areas should receive the first priority for undertaking drainage measures.

Once the quantity of recharge to the groundwater table and the requirement of plantation area for biodrainage are estimated, it would be advisable to divide and sub-divide the area for deciding on the most suitable locations and sizes of biodrainage plantation plots. Large number of small plantation plots, if feasible, should be preferred over less number of big-sized plantation plots. They would be more effective because the requirement of draw-down of the groundwater table underneath the plantations would be relatively less enabling the trees to transpire water at a higher rate with lower root depths.

In flat irrigated command areas, lands along the periphery or boundary of small units of irrigated areas, outside the command areas, can be selected for biodrainage plantations.

There can be considerable flexibility in choosing and earmarking plots for biodrainage plantations. Technical consideration of providing greatest possible effectiveness in drainage with the smallest land area requirement is important, but adjustments to suit the prevailing site conditions are possible.

PATTERN OF BIODRAINAGE PLANTATIONS

Plantations along highways and canals are recommended for biodrainage because of the easy availability of land and convenience of maintenance. The width of the planting strips can be kept as large as possible.

Long parallel strips of, say 100 m width in the irrigated areas can be adequate. The spacing between the strips would depend upon the area needed for plantations and the hydraulic conductivity of the sub-strata. Where a 10 per cent area is needed for biodrainage plantations, the spacing between parallel strips can be about 1 km. In case less land, say only 5 per cent is needed, the width of the plantation strips can be reduced to 50 m.

In case it is not possible to arrange lands in parallel strips, block plantations can be adopted. These should be scattered and distributed in such a manner that they can provide effective drainage to all parts of the irrigated area. Their total area should not be less than what is needed to transpire the volume of water equal to the net recharge to the groundwater table.

Plantations along field boundaries and within the fields can be considered and accounted for while estimating the balance need of area for the biodrainage plantations.

AGENCY OF RAISING PLANTATIONS

Plantations along highways and canals can be raised and maintained by the agencies in charge of the highways and canals. Those along field boundaries and within the fields can best be left to the care of individual owners. Block and strip plantations should be raised and maintained by communities or groups of people who have a direct stake in the plantations. The benefits of plantation produce should also go to the community or the group as a whole to be shared and divided amongst the individuals constituting the group. Where, for any reason, it is not possible for the community to accept the responsibility, the project operating authority or the government should undertake the work of raising and maintaining plantations, till suitable arrangements of transferring the responsibility to the local community are in place.

The general experience is that when the government or public administration is involved, the costs are high and benefits small. Community participation is also necessary to prevent unauthorised trespassing and damage to the plantations. The success of biodrainage schemes would greatly depend upon the collaboration between the Government, project authority, the local community and individual farmers. This cannot be brought about without good management.

In Jaloi village of Jaipur district, Rajasthan, India *Dalbergia sisso* trees were planted over an area of 90 ha., on pasture land during the period 1994–98. The villagers were initially reluctant to grow trees on pasture land. But they discovered very soon that grass could be grown under the trees to feed the animals. The pruning of trees provided additional feed for the live stock. The grown up trees are now a valuable asset to the village. Experience on planting of *Acacia nilotica, Prosopis coneraria, Eucalyptus* and other species of trees in adjoining village was similar. As a result the villagers are now very possessive and protective of the planted trees. The barren land under the trees has turned productive. The soil erosion has been checked. *Rajasthan Patrika* in a series of articles published (2001) reports that about two million trees are now growing over about three thousand hectares of land in the area and this has significantly improved the economic conditions in the villages.

REVIEW OF LAWS

There should be a general review of all public and local administrative laws and regulations, particularly those relating to land water use and afforestations. Amendments that would promote the better use of land, water and forests should be considered and enacted in the overall larger public interest.

The administration should be decentralised to transfer responsibility alongwith the authority to local communities to the maximum extent possible. The effective administration capable of ensuring a fair distribution of resources on the principles of equity, which is responsive to public needs and requirements and ready to settle disputes quickly, is needed to obtain best results.

INTERACTION

There can be no substitute to good interaction between managers and those who stand to gain by good management. The absence or inadequacy of interaction is very often an important cause of mismanagement. Local people are best aware of their own difficulties. They may not be knowledgeable about the best technologies and the most appropriate solutions to their problems, but they would understand if these are explained to them. They would be able to offer comments and suggestions. There should be a forum for such interaction. There also ought to be a system of regular and frequent interaction where managers and the beneficiaries can exchange views, experiences and suggestions on various programmes and activities and their implementation.

STAGES OF PLANNING

The principles of planning for biodrainage have been described in an earlier chapter. The procedure to be followed in planning would depend upon the prevailing circumstances and conditions at the location under consideration. These may vary significantly from place to place and would also depend upon the stage at which biodrainage is to be undertaken at the pre-project or pre-irrigation stage or at some post-irrigation stage. In the latter case, it would be important to consider whether the groundwater table has already risen sufficiently and caused waterlogging or it is likely to pose a threat of waterlogging soon, or whether sufficient time is available before the problem of waterlogging is likely to occur. It may be convenient to classify the prevailing conditions under the following broad categories.

Humid Region with Groundwater of Good Quality

Generally, humid regions have soils and groundwater of good quality. The natural precipitation washes down the salts in soils which are drained away naturally. The top soils and the groundwater are of a reasonable good quality to enable irrigated agriculture.

Natural precipitation may not occur uniformly throughout the year to match with the crop water requirements and therefore supplemental irrigation may be needed. This can be done by storing surface water and making it available for irrigation when needed or by pumping out groundwater for agricultural use.

In case planning is being done at the pre-project stage, it would be advisable to carry out advance water balance studies. If the surface water potential exceeds the crop water requirement of the region, the boundaries of the project area can be enlarged to make full use of the available water potential, subject to the limit of economic viability. The groundwater potential, including that due to estimated recharge from future irrigation should also be accounted for. The objective should be to limit the quantity of surface water to be used and plan its conjunctive use alongwith the groundwater to enable irrigation on the maximum possible area without the threat of the water table rising to unacceptable levels. In case there is still some surplus water, it may be allowed to flow downstream in natural streams.

Agro-forestry under such conditions may be needed only to meet the requirements of timber and fuel based on economic considerations and to meet the environmental needs. Biodrainage can be practised in case there is no need for surplus water in the downstream area.

In operative projects, if and when the rise of the groundwater table due to irrigation, is already posing a threat of waterlogging, it should first be considered whether and to what extent the use of the surface water can be reduced and substituted by the use of the groundwater to stop the rising trend of the groundwater table. The saved surface water can be used by extending the size of the area under irrigated agriculture and if this is not feasible, then the excess water should be allowed to flow down into the natural streams. If the combined surface and groundwater resource is in excess of the irrigation requirement, the role of agro-forestry would be limited to providing the requirements of forest produce. Biodrainage can be practised if the surplus water cannot be put to more gainful use in the downstream area.

In case of humid regions where the areas have become waterlogged due to excess precipitation and irrigation, biodrainage can play an important role in the reclamation of the waterlogged areas. The plantation of trees in strips along the periphery of the waterlogged areas, would increase the rate of withdrawal of water from the ground due to transpiration. With the growth of trees, the water level in the waterlogged area would gradually go down. The water on the surface would recede and the waterlogged area would shrink in size with the passage of time. Tree plantations can then be taken up on the land from where the water has receded. This would further speed up the transpiration from trees and the reclamation of waterlogged areas.

Dry Arid Region with Good Quality Groundwater

In dry arid regions, the water is generally scarce and irrigation is often practised by bringing surface water from outside regions. Recharge from irrigation disturbs the previously existing groundwater balance and the groundwater table starts rising. If the groundwater is of good quality, this should pose no problem. The groundwater can be pumped and put to conjunctive use alongwith the surface water for irrigation. This can enable a lowering down of the groundwater table to acceptable levels. Under such conditions, planning at pre-project stage, should provide for the conjunctive use of groundwater alongwith surface water so that the recharge of groundwater due to irrigation is balanced by groundwater withdrawal by pumping.

However, there are examples galore where inspite of the above explained position, waterlogging occurs in dry arid regions on the practice of irrigation even where the groundwater is of good quality. This is due to the excessive use of surface water only for irrigation and little use of groundwater in spite of its good quality. Such a situation develops and prevails in the early operative years of a project when surface water is surplus and is used in excess. Very often the cost of surface water to farmers is subsidised and is much less than that of groundwater which is required to be pumped. The farmers are therefore reluctant to use groundwater, more so when waterlogging adversely affects farmers other than those who benefit by the use of excess water. The waterlogging problem in such situations is the direct result of wrong policies and mismanagement. Providing irrigation water to farmers at a price which is hardly a fraction of the actual costs acts as a strong disincentive for using groundwater. Very soon, vested interests develop which resist any reforms or corrective action. Mismanagement and lack of farmers participation in the governance and distribution of water magnify the problem.

The first priority under such conditions, should be to reduce the use of surface water for irrigation and to promote the use of groundwater. This can almost wholly take care of the problem. Agro-forestry should be practised to provide timber and fuel and to improve environment. Biodrainage can be practised to reclaim the area that are already waterlogged.

Dry Arid Regions with Groundwater of Poor Quality

The requirement of irrigation is maximum in the dry arid regions, where the groundwater is generally saline because of inadequate leaching and washing out of salts. The waterlogging and secondary salinisation problem is most common in the dry arid regions brought under irrigation by bringing water from outside regions. Because of the poor quality of groundwater, the remedy for the problem is quite difficult and problematic. The presently practised remedial measure is to provide subsurface drainage by installing perforated underground pipelines to intercept and remove the subsurface saline water and maintain the groundwater table below the root zone of the crops. This method is quite effective in controlling the rise of the groundwater table and to maintaining it at a safe level, but it is quite expensive. Moreover, the discharge of saline and sometimes toxic drainage water pollutes the water of lakes or rivers into which the drainage water is discharged. This can adversely affect the health of human beings, animal and aquatic life and degrades the environment. There is an urgent need to find better alternatives.

Planning at the pre-project stage should take full cognizance of the problem. Enlarging the size of the area to be irrigated by adopting low intensity of irrigation, would ensure more efficient use of irrigation water and a slower rate of rise of the groundwater table. Generally, the quality of groundwater improves due to the dilution effect by good quality irrigation recharge water and wherever possible, provision should be made for the conjunctive use of usable groundwater alongwith the canal waters. The feasibility of subsurface drainage and disposal of drainage water should be studied. Biodrainage can be effective and most suitable remedy and this should be considered and provided for. Un-irrigated land in the dry arid regions has very little value and it should be possible to select, earmark and reserve lands for biodrainage plantations at the pre-project stage.

Attention to the problem of waterlogging and secondary salinisation is generally drawn at the post-project or operative stage when some areas, after productive use for many years, get waterlogged and become unproductive and when the threat of waterlogging looms large on other areas. This causes great hardship and misery to the affected farmers who are deprived of their only source of livelihood. Subsurface drainage, through underground perforated pipelines, can be a remedy if it can be achieved at acceptable costs and if the disposal of saline drainage water does not create unacceptable health and environmental hazards.

Biodrainage can be an effective remedy under such conditions. Tree plantations can transpire excess water from the ground to lower down the groundwater table. Timely action would prevent waterlogging. The waterlogged lands can also be reclaimed. Biomass produce from tree plantations removes some minerals from the ground and if the irrigation water is of good quality, biodrainage plantations can help achieve both water and salt balance and enable sustained irrigated agriculture. Biodrainage is likely to be the least expensive and most beneficial remedy under such conditions. Its feasibility should be carefully considered at the planning stage and necessary provisions be made for its implementation.

DESIGN FEATURES

Good planning and design of drainage measures would require that the features that cause or aggravate the problem of waterlogging are first identified and taken care of. Remedial measures should then be planned to provide best protection at the least cost. For this, the following aspects need careful consideration.

INTENSITY OF IRRIGATION

Irrigation can be practised in an intensive manner to yield the maximum possible produce from a unit of land. Two or more crops can be taken in a year from each piece of land. Drainage requirements would be maximum under such a practice. Alternatively, the available water supply can be allocated to a large area so that irrigated agriculture is possible only on a part area during a crop season. The ratio of land on which irrigated agriculture is actually practised to the total irrigable area is generally called 'intensity of irrigation'. In case irrigated crops are grown on a 50 per cent area in winter and on a 40 per cent area in summer, the annual intensity of irrigation would be 90 per cent. An irrigation project can be planned to provide intensity of irrigation in a range of say 60 to 200 per cent. In a water-deficit region, the total agricultural produce and benefits largely depend upon the total volume of water available for irrigation and would therefore be nearly equal regardless of whether the practised intensity of irrigation is 60 or 200 per cent. The quantum of benefit to individual farmers would be less with a lower intensity of irrigation but the total benefit from the project would be no less because a larger number of farmers and a larger area would benefit. On the other hand, the threat of waterlogging and the need for drainage measures would be minimal with the lowest possible intensity of irrigation.

Therefore, it may be preferable to plan irrigation projects, particularly in the dry arid regions and water-deficit areas, with as low an area intensity factor of irrigated agriculture (A_F) as possible. This would enable more sustainable irrigation.

CROP WATER REQUIREMENT

The water requirement for crops largely depends upon climatic conditions, crop species and soils. The water requirement for winter crops is less than for summer crops. Therefore, under water-scarce conditions, the growing of winter crops should receive preference over summer crops. Some crops are more sensitive to water stress conditions than others and their yields diminish significantly even when there is a slight deficit in water supply. Paddy is one such crop. It may not be appropriate to plan the growing of such crops in dry arid regions. Crops that have a high requirement of water for growing also contribute to high percolation losses and recharge of groundwater and therefore drainage requirements for such crops are more than in the case of less water-consuming crops. This aspect is to be taken into consideration while planning and designing irrigation projects. An attempt should be made to maximise the agricultural produce value per unit of water and the allocation of irrigation water per unit area (I_R) should be made on this basis.

WATER USE EFFICIENCY

A significant quantity of water is lost by evaporation, seepage from water conveyance systems and during irrigation application in the field. In surface irrigation systems, the range of this loss may vary from 30 to 60 per cent. Drainage water requirements are directly in proportions to the deep percolation losses from seepage and infiltration. It is desirable to reduce such losses to the minimum extent possible. Piped conveyance systems and the use of sprinklers or drips can greatly reduce the losses as well as drainage requirements. But these require high capital investments, good technology and supportive markets, which may not be feasible in large areas in the developing countries for quite some time to come. Therefore the emphasis should be on the lining of canals to reduce seepage losses and to improve field application efficiencies. The recharge factor (R_F) i.e., the ratio of net recharge of groundwater to the total irrigation water supply, should be as small as possible.

TREE WATER USE

Biodrainage plantations should be planned to transpire all excess water from the ground-water table. The area required to be brought under he plantations would depend upon the transpiration rates of the trees. Some species of trees like the Eucalyptus are known to consume large volumes of water. Some researchers have reported transpiration rates as high as three times of A_{pan} and more where A_{pan} is the measured annual evaporation from a standard pan. It would be reasonable to assume, at the present stage, that the biodrainage plantations can transpire as much quantity of water as is evaporated on a standard pan. It is estimated that with irrigation water application (I_R) of 600 mm, and area intensity factor of irrigated agriculture (A_F) as 1.0, the ratio of net recharge (of groundwater) to total irrigation water supply (R_F) as 0.3, and A_{pan} equal to 1500 mm, 12 per cent of the culturable area may be needed for biodrainage plantations. If the transpiration rates are higher, the need for the area for biodrainage plantations would be proportionately reduced. In desert areas, annual transpiration rates exceeding 3000 mm have been estimated and this can reduce the requirement of land for biodrainage plantations to just 6 per cent.

Enough data is not available on the maximum water use capability of trees of different species. It is likely that many species of trees, other than *Eucalyptus*, are capable of transpiring enough water, to be suitable for growing for biodrainage.

SALT TOLERANCE OF PLANTATIONS TREES

Plantations for biodrainage should be capable of tolerating salinity in groundwater. Tree water use and transpiration are greatest when the water is non saline (EC < 4.0 ds/m). Transpiration rates reduce with an increase in water salinity. In case of *Eucalyptus* reduction in tree water use to about one-half has been reported at groundwater salinity EC equal to 12.0 ds/m. It follows that the capability of tree plantations to provide biodrainage would depend upon the salinity of groundwater. This factor is to be accounted for while planning and designing biodrainage plantations. Fortunately, the groundwater salinity levels in irrigated areas, show a declining trend with the passage of time and are generally less than 12 ds/m at the time when the groundwater table has risen and come close to the ground surface. In areas with groundwater of higher salinity, special species of trees with higher tolerance to salinity, would be needed for biodrainage.

BIOMASS PRODUCE

Biomass produce from trees that is removed from the ground and utilised, results in the net removal of minerals from the soil equal to the mineral content in the biomass. Trees that yield high utilisable biomass produce and those in which mineral content is high would be more suitable for biodrainage plantations. They would remove excess water as well as minerals from the ground, whose quantities can be estimated.

The biomass produce from tree per unit area would depend upon the tree species, density of plantation (i.e., number of tree per ha), stage of growth and whether adequate water supply is available for the trees or not. The selection of tree species to be planted should be made from the species that are known to grow well in the area without irrigation. Deep rooted trees should be preferred. Plantations at spacing of 3 m × 3 m with a density of about 1100 trees per ha are quite common. But more dense plantations of 2500 trees per ha or more can be adopted if they yield a higher biomass produce per unit area. The spacing that gives the maximum biomass produce should be adopted as this would also correspond to the highest rates of transpiration and mineral removal from the ground by the plants. The utilisable biomass produce, on a dry basis, of up to 30 tonnes per ha per annum from mixed plantations has been reported. The available data and information should be carefully analysed to estimate the expected utilisable biomass produce in the area under consideration and a reasonably achievable target should be adopted for planning and design.

ECONOMICS

All developmental programmes aim to maximise benefits. Most benefits can be measured and quantified in economic terms but some social benefits and those relating to environment cannot be quantified easily. There are some activities that yield high economic returns for some time but which are not sustainable in the long run. In the case of irrigation and drainage, the following results should be ensured:

- The project should give the highest possible return from the total available land or water whichever is relatively scarce;
- The irrigated agriculture should be sustainable on a permanent basis;
- The project should contribute to the general improvement of the environment. Besides, it should not cause degradation of land and water resources.

It would be advisable not to look at the irrigation and drainage project in isolation. The status of other infrastructural works, marketing conditions, the availability of credit and input requirements, administrative laws and regulations and, social structure and behaviour – all have an important bearing on the results.

HUMAN RESOURCE

While the world population continues to grow, the land and water resources remain static. In the overall reckoning, there is no shortage of human resources. But human population is not evenly distributed and there are some regions and countries where the lack of population is a constraint in developing available land and water resources. At the same time, because of the low population and low demand, there is no pressure or urgency to rapidly develop the resources to their full potential. On the other hand, in densely populated countries there is no dearth of population. But even in densely populated countries, there are deserts and thinly populated areas, the primary cause of low density of population being lack of water. The desert lands can be developed by bringing water from outside. Here, new settlements would be needed, and it would be advisable to plan new settlements properly so that the human resource, both in quantity and quality, is adequate to meet the development challenges.

REVIEW AND MONITORING

No planning or implementation can ever be perfect. Assumptions and estimates made at the planning and design stages can go awry. Hurdles and problems, not anticipated earlier, can crop up. Unforeseen difficulties can slow down the implementation. The actual results and benefits may be at variance from those that were initially accounted for. Some planned activities may not be found feasible while others, not thought of earlier, may have to be incorporated. Continuous review and monitoring, is thus needed to facilitate adjustments for resolving such problems. Irrigation and drainage projects can be no exception. In the case of biodrainage, which is a relatively new science, much is yet to be learnt by research and experience and the need for review and monitoring would, therefore, be all the more important.

RESEARCH

Some research work is needed to fine tune and fully develop the science of biodrainage. There are some gaps in information. In some matters, the reported results and findings show great variance. New studies are needed on a few aspects. More extensive field trials are needed. The important subjects on which research work and studies are needed are as follows:

- Rates of transpiration and tree water use by different species of trees growing with the help of sub-irrigation from the groundwater table, with varying depths of the groundwater table and with water of varying degrees of salinity;

- Root depths of trees of different species and their capability to draw water from various depths;
- Mineral contents in crops and trees – total as well as element-wise; information on how the mineral content in crops and trees is related to the mineral content in the soil and water at the place where they are growing;
- Biomass produce from different crops and trees, under different conditions of growth; density of plantation and practices to maximise biomass produce;
- Draw-down of the groundwater table underneath the plantations and groundwater movement towards the plantation area and its profile; applicability and validation of generally accepted theories like the Donnan equation;
- General change in groundwater salinity in irrigated areas with the passage of time; and
- Change in soil and groundwater salinity underneath the plantations, with the passage of time.

CONCLUSIONS

Good management is vital for the success of any scheme. Drainage measures including biodrainage are no exception. Land and water resources should not be allowed to degrade while being put to productive use. Proper surveys, investigations, planning and design can help in finding the best suited measures for implementing and operating irrigation and drainage projects. These would vary from location to location depending upon the topography, climate, soils, groundwater and a host of other prevailing natural conditions.

Biodrainage can be the most suitable drainage measure under certain conditions to overcome the problems of waterlogging and salinisation particularly in the dry arid regions. To succeed it would need efficient management.

In addition, careful review and field trials are needed to fully develop the science of biodrainage.

REFERENCE

Rajasthan Patrika (2001). Articles by Ashok Sharma in the Daily Newspaper Issues of 8, 10 and 16 January.

Epilogue

The author had the opportunity and privilege of working for a long period in building the IGNP, a major irrigation project in the Thar desert of India. Canal irrigation brought about a remarkable transformation of the land and ushered in an era of prosperity. Sadly, within a period of three decades, some parts of the irrigated lands were rendered unproductive due to waterlogging and others were threatened.

The plantations had been raised along the canals primarily to prevent embankment erosion and silt deposit in canals due to strong winds that prevail in the area. It was discovered that these plantations had a salutary impact in depressing the shallow groundwater table in the adjoining areas. This prompted a study to find whether biodrainage could be an answer to the waterlogging and salinity problems. Field observation, study and research led to the writing of a few papers that were presented at national and international fora form time to time. They were received enthusiastically by some and disbelievingly by others. A few scientists and researchers rejected the concept outright, almost scornfully. This spurred further study and research as well as interaction with many knowledgeable engineers, hydrologists, soil scientists, plant physiologists, foresters and researchers. Although not intended in the beginning, all this finally resulted in the writing of the book 'A Biological Option for Controlling Waterlogging and Salinity (2001). The subject of biodrainage received special attention at the 8th International Drainage Workshop at New Delhi (2000). The International Commission of Irrigation and Drainage (ICID) took note of the matter and recommended more research. IPTRID (FAO) prepared a synthesis report on the subject (2002) and recognised the potential of biodrainage. As a follow up to these recommendations many research programmes were undertaken in India under different climatic and physiographic conditions. These have produced very promising results and the studies and their results are incorporated in a new chapter in this book.

There are some gaps in our knowledge and some shortcomings would be found in the descriptions presented in the book. There is need to improve and find more appropriate answers through field trials. The author hopes that this book would promote the adoption of biodrainage and the combined technological strength of scientific institutions would be applied to further develop this science and its use for the benefit of humanity and the protection of the environment.

Index